INVERTEBRATE RELATIONSHIPS

Patterns in animal evolution

PAT WILLMER
Department of Biology and Medicine
University of St Andrews

Published by the Press Syndicate of the University of Cambridge
The Pitt Building, Trumpington Street, Cambridge CB2 1RP
40 West 20th Street, New York, NY 10011–4211, USA
10 Stamford Road, Oakleigh, Melbourne 3166, Australia

First published 1990
Reprinted 1990, 1991, 1994, 1996

Printed in Malta by Interprint Limited

British Library cataloguing in publication data

Willmer, P.G.
Invertebrate relationships: patterns in animal evolution.
1. Invertebrates—Evolution
I. Title
592′.038 QL362

Library of Congress cataloging in publication data

Willmer, P.G. (Patricia Gillian), 1953–
Invertebrate relationships
Bibliography: p.
Includes index.
1. Invertebrates–Evolution. 2. Invertebrates.
Fossil. I. Title.
QL362.75.W55 1990 592′.038 87–32019

ISBN 0 521 33064 5 hardback
ISBN 0 521 33712 7 paperback

INVERTEBRATE RELATIONSHIPS
Patterns in animal evolution

To Kate and Janet,
who taught me

Contents

Acknowledgements

Even more than most books aimed at a student market, this book owes its very existence to those undergraduates I have taught, originally at Cambridge and more recently at Oxford. They made me aware of the need for such a book, both by their enthusiasm for delving into these more esoteric areas of invertebrate zoology and by their complaints about existing literature; they gave me some of the ideas it contains, through many hours of tutorial discussions; and some of them helped persuade me to write it. Many have read chapters for me and pointed out explanations that were less than clear. To all of these students I owe an enormous debt.

More specifically, colleagues old and new have helped me by their willingness to read and comment on specific chapters, and to discuss ideas and make me aware of facts that I had omitted or misinterpreted. It is a privilege to thank Drs Matthew Holley, Henry Bennet-Clark, Peter Holland, Anne Keymer, and Cathy Kennedy, for reading parts of the text; and Dick Manuel, who has encouraged me throughout, and whose vast practical knowledge of the living invertebrates has helped in so many ways. My greatest debts are to Dr Jonathan Wright, who has patiently allowed me to inflict the whole book on him and has provided kindly critical expertise even when he disagreed with my approach and conclusions; and above all to Dr Janet Moore, who first inspired me with enthusiasm for the invertebrates and who continued to sustain and direct my interest with her comments on each successive chapter as it was written.

Despite the best efforts of all these people, errors and inconsistencies will surely remain to be found by others in a book of this kind; these of course arise entirely from my own misconceptions, and I hope readers will enjoy tracking them down.

Finally I should mention Bridget Peace, who drew the illustrations that precede each chapter in part 3 of the book and which provide such a pleasant contrast to my own functional and remarkably non-artistic text figures.

Oxford, July 1987.

PART I

Introduction to animal phylogeny

In the first part of this book, some of the principles and problems of phylogenetic study are considered. Many of the classical morphological characters that have been used to classify invertebrates are critically examined, and the schemes of invertebrate relationships traditionally derived from interpretation of these characters are compared.

1 Approaches to animal phylogeny

1.1 Introduction

Phylogenetic study is important, and is possible. There is a right answer to the question of how all the familiar living animals are related to each other; they do, at some point in the past, share common ancestral forms. And looking for that right answer, though it is surrounded by much theoretical debate, is not purely a matter of academic speculation. The search in itself is feasible, for there are many clues and possible sources of evidence; furthermore, the search is informative and, above all, interesting. This book is written to try to convince both students and teachers of these views, and to introduce new insights and sources of evidence that bear on the problem, particularly from the application of technological innovations.

However, there are several other related reasons for a book of this kind. Firstly, phylogenetic study is in itself a serious and stimulating challenge for any zoology student seeking to understand the patterns and principles of animal evolution. Very few courses in vertebrate zoology are conducted without a sound phylogenetic framework, yet it seems that when the other 99% of animals - the invertebrates - are at issue we are only too often quite happy to leave phylogeny aside. Many courses now consider the invertebrates independently of their evolutionary framework, or with such a meagre (and quite probably incorrect) treatment of the degrees of relationship that little or no underlying pattern can be perceived. The result can too often be a series of unsatisfactory and disconnected examples of how invertebrates are built and what they do; for without an understanding of the relationships between animal groups, much of the teaching of physiology, mechanics and adaptation loses all structure and is conducted in a vacuum. The very diversity of animal design, and its effects on ecological diversity, are readily lost when only a handful of the larger and more obvious phyla are mentioned. Phylogeny gives a structure and context to the currently unfavoured 'middle ground' of zoology, between the rapidly advancing intricacies of molecular biology on the one hand and the equally fashionable overviews of behavioural, ecological and evolutionary theory on the other. If we lose sight of the structure of the middle ground, the two ends of the zoological spectrum may eventually risk losing sight of their links and interdependencies.

Secondly, the literature on recent developments in phylogeny is scattered, and reviews are few and far between. The point has been reached when students asked to consider the possible origins and affinities of a group of invertebrates will often dismiss the problem as hopeless. Or, more importantly, they will cheerfully set out to find an answer but fail miserably to discover any appropriate or helpful sources, or will find just one of several possible views and swallow it uncritically. In any event, faced with phylogenetic literature at least twenty or thirty years old, they may feel quite justified in abandoning the subject as outdated and irrelevant. Yet the literature is there, albeit in dispersed and sometimes inaccessible form. In particular, there has been a considerable advance in the basic descriptive literature of smaller invertebrates, concomitant on improved techniques in microscopy. This literature is disjointed and specialist, and much of it has yet to be put into context where its relevance becomes apparent. Some of it particularly concerns the interstitial fauna that was rarely appreciated by phylogeneticists before the 1960s but that has forced some re-evaluation of central issues. A book bringing together all this new literature is long overdue.

Thirdly, though we live in an age of genetic, molecular and ultrastructural research, there is surprisingly little appreciation as yet (especially in English-speaking countries) of the implications these new disciplines have had for animal phylogeny. Invertebrates have always been important tools in such research, but only too often new findings in these fields are subjected to quite outdated comparative treatments. Much of the material discussed in chapters 4, 5 and 6 is of this kind; genetic, developmental or ultrastructural findings are reported for a few species (characteristically an insect, a sea urchin, and a nematode!), and are then fitted to a simple (and usually outdated) phylogenetic scheme drawn from the nearest textbook, in support of which scheme they are thereafter cited. There can be little doubt that molecular approaches to phylogeny will in the future be the major source of progress in invertebrate biology; so it will, I hope, be useful to have available a modern review of the principles, problems and possibilities in invertebrate phylogeny, culled from other sources of evidence.

There is one further reason why a book of this kind may be appropriate at present. The whole idea of phylogenetic analysis as an academic discipline has been subject to something of a revolution in the last twenty years, and there is profuse debate as to the proper methodology for such analysis; useful introductions to the literature can be found in Joysey & Friday (1982). The terminology to be used has certainly been clarified and made more rigorous by advocates of the Hennigian phylogenetic systematic (or 'cladistic') approach (Hennig 1966; Eldredge & Cracraft 1980; Wiley 1981). As yet, these techniques have largely been applied to controversies about vertebrate relationships, or to invertebrate problems at lower taxonomic levels though Ax (1987) has attempted a cladistic analysis of the phyla of lower invertebrates. There is no doubt that cladistic analyses of the whole of invertebrate phylogeny would be greatly welcomed in some quarters. It should therefore be made quite clear at the outset that this book is *not* attempting to fulfil that role.

Avoidance of rigorously cladistic techniques is a consequence of a personal view that cladistic treatises are generally rather unreadable for non-specialists; for a book of this kind, more informal language and argument are needed. More seriously, two major premises on which cladistic theory and practice rely do not seem 'appropriate' to invertebrate phylogenetics, at least in their present state. Firstly, the cladist precludes the use of adaptationist arguments, which are dismissed as 'story-telling'. In contrast, traditional evolutionary taxonomists often support their ideas with a consideration of functional continuity in changing environments (see Bock 1981). Thus cladograms, though rigorously derived from clearly defined character states, may be further removed from reality (how the evolutionary changes could actually have occurred, due to particular ecological, physiological or mechanical constraints) than are evolutionary trees (see Hull 1979; Charig 1982). Secondly, cladists are required by their methodology to assume that arrangements involving minimum convergence are correct - that convergence has not occurred unless it can conclusively be shown to have occurred (see Patterson 1982; Cain 1982; Charig 1982). For relationships at lower taxonomic levels this may be fine; there is no disputing that cladistic analysis can then be both useful and highly instructive. But it does not work at the phyletic level with which this book is concerned, where characters are so often of an apparently 'simple' nature and where (as should become clear) convergence seems to have been far more common than a practising cladist would care to admit, or can bear to contemplate.

However, in deference to the cladists who will certainly disapprove of the informality of arguments used here, perhaps it will prove that the chapters that follow do at least bring together much of the evidence and raw material on which others, more skilled in the necessary formalities, can build a cladistic analysis to their taste.

Probably by the choice of methodology this book will inevitably have some tendency towards a polyphyletic phylogeny; for it is noticeable that cladists commonly believe groups to be monophyletic where evolutionary taxonomists detect several convergent sub-groups and believe in polyphyletic origins. We should therefore begin with a review of the existing schemes of invertebrate relationships, to see how conceptions of the overall shape of the animal kingdom have varied according to the methodology and premises that different authors have adopted.

1.2 Invertebrate phylogenetic schemes

There is fortunately a reasonably good general consensus about the rank of phylum in the animal kingdom; it is probably the most satisfactory taxon after the species. Each phylum (and there are around 32-6 in all) is recognised as including all those animals sharing a particular body plan (or 'Bauplan'), so that, however much the group has diversified, each of its members can be derived from an ancestor sharing certain diagnostic features. Although it is often very hard to define a phylum satisfactorily, nevertheless its limits can be recognised, and there is usually little dispute as to whether a particular taxon *is* of phyletic status.

This is not always true, of course; some authors prefer to keep the numbers of groups accorded phyletic status low, and will unite groups elsewhere regarded as phyla in themselves. For example the lophophorate groups (bryozoans, brachiopods, phoronids) may be seen as one phylum, instead of three; or the pseudocoelomate groups may be amalgamated as the phylum Aschelminthes. At the same time, others try to split up traditional phyla; the clearest example in recent times is the tendency to regard the arthropods as three (or more) independent phyla. Sometimes these arguments are dismissed as trivial, merely a reflection of the particular author's 'splitting' or 'lumping' tendencies; but they are significant in that they make assumptions about common ancestries that should be properly examined. A phylum strictly should contain all the descendants of the first animal to have that ground plan, and should not contain other animals. Lumping or splitting traditional phyla therefore requires careful argument; similarities previously recognised as homologous must be justified (functionally, and/or by new evidence of actual difference) as merely convergent. Some of the chapters that follow will examine problems of this kind.

However, the major issues in invertebrate phylogeny inevitably concern the super-phyletic groupings (assemblages in which several of the agreed phyla are united with an assumption of common ancestry), and their sequential relationships. These 'super-phyla' will be few in number, and should provide a structure to an overview of the animal kingdom; they may also provide useful receptacles for early fossil groups that may be impossible to place in extant phyla. Since these groupings naturally concern the links between rather different body plans, the evidence on which they are based is bound to be more controversial, and it is no surprise to find intense argument about the criteria involved and the resultant shape of the 'phylogenetic tree' of invertebrates. The starting point for this book must therefore be a quick survey of some current theories about invertebrate super-phyla and their interrelationships.

Traditional schemes

Reference to almost any range of English-language textbooks of zoology is likely to reveal a rather standard underlying phylogenetic approach. The protozoans may or may not be included as a first chapter; thereafter the cnidarians and ctenophores will be covered; then the acoelomate flatworms and nemerteans will appear, with nematodes and their relatives as a sideline, followed by annelids, molluscs and arthropods; and the echinoderms will round off the book. Smaller phyla are slotted in at standard places through the text. If the book contains a chapter covering phylogenetic matters in their own right, it will probably justify this sequence in terms of a 'planula'-type ancestor for the metazoans, with diploblasts preceding triploblasts, and acoelomates giving rise to pseudocoelomates and coelomates. The latter are split into two major groups, the **Protostomia** including the three large phyla Annelida, Mollusca and Arthropoda, and the **Deuterostomia** incorporating echinoderms and a few minor phyla together with the chordates (fig 1.1). A phylogenetic tree will generally be given, similar to the format of the examples shown in fig 1.2. Such schemes originated in the late nineteenth and early twentieth century, with zoologists such as Haeckel, Grobben, and Hatschek; but more recently they owe an enormous debt to the works of Hyman (1940, 1951*a*,*b*, 1955, 1959), reviewing most of the animal kingdom in terms of structure and function and giving the most recent really intensive analysis of the phylogenetic status of each invertebrate phylum.

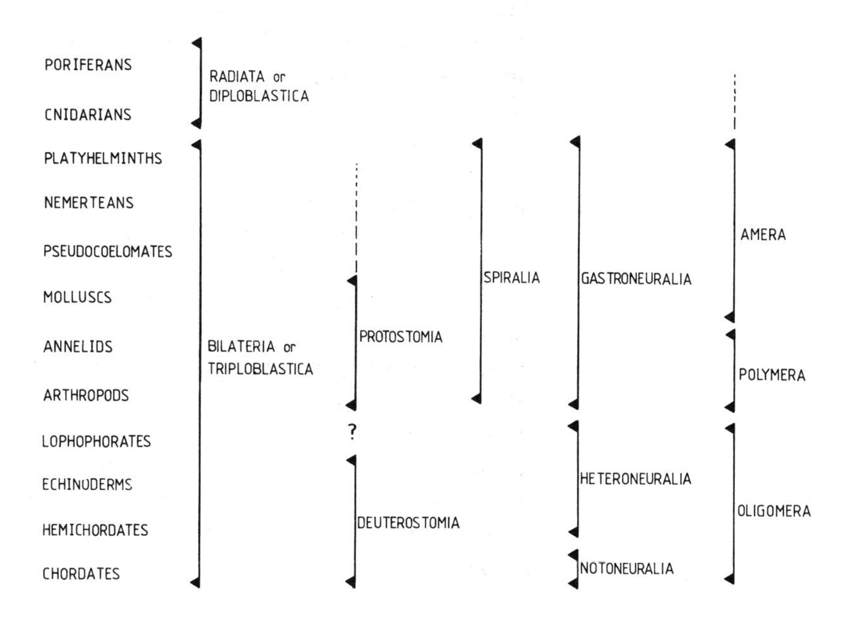

Fig 1.1. Terminology of invertebrate 'super-phyla'.

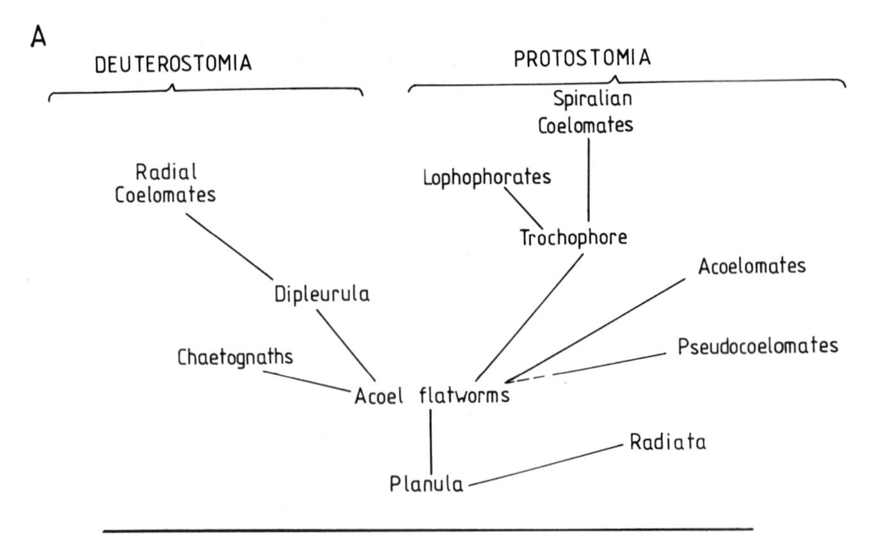

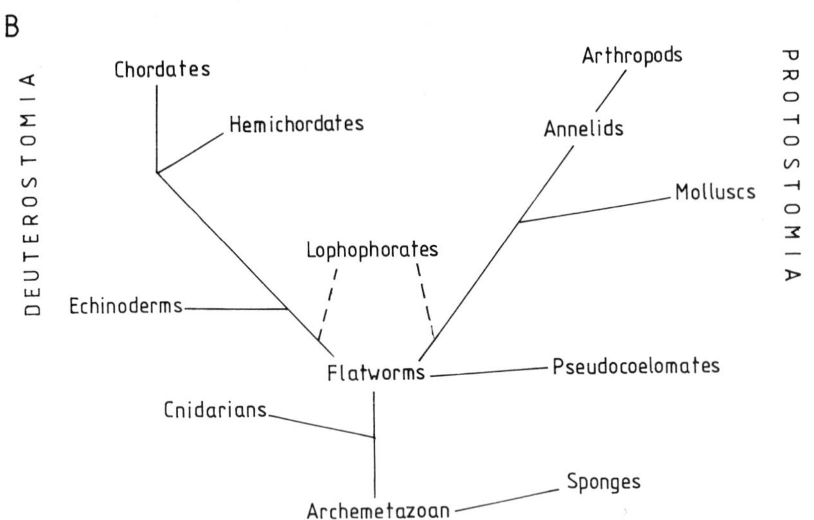

Fig 1.2. Traditional dichotomous phylogenetic trees of the invertebrates, with a major split between protostomes and deuterostomes. *A*, based on Hyman (1940); *B*, based on Barnes (1987).

In essence, the traditional views on either side of the Atlantic have followed or been reinforced by Hyman's synthesis and have therefore been of this dichotomous form. There may be contention about the exact whereabouts within this dichotomy of certain phyla, and even major arguments about the status of some of the larger phyla. Indeed two important changes are now quite frequently met with; molluscs may not really be coelomate protostomes, and arthropods may be an amalgam of at least three independent lines. Nevertheless, the essential structure is given in a rather uniform fashion, and though many caveats may be presented about details and the notorious problems of invertebrate classification, authors will generally state or imply that there can be little doubt about the major distinctions they have drawn.

Archecoelomate views

By consulting other sources, it quickly becomes apparent that not all zoologists are quite as convinced of the essentially dichotomous pattern of animal relationships, with an orderly progression through the grades of acoelomate, pseudocoelomate and coelomate, as is implied above. In particular, there is a long-standing tradition in Europe (and particularly in the German-speaking countries) that has a very different concept of the origins of metazoans and the evolutionary sequences linking major groups. A characteristic German phylogeny based on Siewing (1976, 1980) is shown in fig 1.3*A*.

Rather than taking an acoelomate 'planula'-type creature as the ancestral metazoan, schemes of this type believe the archemetazoan to have been of coelomate organisation, derived from something like a cnidarian and having (usually) three pairs of body cavity pouches budded off from the gut ('enterocoely' - see chapter 2). Since most of the phyla identified as 'deuterostomes' are also of this tripartite enterocoelic coelomic design, they (and perhaps also the lophophorates) take up position near the base of the invertebrate family tree as close relatives of the 'archecoelomates'; and all the remaining groups are derived from them. These derived, traditionally 'protostome', groups are usually denoted by the alternative name **Spiralia** by authors of this school; this refers to their spiral cleavage, another of the

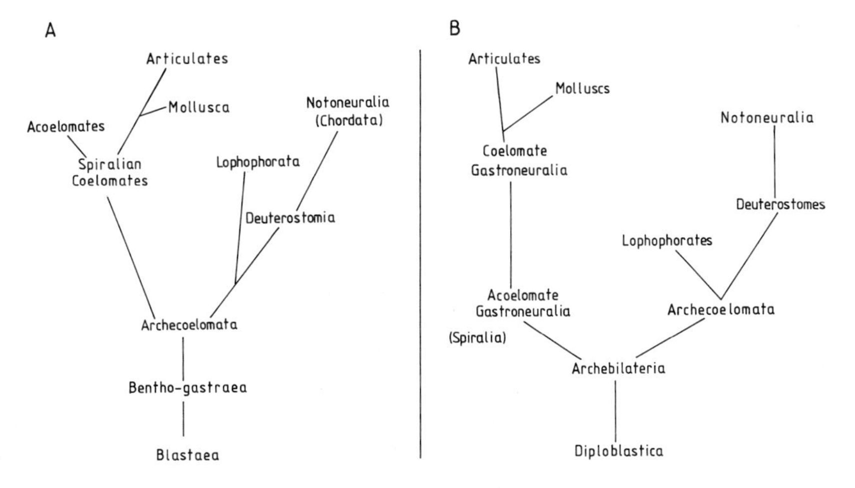

Fig 1.3. Archecoelomate theories concerning invertebrate relationships, with a trimeric coelomate near the base of the animal kingdom; *A*, based on Siewing (1976, 1980); *B*, modified version by Reisinger (1970, 1972).

embryological features (discussed in chapter 5) that are used to define the protostome/deuterostome super-phyla.

For most archecoelomate theorists, then, the tripartite coelomates (deuterostomes) precede and in some sense give rise to spiralians (protostomes). The latter are both of primitive coelomate and also of secondarily acoelomate form; and it is the suggestion that acoelomates (animals with a solid form and no body cavity, such as platyhelminths) are a regressive group, derived from larger coelomate animals with a pronounced body cavity (such as annelids) that some zoologists find hard to believe. Though in these archecoelomate schemes the animal kingdom is still somewhat dichotomous, one branch is an offshoot of the other rather than each diverging from a common trunk. Schemes of this sort often use a whole new set of terminology; the **Gastroneuralia** are roughly equivalent to the protostomes (having a ventral nervous system), the **Heteroneuralia** or **Oligomera** are the tripartite groups, and **Notoneuralia** are the higher deuterostomes including chordates, with a dorsal nervous system close to the notochord; again, the terminology is clarified in fig 1.1.

There is an alternative version of the archecoelomate view, promulgated by Reisinger (1970, 1972), which retains the concept of trimeric archecoelomates, but rejects the more controversial idea that acoelomates are secondary to coelomates (fig 1.3*B*), invoking an ancestral acoelomate form as the starting point in which enterocoely arose. Though Reisinger maintains that archecoelomates are a vital part of his scheme and close to the origin of metazoans, his scheme in fact looks much more like the traditional views of fig 1.2 than like other archecoelomate phylogenies.

Spiralian ancestry schemes

As might be expected, there are also views that take the protostomes or spiralians to be the more primitive of the super-phyletic groupings, with deuterostomes as a derived offshoot. Nichols (1971) developed one such scheme (fig 1.4*A*), with sipunculan worms as the link group between the two traditional branches. Another example is shown in fig 1.4*B*, based on the writings of Salvini-Plawen (1982*a*, 1985), again having the unusual feature of dispensing with a dichotomous tree and making the animal kingdom linear instead. Spiralians are derived from acoelomates, and in their turn give rise to non-spiralian groups; though the new terminology popular with other Germanic writers is adopted, with Notoneuralia as the endpoint of the monophyletic lineage. Salvini-Plawen refers to these categories as 'mega-groups' rather than super-phyla, but the principle is clearly the same as in other schemes.

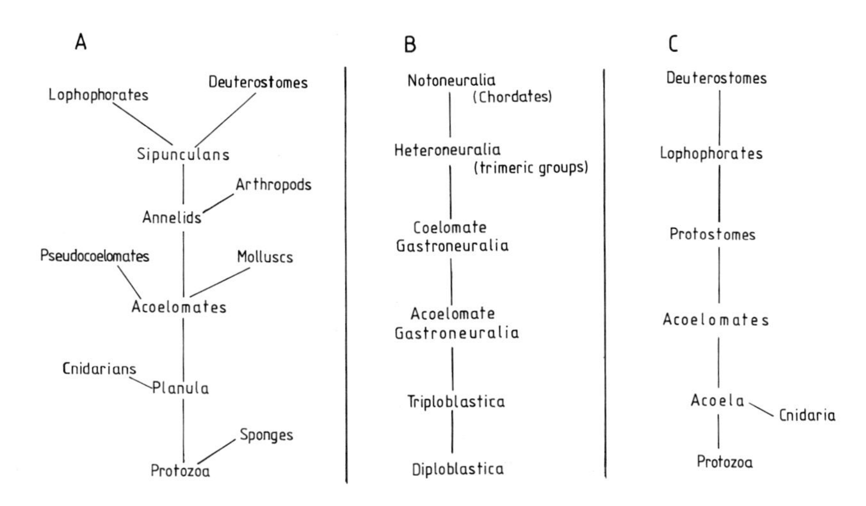

Fig 1.4. Linear phylogenetic trees in which the protostomes (spiralians) precede the deuterostomes. *A*, Nichols (1971); *B*, Salvini-Plawen (1982*a*, 1985); *C*, Hadzi, (1953, 1963).

Hadzi (1953,1963) developed another linear scheme in which protostomes precede deuterostomes, shown in fig 1.4*C*. However, it has the odd feature of commencing with a ciliate protozoan and proceeding directly to a flatworm stage. The reasons for this are discussed further in chapter 7; it certainly leads to a very non-traditional view of relative status amongst the lower metazoan groups.

Yet another scheme having some resemblances to that of Salvini-Plawen, but even more heretical in its approach, has been developed by Gutmann and a team of co-workers (Gutmann 1981; fig 1.5). In effect it retains the central distinction between traditional 'protostomes' and 'deuterostomes', but derives both from a fully segmented spiralian coelomate group (Polymera), reverting to an older idea that chordates were directly evolved from an annelid-like form. (It therefore runs the evolutionary sequence of deuterostome groups 'backwards' relative to most other schemes; chordates as the stem, with echinoderms and hemichordates as later developments - see chapter 12).

One more scheme of this kind has recently been presented by Jefferies (1986), and is reproduced in fig 1.6. It differs from other phylogenies in giving precise and explicit indications of the characters used to define the branching pattern, for it is in fact a cladogram rather than an evolutionary tree, and virtually the only true cladogram yet produced at the level of invertebrate phyla. The characters used to erect the dichotomies in this cladogram are sometimes very traditional ones, common to other schemes (though not necessarily very valid ones, as future chapters will show). But in other cases they are exceedingly peculiar - for example using the presence of an anus to characterise all groups above flatworms, or of a blood vascular system for all

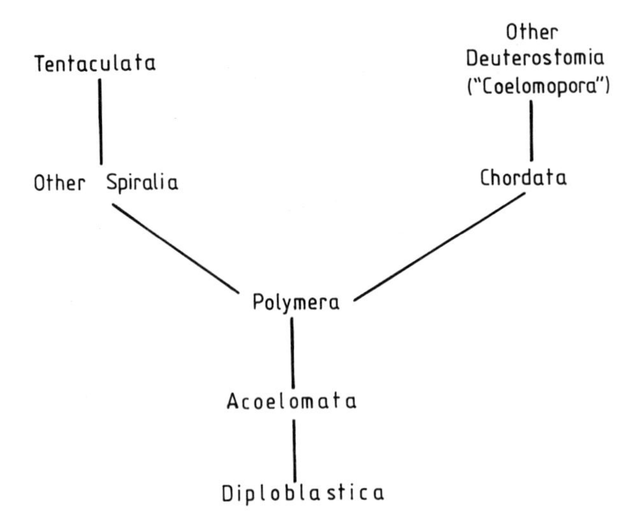

Fig 1.5. The phylogenetic tree devised by Gutmann (1981) with a segmented worm (Polymera) as the rootstock for all coelomate groups including chordates.

groups at or above the nemertean level. Both of these are characters that can be demonstrated with considerable conviction to have been developed repeatedly and convergently (and generally both must go together, as will also be shown). No modern evolutionary taxonomist of invertebrates would use such characters or accept such a scheme. The whole monophyletic sequence that Jefferies puts together over-simplifies the situation of invertebrate relationships

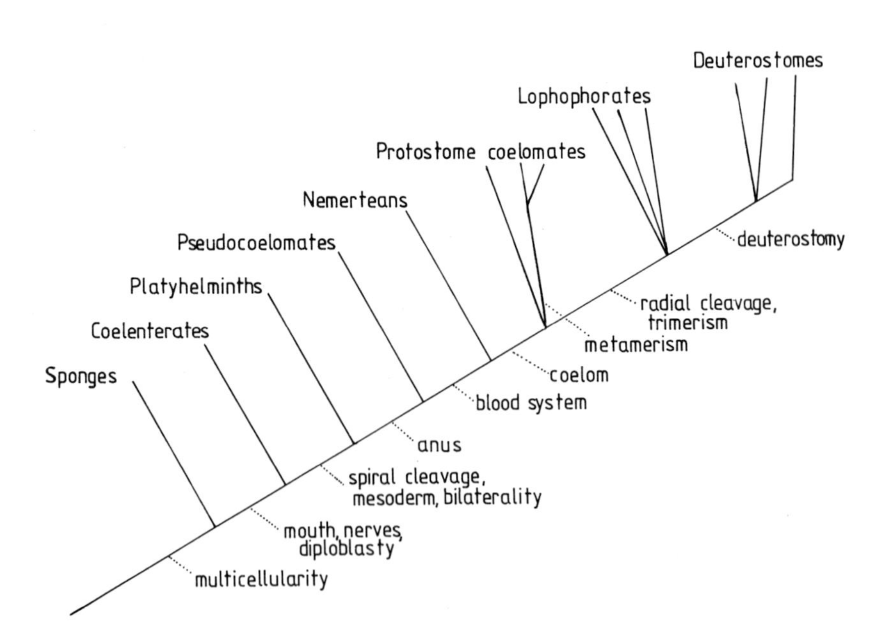

Fig 1.6. Jefferies' (1986) cladogram of the invertebrates.

in a somewhat bizarre fashion; hardly a recommendation for others to use cladistic methodology.

Some alternative schemes

Apart from all the above views, most of which still retain the essential idea that there are two major coelomate super-phyla and a few smaller satellite groups, in whatever arrangement these may be given, a number of less classical approaches to invertebrate phylogeny have been proposed.

A modification of the traditional dichotomous views, much quoted in English writing, is that of Clark (1964, 1979), as shown in fig 1.7. The primary division between protostomes and deuterostomes is retained in that the latter group are specifically separated, as part of a trimeric coelomate radiation. But Clark introduces other groups of equal status, based on the idea that the coelomate groups represent a polyphyletic assemblage in which the coelomic cavity was invented more than once and radiated in various directions largely dictated by mechanical needs; the major divisions are between metameric, trimeric and unsegmented coelomates. This whole scheme owes much to a mechanical/functional analysis of invertebrate design.

Valentine (1973*a*) erected five coelomate super-phyla (fig 1.8) with deuterostomes and lophophorates being joined by three new groups split up from the usual protostome assemblage. This, like Clark's view, was largely based on the segmentation involved; Metameria or Polymera included annelids and arthropods, with numerous segments, Sipunculata or Amera included only

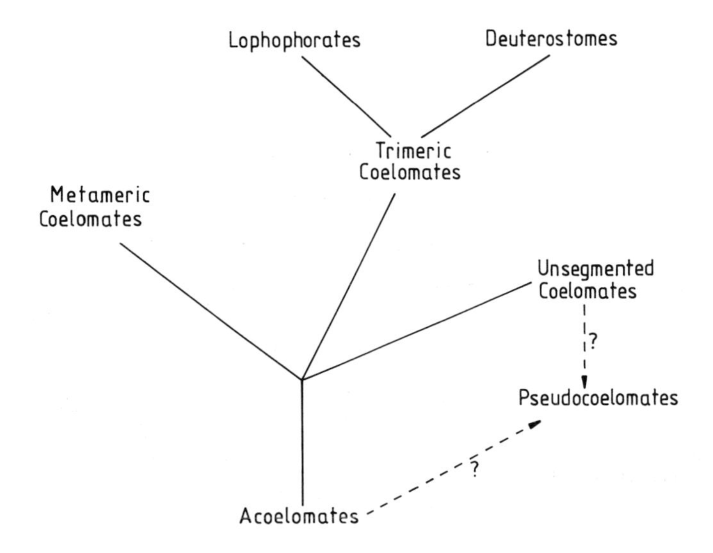

Fig 1.7. A scheme involving multiple branching of coelomate invertebrates, based on Clark (1979).

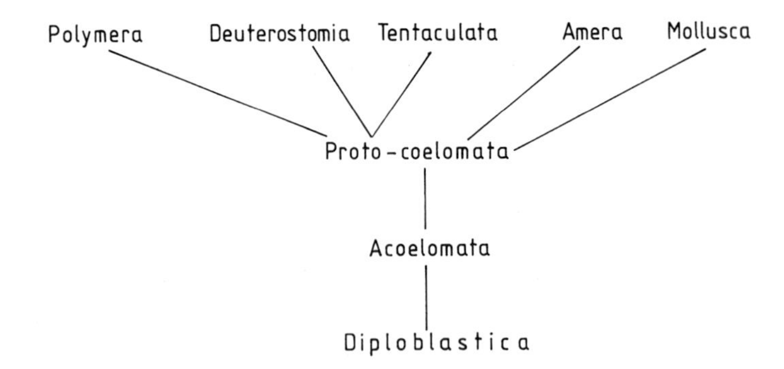

Fig 1.8. Polyphyly of the coelomate condition, with five coelomate super-phyla (Valentine 1973*a*).

the unsegmented infaunal sipunculan worms, and Molluscata encompassed creeping animals with 'seriated organ systems but an unsegmented coelom'; in effect, only the molluscs. However, this scheme is intended to focus attention on adaptive radiations and ecological constraints on building plans, rather than on which groups arose from which others; Valentine maintains that this should be the purpose of establishing super-phyla, rather than merely using traditional morphological criteria.

Polyphyletic views may also be taken to rather greater extremes than has yet been discussed. Several authors have proposed that the metazoans as a whole are polyphyletic, and it could therefore reasonably be proposed that small groups of phyla arose independently from protist stock. One theory of this kind was presented in some detail by Inglis (1985). Another highly polyphyletic invertebrate scheme, advanced by Nursall (1962), is shown in fig

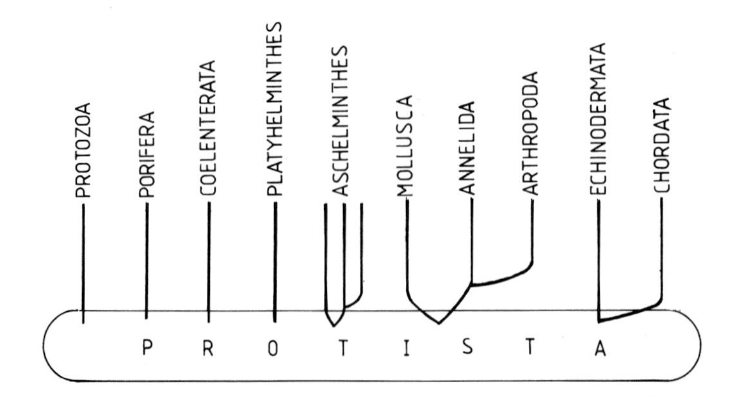

Fig 1.9. A scheme in which most invertebrate phyla are derived independently from the protistan stage, based on Nursall (1962).

1.9; with so many separate origins proposed, it hardly requires the use of 'super-phyla' at all, and the traditional groupings are largely rejected.

1.3 Conclusions

Existing phylogenetic schemes clearly vary enormously, and range from the linear and monophyletic, resembling a mono-axial vine; via dichotomous traditionally 'tree-shaped' schemes, and single-stem but highly radiating 'bush-like' schemes; to the extremely polyphyletic, multiple-origin schemes best likened to a field of grass. An overview comparison of these possibilities, and their consequences for the relative status of major groupings, is shown in fig 1.10. Clearly they are wildly different in their implications, and the very fact that so many opposing views can be held by eminent and knowledgeable zoologists testifies to the problems of phylogeny; they are all attempting to reconstruct a tree with evidence only from the most recent leaves and a few twigs and branches, and inevitably their views of major branchings differ. There can be no absolute proof, or absolute refutation, of any one of these schemes. However, many of the essential morphological features on which all these schemes depend are critically discussed in the next chapter, and it is to be hoped that the further evidence reviewed in parts 2 and 3 of this book will give some clues to choosing between such schemes.

Given the various problems of interpretation and representation in taxonomy, an analysis of invertebrate evolution that takes due cognisance of ecophysiological and mechanical constraints seems, to me at least, to be more useful at present than a potentially more rigorous but probably less readable

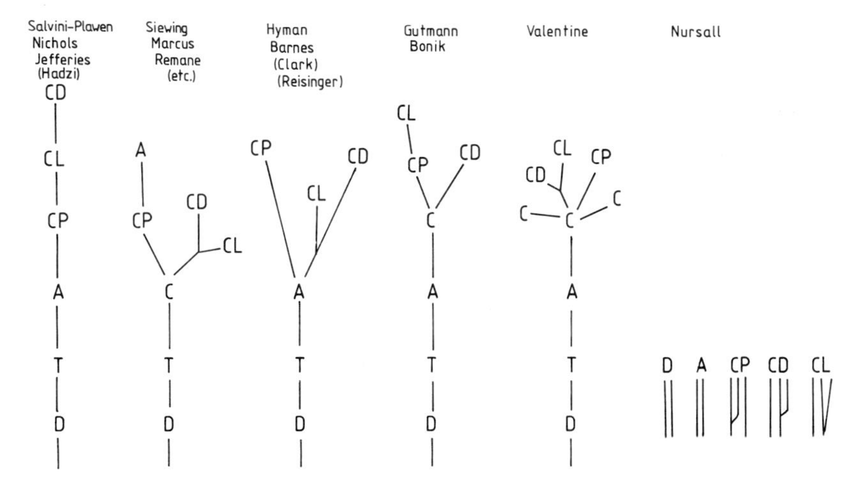

Fig 1.10. An overview of various invertebrate phylogenetic schemes, ranging from the linear 'vine'-like monophyletic sequences through dichotomously branched 'trees' to polyphyletic 'fields of grass': D, Diploblastica; T, Triploblastica; A, Acoelomates; C, Coelomates; CP, Protostomes; CD, Deuterostomes; CL, Lophophorates.

cladistic treatise. Certainly polyphyletic groupings will be avoided, and shared derived characters (synapomorphies) will be used rather than primitive or negative characters; to this extent, the treatment here will be cladistic. But the traditional 'phylogenetic tree' or radiation diagram, where multiple branching is permitted and some indication of timespans may be included, can still aspire to convey information to the reader (and indeed make the reader aware of the gaps in our information!) far better than a formalised cladogram. So this book hereafter adopts a pragmatic approach, permitting a number of features that would be anathema to a cladist (supra-specific ancestors, paraphyletic groups, multiple non-dichotomous branchings, and adaptationist 'story-telling' arguments), in the hope of distinguishing sensibly between all the different possible schemes of invertebrate relationships outlined above.

2 Morphological patterns and traditional divisions

2.1 Introduction

Certain features of basic animal design are critical to all schemes of invertebrate phylogeny and thus underlie, explicitly or otherwise, all zoological textbooks and courses. These are features that contribute to our conventional hierarchical view of animals, giving the sort of schemes summarised in chapter 1: the planes of symmetry that distinguish radiate phyla from 'higher' bilateral groups, the diploblastic nature of 'lower' animals and the added mesoderm giving triploblasty, the presence of varying types of body cavity, or of serial repetitions of structure, and the use of differing systems of skeletal support. Unless we consider the validity of these characters, we can get nowhere in a review of invertebrate phylogeny - they are the very stuff of which theories have been made, the classic comparative morphology against which all other criteria will be judged. This chapter is therefore devoted to a fairly lengthy examination of these topics, and forms both the starting point and in some sense the core of this book.

2.2 Patterns of symmetry

Animals are frequently divided in textbooks into the Radiata and the Bilateria, with the implication that as 'lower' animals are radially symmetrical so the earliest animals must have begun with that pattern. For the present this latter point will be left aside, as the symmetry of the first metazoans will form a central issue in chapter 7. But it is important here to decide if there is a fundamental difference between the so-called radiate animals and the rest of the animal kingdom.

Beklemishev (1969) produced a comprehensive analysis of symmetry in animals, in which he stressed that all metazoans share a fundamental monaxonal symmetry with an axis passing from the oral to aboral poles (fig 2.1*A*) and so giving an element of radial symmetry, this being particularly evident in embryological stages. Even the early blastula exhibits this axis, the animal/aboral pole soon becoming distinguishable by its tendency to lead during locomotion and to adhere when the animal becomes sessile; it is here that the ectoderm will normally form and the nervous system will be centred. The modern radiate animals retained this embryological and larval axis of symmetry (Hatschek suggested the name 'Protaxonia' for them because of this), while all subsequent groups had a new adult axis of symmetry making them bilateral (hence 'Heteraxonia'). Beklemishev believed that whilst all groups at or above the flatworm grade do have adult bilateral symmetry secondarily, many of them still show elements of their primary radiality - as evidence, he cited the nervous systems of flatworms (where there are several separate longitudinal nerves (fig 2.1*B*), rather than a pair ventrally as in most other metazoans), and those of nematodes and gastrotrichs, where a 'head-on' appearance of tri- or hexa-radiality is fairly common (fig 2.1*C*). His view therefore supported strongly the traditional thesis that bilaterality gradually succeeded radiality in the history of animals.

Closer examination of cnidarians and ctenophores reveals extensive manifestation of biradiality, however, as Beklemishev recognised. Anthozoan sea anemones are particularly endowed with an axis of bilaterality, because of the paired siphonoglyphs and the disposition of muscles on mesenteries (fig

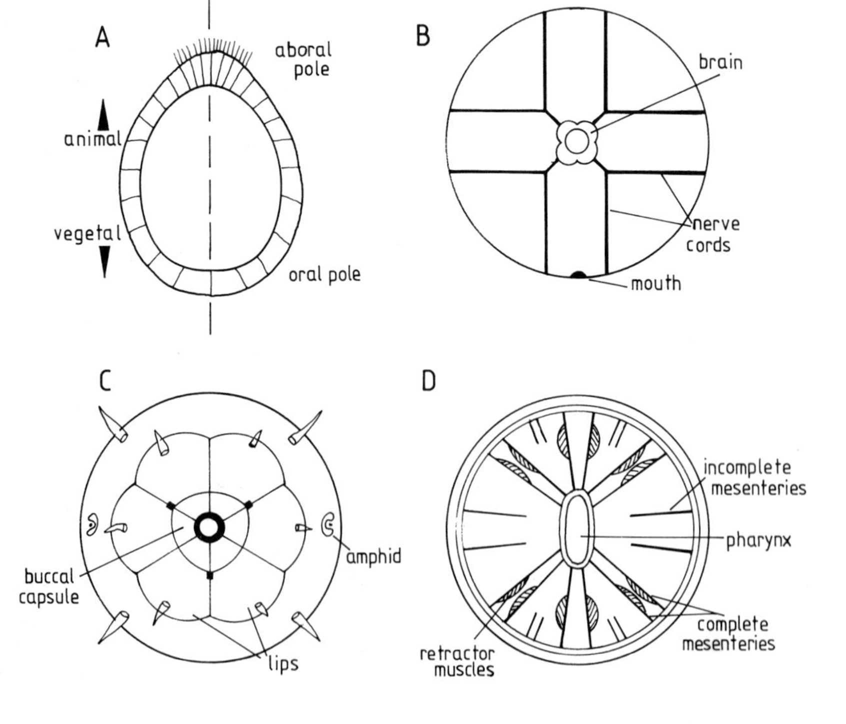

Fig 2.1. Patterns of symmetry in invertebrates, showing the elements of bilaterality present in supposedly radial groups. *A*, the early blastula; *B*, front view of a platyhelminth, showing the arrangement of nerves; *C*, front view of a nematode; *D*, transverse section of an anthozoan sea anemone, with bilateral pharynx and bilaterally disposed muscles on the mesenteries.

2.1*D*). Beklemishev and others have regarded this as a secondary development. But it would be equally possible to interpret the trend as having been from primitive bilaterality (extensively expressed in protozoans anyway, from which descent must be assumed) towards radiality. Full radiality is currently only expressed in some hydrozoans, most obviously *Hydra* itself, which may be the more advanced type of cnidarian (though this is an important issue in its own right, discussed further in chapter 7).

The idea that bilaterality came first would fit well with the ecological reasoning usually given for patterns of symmetry. It is often said that radiality correlates with pelagic life and bilaterality with benthic life, but this is clearly an over-simplification. Radiality is appropriate for a planktonic drifting form, which may be moved in any direction, or for a sessile form which may be assailed by currents or by prey from any direction. Bilaterality is required in an animal that is an active pelagic swimmer or an active benthic crawler across substrates, where a head end normally leads and where sensory capacities are concentrated. Thus either form of symmetry may be appropriate for either pelagic or benthic forms - it depends on how they live, rather than where they live. If cnidarians were initially bilateral their present radiality could readily be interpreted as a consequence of their being sessile as polyps and/or passive planktonic drifters as medusae. Whereas if they were primitively radial it is hard to see why some of their modern forms are often quite markedly bilateral despite their lifestyle. Certainly within the acknowledged Bilateria groups a secondary trend towards radiality on becoming sessile is very common (bryozoans, tunicates and the like; and, historically, the echinoderms, once included in the 'Radiata' but now firmly recognised as secondarily pentaradial). It seems quite likely that the phylum Cnidaria was not primitively radial.

With regard to the other two classically 'radial' groups, sponges and ctenophores, the situation is more clearcut. Sponges are most frequently totally asymmetrical in form (though there is perhaps a radial primary unit associated with each osculum in the simplest types), and their overall growth may be, with about equal frequency, roughly radial (usually in fairly still deep waters), or roughly bilateral (taking on a broad fan-shape when growing across a strong current). They have clear elements of bilaterality during their ontogeny, especially in the larval stages (see chapter 7). It seems illogical to regard them as part of a radial assemblage. Ctenophores are often superficially radial, particularly if they are of the 'sea gooseberry' type (*Pleurobrachia*), but anatomically show significant octoradiate, quadriradiate, and in particular bilateral features, in the disposition of ciliary bands, water canals, sense organs and tentacles (see chapter 7). The animals can best be described as biradiate, and this condition is particularly clearly seen in the more ribbon-like, actively swimming forms such as *Cestus*. This biradiality is established within the very early cleavages during embryology.

It therefore seems unwise to regard the radial/bilateral criterion as a very substantial one in looking for animal relationships. It is quite probable that a biradial condition was primitive, and that transitions between dominant radiality and dominant bilaterality have taken place repeatedly in most phyla. There is little justification for regarding three of the 'lower' phyla as being fundamentally different from all others by virtue of a dubious 'primary' radiality, and indeed we will see in chapter 7 that there is no real justification for believing that any of these three groups - sponges, cnidarians and ctenophores - are in any sense more closely related to each other than to the rest of the Metazoa. The whole concept of the Radiata is unhelpful, as most textbooks now recognise. Therefore the currently prevalent use of the term Bilateria for the other major animal phyla (a trend evident in much of the German literature of the last two decades) must be equally suspect, and we should avoid these terms in anything but a purely descriptive sense.

2.3 Germ layers

The idea that animals are built in a layered fashion has been current for many centuries, but was only formalised in the latter half of the nineteenth century by embryologists such as Huxley, Lankester and Haeckel. The germ layer theory visualises three basic layers of animal tissue, and a correspondence or homology of these layers throughout the animal kingdom; thus it allows direct comparisons between the tissues of all animals, even where layering is obscured in the adult, and this has been extremely influential in shaping our views on animal relationships.

In its usual form, the theory suggests that just two primary germ layers are formed, by the process of gastrulation - an outer ectoderm, and an inner endoderm - and that the mesoderm is subsequently derived from one or both of these layers and lies in an intermediate position to give the typical triploblastic condition (shown in an idealised form in fig 2.2).
Each layer has characteristic properties and gives rise to a characteristic set of tissues or organs, as follows
(*a*) **Ectoderm** is usually derived from the upper (animal) pole in the embryo; it surrounds the outer surface of all organisms, and may line the fore and hind guts and certain other ducts. It has a protective function (and may secrete specific protective layers), but also gives rise to the nervous system and sense organs, and performs a role in ciliary/flagellar locomotion or in producing hydrokinetic effects (creating water currents, for example in filter feeders).

Because of this last function, Beklemishev (1969) gave it the alternative name of 'kinetoblast'.

(*b*) **Endoderm,** deriving from the lower (vegetal) pole in most embryos, forms the gut lining and certain organs derived from the gut; hence Beklemishev dubbed it the 'phagocytoblast'.

(*c*) **Mesoderm** may be derived from either of these two primary layers (and is thus sometimes called a secondary germ layer), but most of it normally comes from the endoderm.

A traditional view (attributed to Hertwig in the 1880s, and discussed by Hyman 1940, 1951*a*) held that a limited amount of loose 'mesenchyme' tissue was usually produced by the ectoderm (therefore called 'ectomesoderm'), and a greater amount of 'mesothelium' tissue was formed by the endoderm (hence 'endomesoderm'); but this view has been questioned by more recent embryological workers, and there are certainly many exceptions to it. A recent critical review of the nature of mesoderm is given by Starck & Siewing (1980). They stress that 'mesenchyme' is the better term for internal tissues of multiple and complex origin, appearing at different phases of development and in different fashions, and plead that the term 'mesoderm' should be reserved for direct derivatives of endoderm. A similar view is taken by Salvini-Plawen (1980*a*), retaining the 'ectomesenchyme' versus 'endomesoderm' distinction.

This involves a quite separate distinction from that of the more familiar (largely vertebrate-related) dichotomy between somatic mesoderm

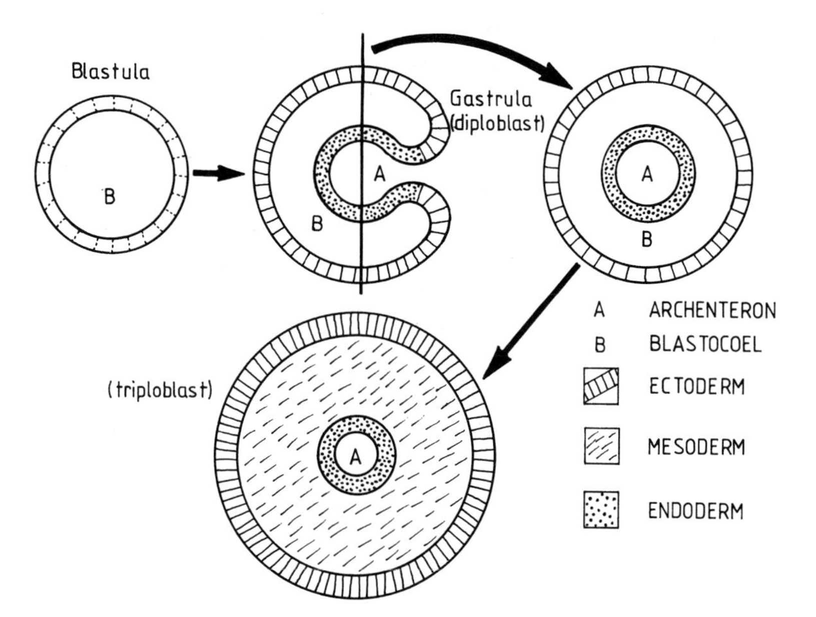

Fig 2.2. Schematic view of the three germ layers and their interrelations with the primary body cavity (blastocoel) and gut (archenteron).

(mesodermal tissues abutting on the ectoderm) and splanchnic mesoderm (abutting on the gut) in those animals where the mesoderm is split by a body cavity.

Mesoderm tissues, whatever their origin, form the bulk of the body material in most metazoans, including connective and vascular tissues, gonads, many excretory systems and most of the muscle; and in some animals mesoderm also forms endoskeletal tissues such as bone.

There are a few obvious problems with this view of animal tissues. Until gastrulation, the cells of an embryo are not noticeably differentiated into types and are not layered, but once the blastula converts into a gastrula there is generally an inner and an outer set of cells, and germ layers should be definable. But actual homology of ectoderm and endoderm is not easy to establish, because of the variety of ways in which gastrulation can occur (discussed in more detail in chapter 5). If the posterior cells invaginate inwards, or the anterior ones grow over them, a reasonable distinction between ectoderm and endoderm occurs and the tissues are spatially homologous in different organisms; but if all parts of the blastula bud off cells into the central space (gastrulation by delamination) then there is no homology and *any* embryonic cell may be producing both ectoderm and endoderm. Calling all these modes of forming the embryonic layers 'gastrulation' at all is a relic of Haeckelian theories of animal development, and calling all the end-products ectoderm and endoderm may be equally spurious - the endoderm of different organisms may have quite different origins. With regard to mesoderm, the situation is even more muddled, since this tissue may be formed in various different ways, and at different times, and may even have multiple origins within one organism (see Hyman 1940; Beklemishev 1969; Starck & Siewing 1980; Salvini-Plawen & Splechtna 1979; Salvini-Plawen 1980*a*). Clearly then, the germ layers cannot be regarded as homologous in their formation or topography. Nor are they very clearly defined as regards potential for differentiation, since recent embryological work demonstrates conclusively that virtually any germ layer, if taken at a young enough stage, will form any organ or tissue when grafted into particular spatial relations with the rest of an organism.

But it remains true that the fates of germ layers are definitive in the course of normal development. In this sense the theory is a useful guide to phyletic relations; in fact it is remarkable that there should be such strong correspondence between the fates of the three germ layers after gastrulation and initial mesoderm formation even in very dissimilar animals.

One remaining issue relating to germ layers concerns the difference between groups possessing mesoderm (**triploblasts**) and those supposedly without it - the **diploblasts**. Conventionally, sponges, cnidarians and ctenophores are accorded diploblastic status (see Pantin 1960), with an acellular mesogloea between ectoderm and endoderm. These groups are often described in elementary texts as being at an entirely separate gastrula-like level of development, which, along with being radiate and being organised only as tissues rather than as organs, leads them to be viewed as evolutionarily 'early' and somehow inferior. However, it is increasingly recognised that this diploblastic/triploblastic criterion leads to a rather artificial division of animals, and that of all these groups only the class Hydrozoa of the cnidarians could in fact truly be described as diploblastic (e.g. Hyman 1940). All the other animals have more or less extensive invasion of the mesogloea by cells (from the ectoderm principally, in most cases); and given the multiple origins of mesoderm in other animals it would be somewhat unreasonable to deny these cells mesodermal status. Some textbooks have therefore dropped the term diploblastic altogether, and perhaps it is no longer a helpful concept, tending to unite in our thoughts three phyla that bear little relation to one another (again, see chapter 7). Rather, we should see the lower metazoans as an assemblage having either relatively sparse or somewhat more obvious complements of mesoderm, probably of an ectomesodermal type in cnidarians and ctenophores and mainly endomesodermal in flatworms. Salvini-Plawen (1980*a*) treats this as a justifiable distinction, but stresses that it should be seen as part of a gradually increasing differentiation of the mesenchyme which is expressed further in the 'higher' worms (see fig 2.10). Therefore, just as there is a continuum of degrees of radiality so there is of triploblasty, and on no logical grounds can the cnidarians and ctenophores be set widely apart from other metazoans.

Thus outdated and excessively recapitulationist views can be misleading in suggesting that there is a basic split between two-layered and three-layered animals. Henceforward the old distinctions between mesogloea, mesenchyme and mesothelium will therefore be largely avoided except as descriptions of different structural types of mesoderm tissues, carrying no embryological implications. Nevertheless it is fair to conclude here that accepting the other concepts and consequences of the germ layer theory seems likely to be a help rather than a hindrance in looking for phylogenetic patterns; and a consistent system for designating the three major tissue types, as in fig 2.2, will be adopted throughout the book.

2.4 Body cavities

The vast majority of animal phyla have acquired some form of body cavity, a fluid-filled space within the framework of cell layers already discussed; and the designation of these has always been a linch-pin of comparative morphology and phylogeny in invertebrates. A simple view of what is meant by the normal terminology of body cavities, and the relationships of each type to the germ layers, is shown in fig 2.3. A brief summary of the conventional status of body cavities in different animal groups is given in table 2.1, to serve as an introduction to what follows.

Functions of body cavities

A fluid-filled cavity that takes up a relatively large space within any animal's body, and within which fluids move and circulate, may serve a number of functions, and have far-reaching effects

(1) The fluid acts as a simple distributive system, supplying nutrients and respiratory gases to the tissues. With a relatively thin outer body wall through which oxygen may reach the body cavity fluid, and a central gut from which foodstuffs are absorbed, all cells of the body are within a short diffusion path of the nutritive fluids, which may be kept in circulation by body movements. Similarly all cells can deposit their wastes into the body cavity.

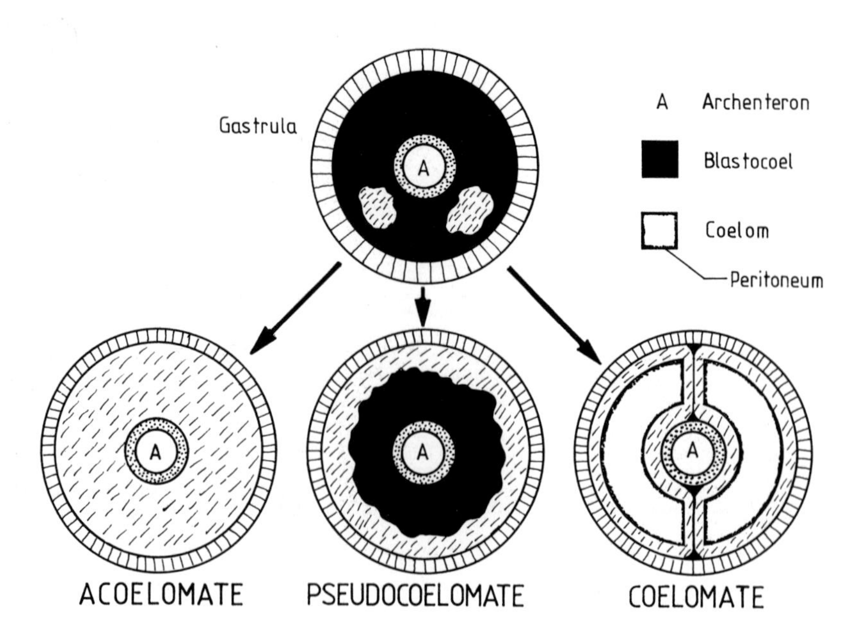

Fig 2.3. The three principal types of animal design in relation to body cavities, and their derivation from an idealised gastrula.

Table 2.1 *Body cavities in Metazoa*

Acoelomate		
	Porifera	Placozoa
	Cnidaria	Mesozoa
	Ctenophora	
	Platyhelminthes	Nemertea
	Gnathostomulida	
Pseudocoelomate		
	Rotifera	Nematoda
	Nematomorpha	Acanthocephala
	Gastrotricha	Kinorhyncha
	Entoprocta	Loricifera
	Priapulida?	
Coelomate		
	Priapulida?	Sipuncula
	Mollusca*	Annelida
	Echiura	Pogonophora
	Tardigrada*	Onychophora*
	Chelicerata*	Crustacea*
	Uniramia*	Pentastomida*
	Phoronida	Bryozoa
	Brachiopoda	Chaetognatha
	Echinodermata	Hemichordata
	Chordata	

*with dominant haemocoel, coelom reduced

(2) Arising directly from these effects, the animal can be of considerably larger size, being freed from the restriction on diffusion paths of the acoelomate condition which generally keeps body dimensions in the millimetre range.
(3) The large fluid-filled space provides a storage area. Complex organs may be suspended within it, and may have independent movement as in the case of hearts. Most obviously the gut is kept separate from the body wall, (particularly important where the two are undergoing peristaltic waves independently and often in different directions).
(4) A further direct consequence of circulating fluids, and the volume they occupy, is that the multiple organs required to serve all parts of the acoelomate

body can be reduced to a single structure (or more commonly a single pair). The most obvious example is the loss of the numerous scattered flame cells of flatworms and their replacement by one or two larger and more complex excretory/osmoregulatory organs, extracting all the wastes dumped into the body cavity from all over the animal.
(5) Independent growth of organs is also possible, especially in the case of gonads which may change size seasonally. Storage of gametes is thus permitted, allowing synchrony of release when suitable conditions occur and/or suitable partners are found. Hence greater fertilisation can be achieved and reproductive success enhanced.
(6) The cavity as a simple circulatory system may allow greater coordination by use of circulating hormones, again important in growth, metamorphosis and reproduction.
(7) The body cavity fluids serve as an efficient hydrostatic skeleton, allowing faster movements and greater shape change than the parenchyma tissues of acoelomate worms. Burrowing can become a major way of life, and swimming and crawling movements are also facilitated.
(8) Eversible structures, such as the proboscis as found in sipunculans, annelids and priapulids, become feasible, and feeding methods can diversify.

Thus body cavities have profound effects on the size, locomotion, and life histories of animals. The hydrostatic effect of a cavity is often taken to be of central and primary importance (Clark 1964, 1969), in particular because it allows much more effective penetration of substrates and the burrowing mode of life. Certainly the greater forces that can be generated, and the greater shape change (allowing efficient anchorage while burrowing) are vital improvements, and acoelomate animals are poor burrowers (though often small enough to be interstitial infauna instead). However, it could reasonably be argued that, rather than the mechanical/hydrostatic advantages of a cavity, the quantum increase in size permitted by body cavities is in itself the most important factor of all; very small animals cannot actually benefit from a hydrostatic skeleton anyway and are adequately served by ciliary locomotion, so that until the size increment is effected the mechanical advantages will be irrelevant. Clark maintains that a body cavity (coelom) could only have evolved in relatively large animals because only then would it be of hydrostatic use; but does not address the problem of how these animals could *be* large without having a body cavity in the first place. In an important sense, then, a body cavity is primarily beneficial by virtue of the size increase it allows (giving less risk of predation by other small creatures, a greater size range of foodstuffs that can be handled, and increased possibilities of homeostatic control and independence of the environment), with other advantages immediately accruing.

All the advantages listed, it must be noted, apply to *any* body cavity, whatever its embryological origin; though these features are often quoted purely as advantages of the coelomate condition. A fluid-filled space inside the body always brings profound advantages, and coeloms have no particular inherent merit; the convention that pseudocoelomate is 'better' than acoelomate, and coelomate 'best' of all, is a particularly inappropriate one. But to prove this, we need to look at the definitions of these conditions and the distinctions between them, and assess the status of the animal groups that, according to table 2.1, belong in each category.

Traditional types and distinctions

In terms of gross definitions, there are three main types of body cavity, distinguished by their embryological origins and site within the body. The distinctions between them were most clearly elucidated by Hyman (1951*a*) and have formed the basis of most invertebrate phylogenies throughout this century; so we should begin by considering this classic position.
(1) The **archenteron** and its derivatives (fig 2.4), formed by intucking during embryological gastrulation, (and hence sometimes called the **gastrocoel**); it opens to the outside via the blastopore. In radiate groups this cavity is commonly termed the **coelenteron**, whereas in bilateral phyla it is the **gut**, and may be blind-ending and diverticulate or a double-ended through-tube. The fluid contained within it is technically part of the outside world, and therefore not a 'true' body cavity; but in most animals it can be temporarily closed off by contraction of sphincters (as in cnidarians). A simple gut cavity may thus perform most of the eight functions listed above.
(2) The **blastocoel** and its derivatives (fig 2.5), the **primary body cavity** formed within the hollow embryo at the blastula stage, having no real lining of its own and no opening to the exterior. This cavity may persist directly as a spacious **pseudocoel**, in those animals where only limited mesoderm is produced to invade the embryological space with cells (fig 2.5*B*). Or more commonly it may be almost obliterated by tissue growth, technically persisting only as the interstitial fluid around the mesoderm cells that have filled in this cavity in acoelomates. There is thus a range of possibilities. A good example of a pseudocoel is found in the nematodes (though some authors describe this cavity as formed of the vacuolated remnants of a few large cells, giving it only a tenuous link with the blastocoel). A good intermediate is the relatively fluid parenchyma of nemerteans, technically 'acoelomate' but with considerable fluid-filled space in practice. And a good 'real' acoelomate is the turbellarian flatworm, with fairly densely packed cellular mesoderm and little evidence of a residual primary body cavity.

In certain phyla this largely obliterated blastocoelic cavity could be said to persist in a more organised state, forming the basis of a **blood vascular system** (BVS). This can be seen particularly clearly by considering annelid development (fig 2.5*C*), where a coelom forms in repeated blocks of mesoderm and grows outwards, progressively obliterating the primary cavity, but leaves vestiges of the blastocoel between the mesoderm blocks to form the characteristic segmentally organised blood vascular system. Most blood vascular systems form in this manner, even in non-segmented groups, and are directly traceable to the blastocoel. (One exception that should be noted is the process giving rise to the anomalous closed vascular system of nemerteans which appears in a manner rather similar to a coelomic cavity - see chapter 8.)

The blastocoelic spaces that form the blood system may reassert themselves even further in some coelomate groups where the coelom is reduced, perhaps secondarily from an annelid-like state, so that a spacious **haemocoel** results (fig 2.5*E*), as in most of the arthropods. Equally, a haemocoel could arise

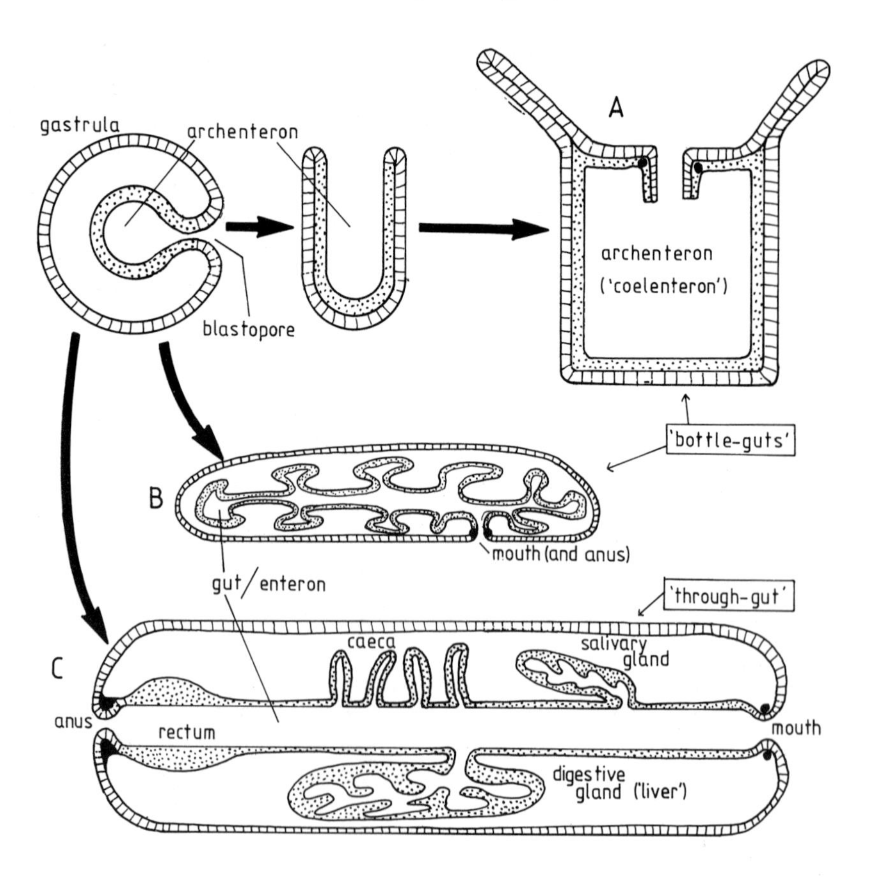

Fig 2.4. The archenteron and its fate in different animals. In cnidarians it remains as a spacious cavity (coelenteron), and in flatworms it becomes the highly branched gut; both of these can be described as bottle-guts. In most other animals it becomes a through-gut with mouth and anus.

directly in acoelomates by retaining more fluid within the developing mesoderm and organising it into a relatively spacious cavity (an extension of the nemertean 'fluid parenchyma' condition). This primary haemocoelic status (fig 2.5*D*) may be found in the molluscs, as we shall see in chapter 10, and it is technically the same thing as the classical pseudocoelomate condition.

(The blastocoelic nature of blood vascular vessels and haemocoels was first discussed by Lang some eighty years ago, but some subsequent authors have maintained that the BVS should more properly be viewed as a de novo development within the extracellular matrix, with primitively unlined vessels (see Ruppert & Carle 1983). Hence the blood vessels are in the site of the embryonic blastocoel, but are new features independent of it, and not strictly a persistent manifestation of it. This is probably a technically better view of the situation, given that some phyla are coelomate but do not retain significant vestiges of the blastocoel as a BVS - for example, the sipunculans. Nevertheless it is useful to realise the spatial relationship between blastocoel, haemocoel and BVS.)

All these terms, then, usually treated quite separately in simple zoology texts, are in fact roughly equivalent and might even be regarded as embryologically homologous. The pseudocoelom of, for example, a nematode (often regarded as some kind of special aberration in the animal kingdom, and carrying the nomenclatural implication of being improper), is technically much the same thing as the haemocoel of an insect or mollusc, the fluid between the mesoderm cells of a nemertean, and the blood within the vessels of an annelid or chordate. More importantly, we should note that *all* triploblastic phyla can

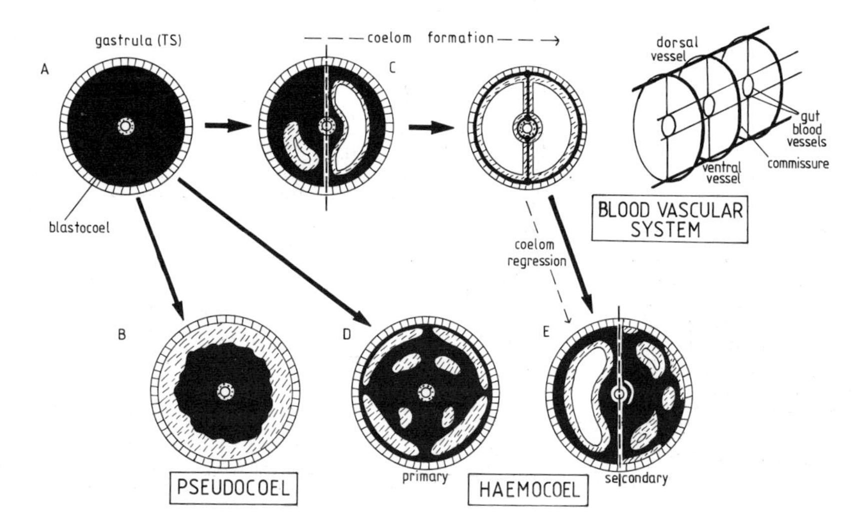

Fig 2.5. The fate of the blastocoel, remaining spacious in pseudocoelomates *(B)* but being progressively obliterated by mesoderm and coelom growth in other groups such as annelids *(C)* to leave only remnants as a blood vascular system. In arthropods the coelom subsequently regresses to produce a secondarily enlarged blastocoel as an open blood system or haemocoel *(E)*; whilst in molluscs the haemocoelic condition may be reached directly *(D)*.

be seen as having some form of this primary body cavity, whether it makes up most of their volume or is virtually obliterated by tissue growth - it is not unusual, or primitive, or inferior. It may, indeed, be 'primary', in its embryological derivation; but even this is open to debate, as some authors regard the pseudocoelom as an example of paedogenesis, the traditional pseudocoelomate phyla being derived secondarily from a coelomate ancestor by retention of the embryonic condition into adult life (see chapter 9).

(3) The **coelom** (fig 2.6), the **secondary body cavity** formed within the developing mesoderm. To qualify for this designation, a body cavity must conform with the minimum definition of a fluid-filled space totally bounded by mesoderm, and lined by a specific cellular mesothelial layer called the peritoneum, which commonly also forms mesenteries suspending the body organs within the cavity. A coelom will often be associated with gamete-producing cells, and usually has a duct ('coelomoduct') communicating with the outside world. Normally it is organised laterally, with right and left halves and a major central mesentery, resulting from bilaterally symmetrical mesoderm growth.

In extant animals, there are conventionally two ways of forming a coelom (described in fuller embryological detail in chapter 5); either by a split occurring within a clump of mesoderm cells (fig 2.6*A* - **schizocoel**), or by pouching from the gut endomesoderm (fig 2.6*B* - **enterocoel**). These methods are part of the means (together with blastopore fate, cleavage patterns, larval forms and some other features) of distinguishing the two super-phyla, Protostomia (Spiralia) and Deuterostomia; matters discussed in

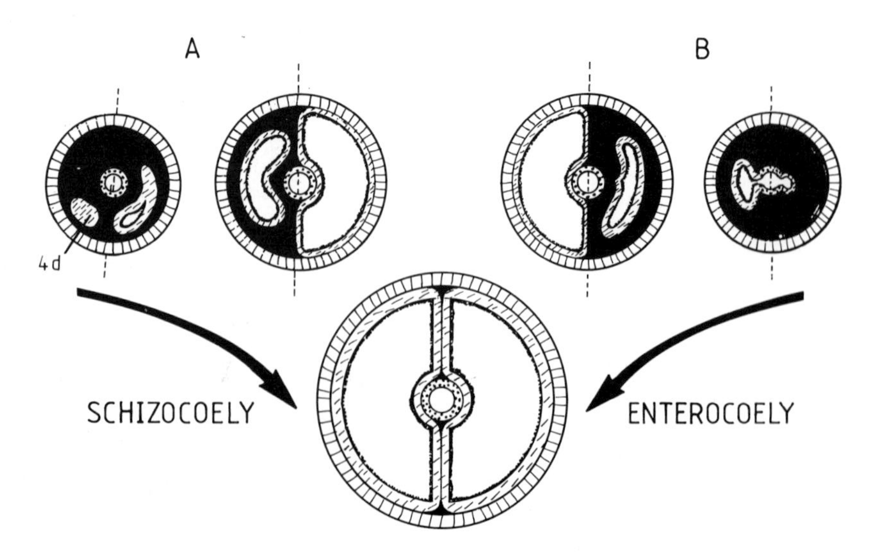

Fig 2.6. The two principal modes of coelom formation, by splitting within the mesodermal bands *(A)*, or by pouching off the gut wall *(B)*. Both produce similar large cavities organised in right and left halves with a central mesentery.

more detail in chapters 5, 8, 12 and 13. Until recently it was widely assumed that these two methods of coelom formation represented a divergence from a single original method (see Beklemishev 1969, or Clark 1964, 1969, for a comprehensive review of these issues). The two most important theories suggested either that the coelom was formed from the cavities of gonads, remaining in an enlarged condition after the shedding of gametes (fig 2.7); or from pouches of the gut that became sealed off (fig 2.8). A brief resumé of evidence for and against these two theories ('gonocoel theory' and 'enterocoel theory' respectively) is given in table 2.2. There is little point in repeating here at great length the controversies that are set out clearly in other texts; useful summaries of the arguments can be found in Remane (1963*a*) and Hartman (1963), and in Clark's papers. The main issue is that either theory has difficulty in accounting for the occurrence of two distinct modes of coelom formation in modern animals. If gonocoely were the primitive mode, why do animals now mainly exhibit schizocoely or enterocoely, where in either case an association with the gonads is not necessarily evident? And if coeloms originated by enterocoely, whether in a cnidarian-like animal with septa or in a diverticulate flatworm form, where did schizocoely come from in modern protostomes? Beklemishev (1969) argues that schizocoelic animals derive their

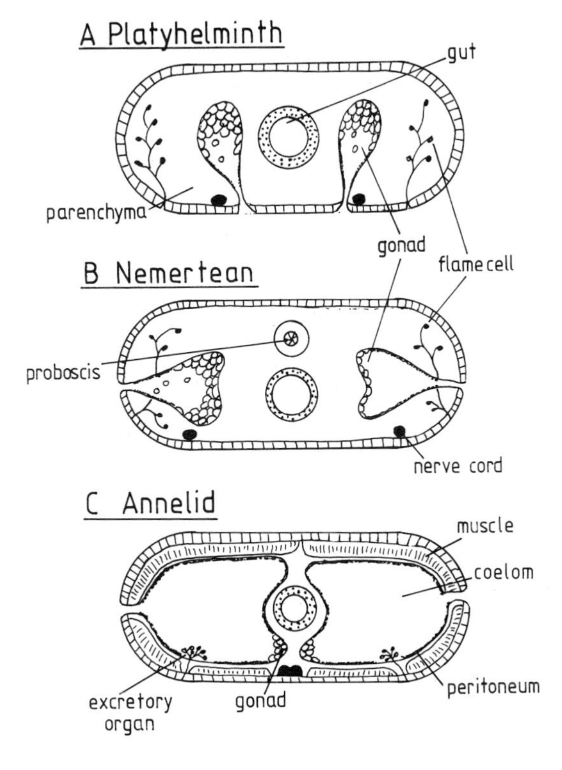

Fig 2.7. The 'gonocoel theory' of coelom origins, where the gonad cavity is thought to persist after the gametes are shed, in a platyhelminth or nemertean; in annelids the gonads are still associated with the coelomic wall (modified from Goodrich 1945).

Table 2.2 *Gonocoelic and enterocoelic origins of the coelom*

A. Problems with gonocoel theories
(i.e. the view that the coelom originated from the serially repeated gonads of lower worms, the cavity of each half of a segment in annelids corresponding to the gonad cavity of a turbellarian or nemertean, left behind and enlarged after gametes were shed; the coelom epithelium, or peritoneum, is the gonad wall, and the coelomoducts are the genital ducts)

(1) Although gonads are commonly associated with the coelom cavity, the coelom usually appears *before* the gonads in ontogeny. In some annelids the genital cells are formed anteriorly and migrate back later into the coelom walls of post-larval segments.
(2) It is implausible that gonads in a small acoelomate would enlarge, and not regress, after shedding of gametes.
(3) Many simple animals die immediately after reproducing, so would never acquire their coelom and body-wall muscles.
(4) Coelom formation is made coincident with segmentation (the gonads of acoelomates being serially repeated) yet there are clearly unsegmented coelomate groups.
(5) Mesoderm formation is supposed to be separate from subsequent cavitation of the gonads; modern enterocoely, where coelom and mesoderm arise together, is thus hard to explain.
(6) Cells from which the mesoderm arises should produce only coelom and gonad, yet in flatworms they also produce parts of the intestine and the mesenchyme that fills the body.

B. Problems with enterocoel theories
(i.e. the view that the coelom formed from pouches off the gut, closing off into separate cavities, either at an anthozoan stage or in a serially repetitive flatworm)

(1) They rely heavily on somewhat improbable comparisons of modern anthozoans or turbellarians with modern segmented coelomate worms.
(2) The anthozoans that best fulfil the requirements of the theory are the cerianthids (where the sequence of septal growth bears some resemblance to the sequence of metameric growth in worms), yet these are so specialised as to be unlikely ancestors of other groups.
(3) It is necessary to assume that unsegmented coelomates are secondary derivatives of segmented forms; and, if the coelom arose in anthozoans, that acoelomate worms are secondary to the coelomate worms.
(4) The selective advantages of sealing off parts of a pouched gut (presumably developed originally to give adequate digestive surface) are difficult to visualise.
(5) The theory is hard to apply to schizocoelic protostomes, where the coelom never bears any relationship to the gut; though some have argued that schizocoely represents proliferation and splitting of gut tissue, rather than pouching of it, and could readily be derived from enterocoely.

coelom from splits in derivatives of the 4d cell that was originally endodermal, so this method could easily be derived from enterocoely - but this seems a rather tenuous and unhelpful notion. More recently Salvini-Plawen (1980*a*) has discussed how deuterostomy could be seen as derived from protostomy, with the lophophorate groups (where in some cases an intermediate mode of formation occurs) as a convenient link. This is perhaps a more attractive proposition than accepting, as so many European authors have done, that enterocoely was the primitive mode and that enterocoelic deuterostomes were the more primitive of the two major 'super-phyla'.

These problems can never be resolved conclusively, and arguments seem rather pointless. Since Clark's analysis of the whole problem was published (1964), the obvious compromise position that he advocated has become much more widespread: if no one primitive mode of coelom formation can account for the currently diverse situation, the sensible and parsimonious view is that coeloms must be polyphyletic, particularly given the dramatic structural and functional differences between extant 'coeloms' and the presumed evolutionary simplicity of acquiring a cavity. Some groups of animals invented coeloms by mesodermal splitting, and others by gut pouching. Others again perhaps used other methods; for example the lophophorates, already briefly mentioned and discussed further in chapter 13; some echiurans and some annelids, that appear to epitheliarise part of the mesenchyme around lacunae (forming a 'syntecocoel', Salvini-Plawen & Splechtna 1979); and perhaps the nemerteans (chapter 8), normally described as acoelomate but with a rhynchocoel operating the eversible proboscis that is by all reasonable definitions a coelom (Remane *et al.* 1980; Ruppert & Carle 1983). In a most important sense, then, coelomates are not a cohesive group (see Salvini-Plawen & Splechtna 1979) and coeloms are no more similar to each other than

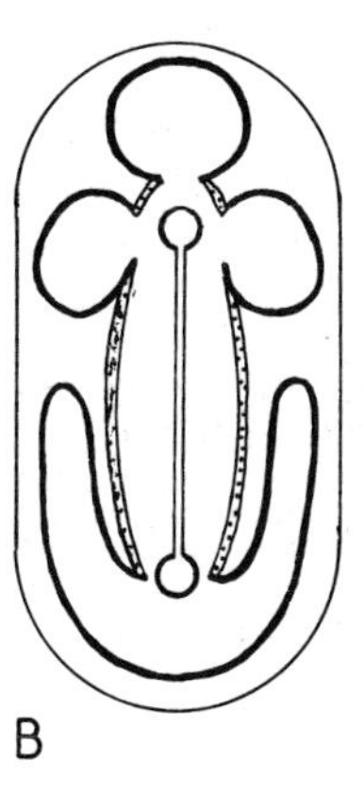

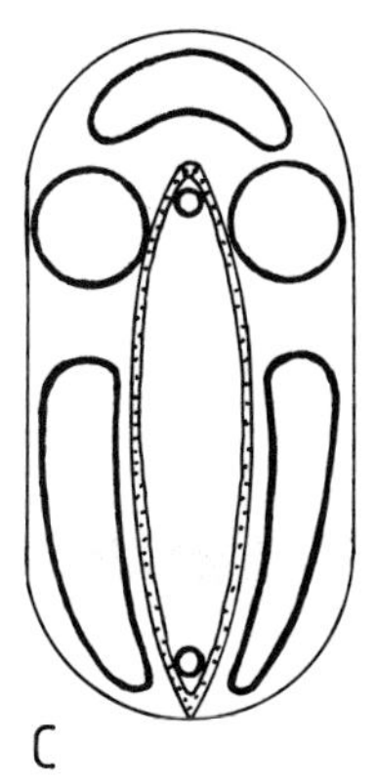

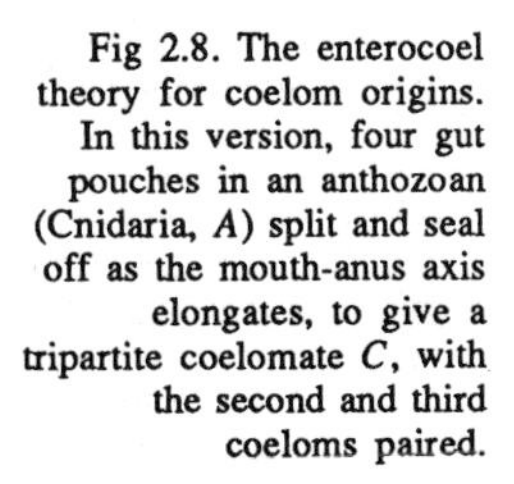
Fig 2.8. The enterocoel theory for coelom origins. In this version, four gut pouches in an anthozoan (Cnidaria, *A*) split and seal off as the mouth-anus axis elongates, to give a tripartite coelomate *C*, with the second and third coeloms paired.

to the various forms of primary body cavity discussed earlier. It is probably wise to split them up into different categories (at least schizocoel and enterocoel), of exactly equivalent status to the haemocoel and pseudocoel already described.

Indeed, given the definition of a coelom, almost any discrete cavity within a mesodermal organ may logically qualify for coelomic status, so that there must be 'gonocoels' and 'nephrocoels' in existence in many phyla - but at this point the whole discussion borders on the ridiculous, and clearly a decision has to be made to restrict the idea of 'coelomates', if it is to be of any use, to those phyla where the secondary mesodermal cavity is spacious during embryonic and/or adult life, and is genuinely a 'whole body' cavity. Beklemishev (1969) and Siewing (1980) even argue that a real coelom should itself be regarded as a large and important 'organ' (it has its own wall, and its own duct); this would be quite consistent with the notion of its having several independent origins, just as many other organs have had.

If coeloms are polyphyletic and the term itself somewhat misleading, there is a further problem in distinguishing between primary and secondary cavities. All groups of animals have retained their primary embryological cavity to some degree; why have some, albeit independently, bothered to invent a secondary body cavity as well? The answer probably lies in a separation of function, leading to greater efficiency of each cavity in a subset of the functions listed earlier. As we have seen, any animal that acquires a secondary body cavity can retain (or re-invent) the vestiges of the original blastocoel as a circulatory blood system, whether closed as in annelids and chordates or in the form of an open haemocoel as in arthropods. Thus many coelomates have the advantage of one 'cavity' performing the roles of circulation and distribution, for which it may become highly specialised by the addition of pumping hearts and carrier molecules, and the other secondary cavity as a spacious hydrostatic system, with body wall and gut highly muscularised. If the animal is segmented as in annelids this may be particularly useful (Ruppert & Carle 1983), for the primary cavity in its reduced state as a blood vascular system actually traverses the whole body, circumventing the effects of septation, whilst the coelomic compartments are isolated and give the worm the merits of hydrostatic segmentation (see below). Thus having a coelom, of whichever type, may give the advantage of division of labour between the two body cavities, and only in this sense can coelomates be said to be 'superior'. Even then, in many groups that develop as coelomates but where hydrostatic effects and/or segmentation are reduced in the adult (for example the arthropods), this separation of function is inappropriate and the coelom may be small and have little role in the functioning of the whole body. And in cases where septate coelomic systems would not help in locomotion, or would preclude necessary volume changes (e.g. leeches, and some interstitial annelids - chapter 8), the 'two system' arrangement may break down with coelom sinuses, or rather solid mesenchyme, occuring throughout the body.

These points about division of labour within body cavities also underline the issue of polyphyly of the coelomic state. As there is a good reason why coeloms might give a selective advantage to some groups, they could reasonably be expected to have evolved more than once. As such they reflect the essential polyphyly of body cavities as a whole; from the analysis given here, there cannot be any doubt that all kinds of body cavities have developed in different fashions, in different groups of invertebrates, using the initial material provided during embryological development in a variety of ways according to lifestyle, size or habitat. Given the enormous advantages accruing from possession of a body cavity, it seems only logical to assume that they should have been invented, retained, and perhaps lost, repeatedly.

An alternative view of body cavities

The analysis given above does suggest an element of hierarchy of body designs from the acoelomate condition, via pseudocoelomates retaining or re-establishing the primary embryological blastocoel, to coelomates acquiring a secondary cavity. However, as we saw in chapter 1, a quite different interpretation of evolutionary sequences has been current particularly with German authors for several decades, and has been advocated strongly in recent years by morphologists such as Siewing (1969, 1980) and Rieger (1980, 1985).

These workers take the view that the coelomate condition pre–dates other designs in an evolutionary sense, the entire spectrum of bilateral animals being descendants of relatively large ancestors endowed with a spacious coelomic cavity. This view is closely allied with 'gastraea' theories of metazoan origins (see chapter 7), and the associated view that coeloms are of enterocoelic origin. This latter view has long been upheld by eminent European authors such as Ax (1963), Marcus (1958), Remane (1963*a*, 1967), and Jägersten (1959, 1972). The coelom is believed to have first appeared in early radiate invertebrates (specifically in octocorallian cnidarians, according to Siewing (1980)), by sealing off enteric pouches (as in fig 2.8). Thus it arose very early in the history of metazoans, and extensive radiation occurred from these early 'archecoelomates'. Remane, Rieger and Siewing then see the acoelomate condition as a regressive response to reduced size and interstitial habit, occurring convergently in several phyla. The pseudocoelomate condition is also deemed to be secondary, perhaps (as before) a result of paedogenetic effects in embryonic or larval coelomates; though some of the authors who support such theories regard the distinction between coeloms and pseudocoeloms as fairly superfluous anyway, as discussed by Lorenzen (1985).

These views clearly turn upside down the evolutionary sequences entrenched in many English language texts, and are at first encounter unacceptably heretical. Nevertheless they form part of a long and carefully thought out European tradition, constituting a cohesive theory of metazoan relationships, and their supporters are able to cite several lines of evidence in their own defence. Indeed the idea that 'large animals came first' is gaining ground on several fronts, in relation to life history patterns (chapter 5 - release of gametes for external fertilisation, and a planktotrophic larval stage, may be primitive traits and are promoted by storage of gametes in a body cavity prior to release) and to fossil evidence (chapter 3 - notably the occurrence of relatively large burrow traces in the Pre-Cambrian, and the prevalence of large priapulids in early fauna). Rieger (1980, 1985) is one of the staunchest recent champions of the archecoelomate cause, stemming largely from his fine-structural studies (chapter 6). His analyses of the mesodermal tissues of platyhelminths are particularly apposite. He points out (1985) that if acoelomates came first then the endomesoderm of all Bilateria would be a derivative of mesenchymal tissues, and would have a connective tissue type organisation; whereas if coelomates came first, the mesoderm of Bilateria would be derived from the endomesoderm of the ancestral radiate gastral pouches, and would thus have a distinctively epithelial type organisation. In flatworms, he identifies three separate grades of mesodermal/parenchymal condition, the more primitive groups lacking any typical 'parenchyma cells'; while gnathostomulids and nemerteans (both classic acoelomates) have tissues that are different again. This leads him to suggest that as there is no unifying pattern of parenchyma in acoelomates the condition is a secondary one, arrived at convergently by several unrelated phyla. He also points out that a perfectly 'normal' acoelomate condition is arrived at by leeches, some interstitial polychaetes, gastrotrichs, entoprocts, many nematodes and some enteropneusts, where it must clearly be a convergent secondary phenomenon due to habitat and lifestyle.

However, Rieger appears to undermine his own arguments somewhat by stressing that within the flatworms the complexity of parenchyma is largely related to size of organism, so that degrees of cellularity and of matrix could be reinterpreted in functional/biomechanical terms and a 'unifying pattern' might not be expected even if the acoelomates *are* primitive. And the fact that many groups can regress to an acoelomate state, so that for example some annelids appear extraordinarily flatworm-like (Rieger 1980; and see chapter 8), is not incompatible with the basic acoelomate condition (in flatworms, gnathostomulids and nemerteans) being evolutionarily primary. Nor is it apparent why such profound regressive changes as Rieger and others require should have occurred in the many turbellarians and nemerteans that are not interstitial at all. Even some of the key German morphologists who support aspects of gastraea-based theories of metazoan evolution have rejected the idea that coelomate conditions are primitive (e.g. Reisinger, see chapter 1). Ax, who formerly (1963) supported the archecoelomate ideas, points out (1985, 1987) that smaller platyhelminths are unequivocally simpler in all respects than the larger forms, even in features that are unrelated to 'regressive' tendencies due to interstitial habit, such as pharynx structure; thus they cannot be a secondarily small taxon, and are not secondarily acoelomate. It also seems most unlikely that the blind-ending gut of platyhelminths could be a secondary condition; it is surely more likely to be a common plesiomorphy of lower metazoans (Ax 1987), shared with the 'diploblastic' phyla, since loss of an anus can have few advantages (as discussed in section 2.6).

Further criticisms of the archecoelomate idea and the phylogenetic sequences they impose are given by Salvini-Plawen (1980*a*). He attacks the issue on three fronts. First he points out (following Reisinger 1972) that nervous systems show a progressive pattern within the cnidarians, spiralian acoelomates and coelomates (fig 2.9); from an unordered superficial plexus in cnidarians to a concentration as a sub-epidermal orthogonal net of connectives in flatworms, on to increasing ventral plus lateral concentration in molluscs and to purely ventral connectives in the articulate (annelid and arthropod) groups. The lophophorate and deuterostome groups (with intra-epidermal nerves) can be derived from this via the larval arrangements of the spiralians,

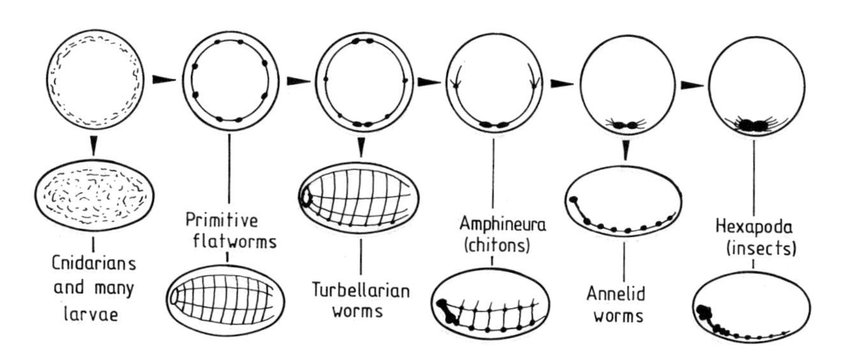

Fig 2.9. A scheme for the evolution of the nervous system, from a nerve net in cnidarians through a primitive orthogonal state in flatworms, and with increasing ventral concentration in higher groups. On this basis, it seems unlikely that acoelomates could be derived from coelomates as the archecoelomate theories require. (Modified from Reisinger 1972).

making these latter two groups somewhat paedomorphic in character; this allows Salvini-Plawen to maintain his view (see above) that deuterostomes are a sort of regressive or paedomorphic protostome. His second point is more directly related to this, in that he considers the site of blastopore relative to mouth and anus to be a continuum in which the protostome condition (mouth formed from blastopore) must come first - we will see more of this issue in chapter 5. Thirdly he maintains, as was mentioned in section 2.3, that progressive differentiation of the mesenchyme occurs (fig 2.10), cnidarians having only ectomesenchyme, flatworms and nemerteans (the 'Mesenchymata') with some additional endomesoderm, and higher spiralians with differentiated and split endomesoderm giving a schizocoel. Enterocoelic deuterostomes are again derived from this latter condition, with the lophophorates as an intermediate, and only have endomesoderm remaining. Salvini-Plawen states that the gap between 'diploblasts' and 'triploblasts' can only be bridged by this route, where acoelomates precede coelomates; to move from cnidarians to deuterostome coelomates, as Siewing and others require, makes the presence of only ectomesenchyme in the former and only endomesoderm in the latter inexplicable. He therefore roundly refutes the whole idea of an 'archecoelomate'.

(The archecoelomate viewpoint will be recurring at various moments in the chapters that follow, as it has important consequences for theories on metazoan origins and relations between lower phyla; reviews of the ideas involved may be found in Schäfer (1973). For now, though, it will be helpful

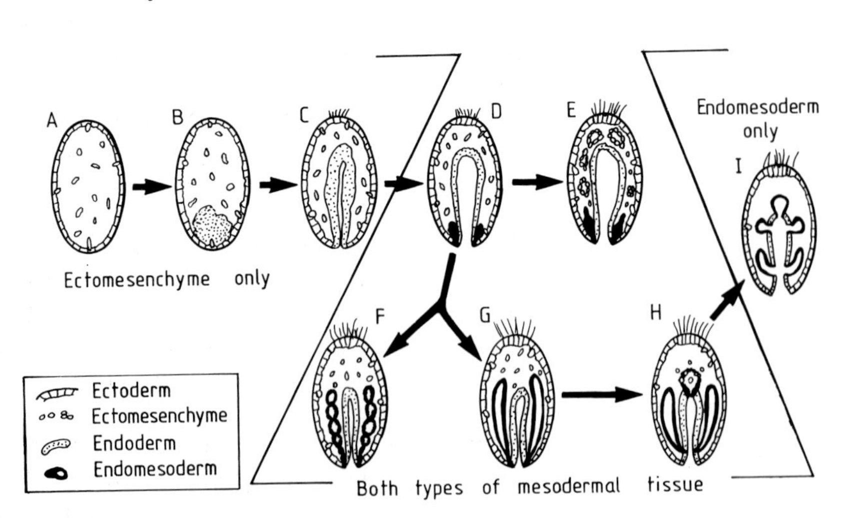

Fig 2.10. The origins of mesodermal tissues in different invertebrate groups (based on Salvini-Plawen & Splechtna 1979). *A*, a hypothetical blastuloid ancestor; *B*, Porifera, disorganised ectomesenchyme; *C*, Cnidaria, with ectomesenchyme organised to form a gastrula or planula; *D*, Platyhelminthes, with endomesoderm forming from the 4d cell; *E*, Mollusca, with ecto- and endo-mesoderm producing organs; *F*, Annelida, with compact mesodermal bands from the 4d cell splitting to give segmented coelomic cavities; *G*, Sipuncula, with an unsegmented coelom in the mesodermal bands; *H*, Phoronida, some Enteropneusta, combined splitting and pouching of the gut wall to form endomesoderm; *I*, some Enteropneusta, Echinodermata, complete enterocoely from the gut wall.

merely to note the vigour of the controversy, and remember that our conventional hierarchy of acoelomate/ pseudocoelomate/ coelomate is certainly an over-simplification (as each type may have been developed repeatedly and from different 'directions') and may even be fundamentally wrong in the overall trend it suggests.)

Whichever view is taken of the evolutionary sequences involved here, quite clearly there are a number of different types of body cavity, independently evolved and all in effect of equivalent status, all performing a variety of functions. How in practice can they be distinguished, and how may they help in deciphering phylogeny?

Practical problems

The body cavities described above cannot in themselves be distinguished, of course, as a cavity is a more-or-less negative feature; their definition relies inevitably on their embryological development, their site within the body, and the nature of their linings. Thus a pseudocoelom should be spatially, and preferably also temporally, contiguous with the embryological blastocoel; should be radially disposed, without interrupting mesenteries, and with organs lying free within it; and should normally have mesoderm only on its outer margin, with no specific epithelial lining (fig 2.5). A coelom should arise ontogenetically 'later', in various fashions such that it is entirely surrounded by mesoderm, normally as one or more pairs of bilaterally disposed cavities thus giving dorsal and ventral mesenteries (fig 2.6); and it should be lined by a continuous cellular epithelium, the peritoneum, which covers and suspends all the organs so that they are 'retro-peritoneal'.

However, in practice it can be nearly impossible to decide just what type of body cavity certain groups of animals have. And it is the last of these distinctions, in theory perhaps the easiest and most conclusive to apply, that has caused most of the problems since the advent of electron microscopy. Some of the controversies are summarised in table 2.3. In essence, it is very difficult to decide whether cavity linings are continuously cellular and nucleate, as required of a coelom, when viewing isolated fixed sections at high power. Where a decision can be made, how should a cavity be characterised if it has a continuous, but syncytial or anucleate lining? (A discussion of one of the most difficult cases, regarding priapulid status, is given by van der Land & Nørrevang 1985, who conclude that the animals are coelomate but have a unique cavity lining - see chapter 9). Worse still, what can be done with phyla where the larval body cavity conforms with one (coelomate) definition and the

Table 2.3 *Characteristics of controversial body cavities.*

Phylum	Cellular cavity lining	Musculature	Designation (usual/possible)
Priapulida	present? but anucleate?	both sides	pseudocoelomate? coelomate?
Chaetognatha			
larval	present	outer only?	coelomate
adult	absent		pseudocoelomate
Bryozoa	present but anucleate?	?both sides	coelomate (?pseudocoelomate)
Rotifera	present but anucleate	?	pseudocoelomate (?coelomate)
Acanthocephala	absent	?outer only	pseudocoelomate - but not persistent blastocoel
Pogonophora	absent?	outer only (no gut)	coelomate (?pseudocoelomate)
Hirudinea	no single cavity	both 'sides'	coelomate
Nemertea (rhynchocoel)	present	both sides	acoelomate (?coelomate)

adult cavity is by all criteria a pseudocoelom? Or with groups such as the enteropneust hemichordates, where there are many different modes of cavity formation within one clearly defined taxon (Remane 1963*a*)?

We are clearly left with a host of difficulties rather than easy answers. Once again, the point that body cavities have multiple origins is underlined; and for phylogenetic purposes the nature of such a cavity may be of very little help. Indeed, given the numerous cases mentioned above in which an acoelomate condition is clearly secondary, from either coelomate or pseudocoelomate precursors, and the apparent ability of a single organism to change its designation as it develops, it may well be that - contrary to most of the simple invertebrate textbooks - the body cavities of animals are amongst the most misleading of all possible taxonomic characters.

2.5 Body divisions - metamerism and segmentation

Occurrence and definitions

Several different phyla of metazoans show a tendency towards repetition of structure along their axis, a phenomenon generally known as metamerism. Examples are shown in fig 2.11; even protozoans may show such longitudinal repetition, particularly in the case of foraminiferans, and in metazoans it occurs in such diverse examples as cnidarians, turbellarians, cestodes, rotifers, kinorhynchs, and of course the familiar annelids, arthropods and chordates. It is therefore possible in all grades of body design, including acoelomate, pseudocoelomate and both schizocoelic and enterocoelic coelomates. Quite clearly, then, a tendency to such repetition is a polyphyletic phenomenon, and early schemes that tried to define a unitary form of 'eumetamerism' that was closely tied in with the evolution of a coelom (itself as we have seen polyphyletic) are unacceptable. They cannot explain successfully either the presence of similar metamerism in both schizocoelic and enterocoelic animals, or, more particularly, the occurrence of non-metameric coelomates - notably the sipunculans.

Nevertheless the idea that 'higher' coelomate animals show a 'true' metamerism that is different in kind from the repetition manifested in, for example, the cnidarians and cestodes, has always been popular - cestodes in particular are regularly described as 'pseudometameric'. The distinction is said to lie in the elements which are repeated, and in the part of the body where growth and repetition occur. Thus real metamerism requires repeated units

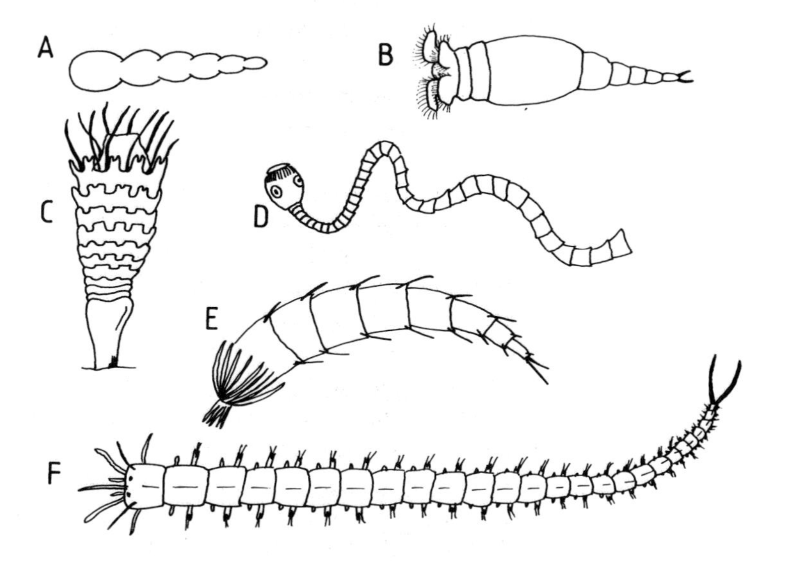

Fig 2.11. Some examples of serial repetition of structure in invertebrates. *A*, foraminiferan protist; *B*, rotifer; *C*, strobilising cnidarian; *D*, tapeworm; *E*, kinorhynch; *F*, polychaete annelid. The appropriate terminology is discussed in the text.

involving every part of the body, and in particular the mesodermal muscles and coelom; whereas animals in the pseudometameric category have only partial repeating sequences, principally relating to the diverticulate gut (in flatworms), or the reproductive apparatus (in cestodes), or the cuticular system (in bdelloid rotifers and kinorhynchs) (see Beklemishev 1969). At the same time, truly metameric animals add new segments sub-terminally, at the distal end of their bodies, while the pseudometameric cestodes add them just behind the head, or scolex.

Cases such as the foraminiferans, cestodes and cnidarians are often referred to by the alternative term 'strobilation', a phenomenon arising from incomplete transverse fission between adjacent units each of which can almost be regarded as an individual animal - in effect, a longitudinal colony is implied. Although the idea of cestodes as colonies was popular earlier this century (with the scolex as the asexual zooid, budding off sexual proglottids) it is now normally accepted that a tapeworm is in fact an individual.

There is then a reasonable consensus that the metamerism of annelids, and of arthropods (which may themselves be an amalgam of three independent groups each of which could have acquired segmentation separately), and of chordates, is different in kind from that of other animals, and these phyla are therefore often dignified with the term 'segmented' instead. This implies a precise and definite repetition of all structures in each segment (now called somites), particularly involving the mesoderm, and with new somites added at the back end of the animal. Note, however, that some texts have used the terms metamerism and segmentation in precisely opposite senses, where segmentation is the general phenomenon of repeated structures, and metamerism the more specific state of complete/mesodermal repetition in annelids, arthropods and chordates - there still seems to be no consensus on this. The term **segmentation** will be used in the more specific sense here, to apply to annelids, arthropods and chordates; the term **metamerism** will be used for the cestode condition and similar strobilised animals; and the term **serial repetition** will serve for non-specific cases of multiple structures on a longitudinal axis, as in flatworms, nemerteans, and (as will be seen in chapter 10) some molluscs. It remains unclear whether some or all of these phenomena have a similar genetic basis (see for example Gerhart *et al.* 1982), though the relevant 'homeotic genes' will be considered further in chapter 4.

Segmentation and its origins

Even the view that complete mesodermal segmentation is special and somehow uniquely definable, and can be applied to a limited range of phyla, has its problems when considering the embryology of segmentation. There are several possible theories as to where 'real' segmentation came from. The 'pseudometamerism' view says that the indefinite repetition in some acoelomates somehow 'crystallised' into segmentation, with muscles becoming segmented to permit serpentine swimming (though many elongate swimmers are perfectly efficient without segmentation). The 'cyclomerism' theory requires the gut pouches of cnidarians to close off forming both coelom and segments simultaneously, as was discussed before (fig 2.8). The 'corm' theory suggests that incomplete transverse division following asexual reproduction gave rise to segments (as happens transiently in some modern turbellarians, and more permanently in strobilised cestodes and scyphozoans). None of these theories, all in effect extending one of the means by which repetition of structure can happen in modern animals, is entirely convincing as an explanation for modern segmented animals - particularly when they require coeloms and segments to go hand in hand.

Beklemishev (1969), following an earlier analysis by Ivanov, suggests that annelids and arthropods in fact show two kinds of segment. There are larval segments, usually 3-13 in number, which begin with a metameric growth of ectodermal structures (particularly the incipient parapodia of polychaete worms), and where the mesodermal band tissues split up subsequently into the same number of pre-determined segments; and there are post-larval segments, growing sub-terminally in front of the anal lobe, and where mesoderm splits up first and other structures follow. He suggests that a cyclomerism-type theory may explain the larval segmentation, uniting it with certain other forms of repetition, but that post-larval segments are more akin to a strobilation phenomenon, caused by incomplete transverse divisions. This view of primary and secondary segmentation fits conveniently with archecoelomate theories, as it allows Remane to suppose that the secondary segments arose later and from sub-division of the third coelomic cavity (metacoel) of his trimeric archetype. Siewing (1976) adopts a similar view, regarding all types of metamery as derivatives of an ancestral trimery. However, Manton (1949) regards the presence of two types of segment in annelids and arthropods as a secondary effect, due to specialised larval needs; the primitive condition was successional formation of identical somites from anterior to posterior.

Nevertheless, the fact that two methods of achieving metamerism can occur in the same animal suggests that there may have been repeated inventions of a metameric state by different animals, using whatever method was embryologically feasible at the time. Metamerism and segmentation are therefore rather a nebulous collection of design possibilities, strongly polyphyletic and unlikely to be of much help in broad phylogenetic terms.

Functional aspects

This view of the problems of repeated structures in animals is further supported by Clark's (1964) analysis of metamerism and segmentation, in which he concludes that the condition has arisen several times. In particular he notes that even the real 'segmentation' of certain coelomate groups cannot be homologous, the chordates having invented it quite separately from the annelid/arthropod lineage. (This is clearly evident from most of the classical phylogenies of the twentieth century shown in chapter 1, where the two groups are widely separate and can share no recent common ancestor that was itself segmented.) He proposes that segmentation is a response to functional needs in each group where it occurs.

What then are the functions and advantages of metamerism that could lead it to be invented independently several times over? There are a number of possibilities.

(1) It may facilitate embryological growth and organisation, as patterns of morphogenesis (and the chemical gradients responsible) can be repeated so that less genetic information is required to produce a certain body size.

(2) It may facilitate and speed up locomotory patterns dependent on neural patterns, as the segmented nervous system can repeat a firing sequence initiated centrally, slightly out of phase in each segment to give metachronal rhythms. This would be relevant in undulatory swimming, in peristalsis and burrowing, in creeping across substrates, and in locomotion based on jointed appendages.

(3) It permits periodic strengthening of the cuticle into rigid plates with intervening softer articulations, relevant to the versatile 'arthropod' designs.

(4) It may be mechanically beneficial in peristaltic/burrowing locomotion to have a septate body where muscular contractions are localised, forces are localised, and energetic cost is reduced.

(5) It may be mechanically beneficial in certain cases for swimming animals, where there is longitudinal stiffening of the body.

Clark's analysis (1964, 1981) particularly stressed these last two points as explanations for segmentation in annelids and in chordates respectively. A septate body cavity is particularly useful to creatures like the earthworm that burrow continuously through relatively hard substrates. Very specific parts of the body can be used as expanded anchors to work against, these being segments where the longitudinal muscles are contracted. Other segments push forward from the anchored regions as the circular muscles contract, giving a strong peristaltic wave down the body. The septa act as bulkheads preventing fluid transmission throughout the body, which would dissipate the force of local contractions, reduce speed, and waste energy. Thus, if body cavities were of use in allowing burrowing at all (section 2.3), segmentation was an additional benefit allowing it to be faster, more forceful and cheaper.

It is worth noting, though, that a great many annelids are successful burrowers despite having lost their internal septation - the oligochaete earthworms are fairly exceptional in the degree to which they have retained full segmentation. So burrowing clearly does not 'require' segmentation. In fact the explanation for loss of the septa probably resides in the two qualifications given above - earthworms are successful as *continuous* burrowers, in fairly ***hard substrates***. By contrast, a great many polychaetes (and other invertebrates) burrow in the softer sands and muds of marine environments, making a burrow to live and feed in and staying there in a fairly sedentary state until the burrow is destroyed. For these animals, of which the lugworm *Arenicola* is a classic example, septation is not very helpful. The energy optimisation required by an earthworm, burrowing relentlessly through the earth, is less relevant to a one-off burrower. And because to burrow in something soft requires a very large shape change of the body to make an adequate anchor (since the substrate tends to give way somewhat as the body expands), septation may be positively disadvantageous. Shape change in a septate body with fixed bulkheads is limited, whereas in a non-septate body it can be extreme, with a bloated bulbous anchor forming at one end as all the circular muscles at the other end contract and force fluid out of their area - an expensive but necessary action. Thus a burrow in soft mud can *only* be made effectively by an unsegmented body, and it is no coincidence that the majority of worms dug up in the littoral zone are in fact unsegmented internally. *Arenicola* is a particularly neat example as it has lost its internal septation in the anterior end - the part that actually forms the burrow - but retained it posteriorly where the body undergoes undulatory movements to ventilate the burrow (though support for the gut during defaecation has been said to be the main reason for this - see Clark 1964).

Hence Clark's view that burrowing is the prime reason for annelid segmentation, set out in an over-simplified form in many subsequent texts, must be qualified according to how and where the burrowing occurs, since more burrowers are actually unsegmented than otherwise. And of course many swimming and crawling polychaetes (and the back end of *Arenicola*) do retain their segmentation in a fairly complete form. Clark demonstrates that segmentation is not mechanically necessary for undulatory swimming motion (as can of course be seen in swimming nemerteans and flatworms), but it may still be true that the better coordination given by a segmented nervous system and musculature was a prime contributor to the evolution of segmentation in annelids as a whole. And it is certainly this aspect of the segmented condition, in conjunction with the bodily rigidity and orderly arrangement of limbs that it permits, which is fully exploited by so many of the arthropods, some or all of which may have arisen from a segmented proto-annelid stock (chapter 11).

The reasons for segmentation in chordates probably do pertain more directly to mechanical swimming needs, according to Clark. Undulatory swimming wherever it is found depends on an appropriate body shape, irrespective of segmentation; but the mechanism for producing the locomotory waves varies between vertebrates and invertebrates. Worms can swim, and move in several other ways, using their circular and longitudinal muscles. Early chordates can *only* swim, and their longitudinal muscles are entirely devoted to achieving this, which they do in a special fashion dependent upon a stiff axial skeleton (the notochord) that can transmit torsional (rather than purely compressional) forces. The resultant locomotory wave requires segmentation of the muscle into specific myotomes.

Thus it seems quite clear that not only were the segmentation patterns of vertebrates and annelids separately evolved, but they arose for quite different purposes. In both cases the segmentation was probably selected for by locomotory need, and has been modified accordingly through each group's evolutionary history.

A logical extension of this view is that other groups have evolved other forms of serial repetition in response to other needs or constraints. Turbellarians have it because their lack of an anus and a body cavity necessitates a diverticulate gut, whose regular lateral branches impose serial order on other organ systems. Nemerteans have serial repetition largely of the reproductive system; and cestodes are serially repetitive for similar reasons. A few groups like kinorhynchs have serial repetition because of cuticular annulation which aids locomotion; and something similar may apply to chitons, where the multiple shell plates permit a firm hold on rocky shores.

As a final point to back up this functional view of metamerism, consider the 'oligomeric' phyla, having only a few partitions to the body (Clark 1964, 1979). Classic examples are the three lophophorate phyla (Phoronida, Bryozoa and Brachiopoda), all with two- or three-part bodies. Though these should technically be considered truly segmented, since the partitioning involves muscles and (coelomic) body cavity with full septation and often visible external divisions, they are clearly unlike the other segmented groups considered above that have multiple repetitions of structures. Their divisions can have little to do with patterned locomotion. Nor are these animals necessarily closely related to either of the two main lineages of segmented animals, either by descent or close common ancestry, as will be seen in chapter 13. But there do seem to be good functional reasons for an oligomeric condition, seen most clearly in phoronids, effectively consisting of two parts (though embryologically tripartite). One section of coelom operates the feeding organ (lophophore), allowing it to extend or retract, while the other permits burrowing or movements within the tube; there are in effect two separate hydraulic systems. In bryozoans, where the rear part of the animal is immovable and permanently encased in a skeletal box, the septum between the two parts is often incomplete and they are functionally one-segmented (though again there are three embryological segments). Hemichordates (chapter 12) are also primitively tripartite; the presumably more primitive pterobranchs retain all three segments, but here the front part produces movement within the tube while the second section operates the feeding tentacles. More advanced enteropneusts use only the front segment for burrowing locomotion and the coelom of the rear segments is largely occluded. Finally the pogonophorans (chapter 8) are effectively oligomeric; though probably related to metameric annelids, they are tube-dwellers using only the front few segments of their body (see Jones 1985) for specific functions and reducing all others to a shrunken buried tail-piece. The oligomerous condition therefore is also probably polyphyletic, arrived at from several alternative 'directions', particularly by burrowers requiring separate parts of their body for feeding, anchorage, aperture guarding and burrowing itself. But whether polyphyly is the case here or not, oligomeric animals certainly underline the independent evolution of different septate conditions for different functional reasons.

Thus all forms of repeated structure have specific functional causes and most of them are undoubtedly of independent origin. It may not be particularly helpful to separate them into categories, and designate only those involving the muscles and body cavity 'proper' segmentation, because these instances are themselves clearly independent of each other, and only appear more profoundly segmented because they reflect a locomotory need and thus primarily involve very 'visible' tissues and organs. But given the widespread recognition of mesodermal segmentation into somites as something rather different in kind from the metameric serial repetition of other groups, the use of distinctive terms will be retained in this book, using 'segmentation' in a specific and restricted sense. However, it is evident that many kinds of metamerism have been independently evolved, for good functional reasons, and that this feature of animal design is rarely likely to be of use for establishing links between phyla.

2.6 Skeletal systems and basic body plans

The times when animals were classified by their skeletons are long past, though elements of these simple early phylogenies perhaps persist in the ubiquitous use of the very term 'invertebrate', as if there were some fundamental split between the bony and the soft-bodied.

However, even within the invertebrates there is a long-standing tradition of dealing with groups in categories that owe much to their skeletons. Most textbooks will first dispense with the fundamentally hydrostatic groups - the cnidarians and the worms - before passing on to the shelled molluscs, the cuticularised arthropods, and finally the peculiar spiny-skinned echinoderms. The implication that soft-bodied forms precede and are 'lower than' those phyla that invented stiff skeletons is hard to avoid. It might be helpful, then, briefly to review skeletal types and their effects on body design, to see if there are real phylogenetic trends here that could be useful.

Traditionally there are three main types of skeletal system, hydrostatic, stiff exoskeleton and stiff endoskeleton, each having its own well-known advantages and drawbacks. But it is probably more useful for present purposes to sub-divide the second group into two further categories, complete exoskeletons as found in all the arthropod groups, and incomplete exoskeletons such as shells, tubes and boxes. For each of the four categories erected (fig 2.12), profoundly different constraints on the resulting animal design are then evident.

Hydrostatic skeletons operate in all the truly 'soft-bodied' animals, where muscles act against a cellular or fully fluid-filled space to bring about shape-change and movement, these changes being limited and directed by the presence of fibrous lattices in the body wall that prevent excessive bulging and reduce the necessity for continuous muscle tone. By far the most effective design for a mobile hydrostatic animal, particularly those that burrow, is the elongate worm-shape. The majority of phyla are entirely made up of 'worms', and virtually all phyla have some representatives that have converged on this form. It is no coincidence that so many early phylogenies classed most of the invertebrates together as 'Vermes'. However, as we have seen, worms come in a great range of internal designs. Some of the smaller types have no anus, and their sac-like guts have to be diverticulate to achieve adequate surface area for unspecialised digestive processes; most have acquired a through-gut (the classic 'tube within a tube' arrangement), giving the advantages of more efficient and specialised sequential food processing and the possibility of continuous ingestion. Worms occur also with all possible versions of body cavities, and with varying degrees of serial repetition or septation. Their external similarity is merely a consequence of the potentials and limitations of the hydrostatic system.

Any worm that becomes entirely covered in a stiffened hard cuticle is rendered immobile, so the complete exoskeletal design option imposes new restraints and requires new inventions. All animals designed in this way must have articulation between hardened plates of the body, and the exoskeleton is thus really only appropriate in segmentally organised animals. In addition articulated limbs become necessary, and new methods of feeding and reproducing and communicating with the outside world are required. Every aspect of the animal's biology is constrained by its cuticle, and design options are severely limited - thus it is no surprise that all animals with exoskeletons again end up looking rather similar. We call them all arthropods, and commonly unite them as a phylum; but as we will see in chapter 11 they, like the worms, may well be a polyphyletic assemblage whose superficial resemblance is a consequence of the convergence imposed by their skeleton.

The option of an incomplete exoskeleton is a much less restrictive one. Size range can be much greater (whereas both worms and arthropods are necessarily quite small), and shape can be enormously diverse. Within the single phylum Mollusca a huge diversity of external morphology, and of habit and habitat, can be seen. The open-ended shell covering a mobile foot gives many of the advantages of the hydrostatic state, but also gives the protection of

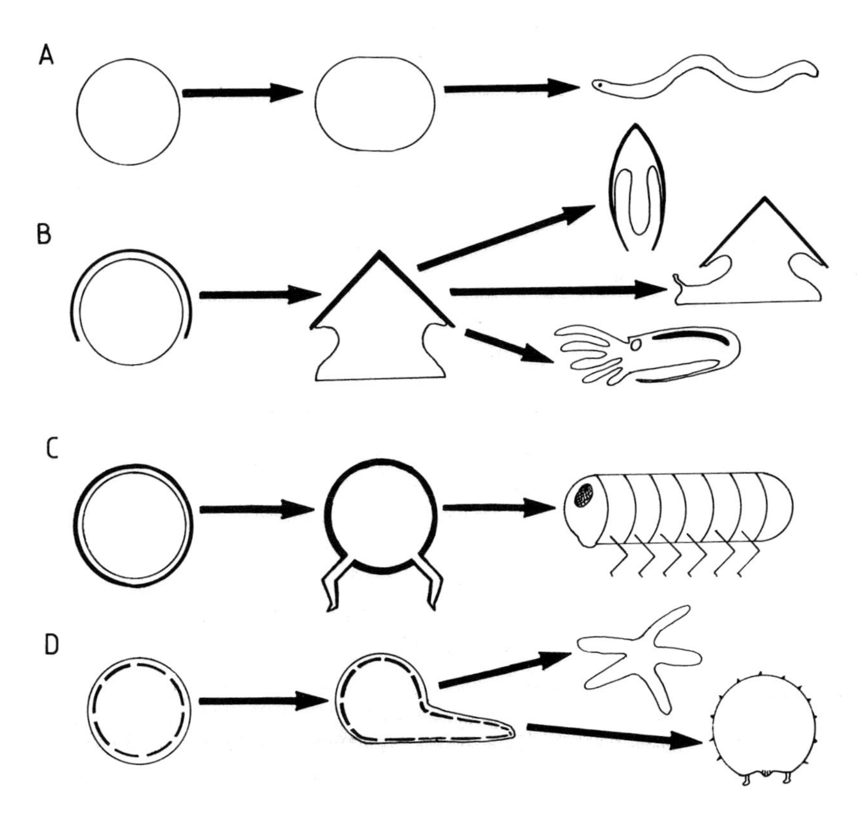

Fig 2.12. The four main skeletal designs found in invertebrates and their effects on form and diversity (see text). *A*, hydrostatic 'worms'; *B*, incomplete exoskeleton, the 'shell' of molluscs; *C*, complete exoskeleton of 'arthropods'; *D*, endoskeleton as internal ossicles in echinoderms.

an exoskeleton against physical and physiological environmental stress and against predators. Given this basic design, different members of the same group may opt for a highly protected enclosed lifestyle, using all the benefits of their shells (bivalves), or may largely lose the shell and use it only as a stiffener or buoyancy aid internally (cephalopods); but it is no surprise that those retaining both the 'hard' and soft' options of their body plan, the gastropods, are the group to show the greatest diversity of form and habit, and the greatest speciation. The incomplete exoskeleton could justifiably be seen as the best compromise in animal design, exemplified by the molluscs but also adopted by brachiopods, many rotifers and protozoans, and in a modified 'fixed' form by many tube-dwelling worms, and box-dwelling colonial bryozoans or corals.

The stiff endoskeletal design, offering less protection and requiring a certain minimum size, is less common in invertebrates - it has mainly been developed within the chordates, where again articulated appendages of some kind are generally required. But most echinoderms (chapter 12), though designed very unlike vertebrates, do have endoskeletons, and achieve a good range of form and size (despite the additional and peculiar limitations of pentamery). Notwithstanding the articulated bone-like ossicles of brittle-star arms, there is clearly less convergence within the endoskeletal assemblage than is evident within the worms and the arthropods. The echinoderm skeletal system is in fact very near the outside of the animal, and is somewhat intermediate between an exo- and endo-skeleton, achieving some of the advantages of each. Paul (1979) likens it to the design of a Mediaeval castle, highly protected but with many slit windows for access and defense; by the same analogy, the molluscan design is a castle with a drawbridge, and the arthropod design a totally sealed castle under siege that happens to move rather well (though a modern tank might be a more appropriate image if an analogy is required from the realms of warfare).

Of the four basic designs, then, two are rather restrictive and cause convergence between all their possessors, even if only very distantly related; and two allow considerable diversity of form, but are evident in relatively few phyla. Nevertheless it is the two structurally restrictive skeletal patterns that have given rise to the dominant fauna of the planet, in both species numbers and biomass; the 'arthropods' and the 'worms' hugely outnumber all other groups and are the major success stories of animal evolution.

Before leaving this topic, it is worth remembering that the four designs outlined here are by no means mutually exclusive, and some of the major successes of particular phyla come from their ability to exploit the advantages of different types of skeleton within one body. Most phyla endowed with stiff skeletons also retain some hydrostatic support from the fluids of their body cavity, and in the arthropods this may be vital in particular adult activities - growth during moulting, leg extension in spiders, protrusion of mouthparts or penes in many insects - as well as in almost all aspects of larval life for the soft-bodied caterpillars and grubs. Exoskeletal arthropods also exploit the principles of an endoskeleton in their endophragmal skeletal elements, giving invaginated internal surfaces such as apodemes for muscles to work against. A great many soft-bodied worms use elements of stiffened material in their mouthparts, or as protective spicules or chaetae. And echinoderms, of course, use their internal ossicles very much as a protective layer, and combine them with hydrostatic tube-feet to effect locomotion.

Skeletal design problems inevitably leave it difficult to draw phylogenetic implications. Each system has clearly been evolved repeatedly, and for stiff skeletons the materials used give little help in sorting out relationships, some being ubiquitous and others highly sporadic as will be seen in chapter 3. Some skeletal systems cause extensive convergence that could seriously obscure phylogenetic differences; others allow great diversity that could be equally confusing; and overlap between the designs adds further possibilities for misinterpretation. However, it should at least be helpful to be aware of the types of problem created by each kind of skeletal design, as we come to examine particular phyla and their relationships in the third part of this book.

2.7 Conclusions

Basic morphological features, of the kind that form the backbone of so many simple phylogenies, turn out on examination to be dangerously misleading. This has been demonstrated in relation to every one of the characters examined in this chapter. The old divisions of radial/bilateral animals are already losing support and have no good logical basis, so the terms Radiata and Bilateria should be avoided unless carefully qualified. The idea of diploblastic and triploblastic body designs is also falling from favour, so that cnidarians and ctenophores can no longer be set far apart from other metazoans. However, the recognition of three major tissue layers - ectoderm, mesoderm and endoderm - remains a useful comparative tool in interpreting later stages of embryology, even though these layers (particularly the mesoderm) may be non-homologous in view of their varying derivation at gastrulation. Grades of body cavity can be usefully distinguished in theory, but in practice their characterisation is fraught with difficulty, and the order in which the various conditions may have been 'invented' during evolution remains contentious. The coelomic body cavity presents particular problems, and its multiple origins will necessitate some rethinking on the issue of the protostome/deuterostome dichotomy. In addition, since body cavities share with the remaining two characters considered here an undoubtedly polyphyletic nature, an understanding of cavity origins, type of segmentation, or site and nature of skeletal tissues is never likely on its own to be a great help in deciding phylogenetic issues.

Unfortunately, then, it will be necessary to treat all these conventionally diagnostic features with a good deal of caution, even suspicion, in the succeeding chapters. It should already be evident that some of the phylogenies outlined in chapter 1 may be relying on rather dubious or outmoded concepts of invertebrate relationships.

PART II

Sources of evidence in invertebrate phylogeny

The chapters in this part of the book concern a number of possible types of evidence that might be of use in deciphering relationships between different groups of invertebrates. Phylogenetic characters could be of several kinds, each of them interrelated; but broadly speaking, the evidence used here and elsewhere is generally of a morphological and biochemical nature, and characters of a physiological, ecological or behavioural type are not employed.

In chapter 2 some of the gross morphological features at the level of the whole organism were considered, representing the adult phenotype; and in chapter 3 the morphology of extinct organisms, left as fossil traces, are discussed. The relations of adult morphological characters to other properties of the organism are shown diagrammatically below, and in chapters 4, 5 and 6 the uses of characters at the successive levels of increasingly integrated biological phenomena shown in this diagram are explored as sources of phylogenetic reasoning. Chapter 4 concerns the genetic material itself, its primary protein products, and other biochemical derivatives. Chapter 5 looks at the processes of embryology and morphogenesis, brought about by interaction of these biochemical products. Chapter 6 examines the resultant morphology at the level of ultrastructure of the cells, cellular products, and tissues; the elements that constitute an adult organism.

The character distributions outlined in these chapters form the basis for much of the analysis of particular phyla and super-phyletic groupings discussed in Part 3.

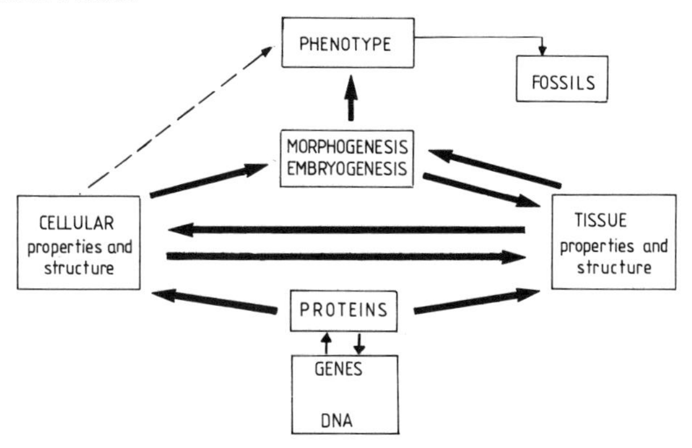

3 Evidence from the fossil record

3.1 Introduction

The fossil record is the most obvious and tangible evidence of the past history of animals, and ideally it should be the final arbiter in deciding between opposing theories on major issues of phylogeny that have been derived from living fauna. This, at least, is the conventional view, and it is only in the last two decades that an increasingly theoretical approach to phylogenetic study has brought into question the role of palaeontology as a superior and ultimate guide to relationships (e.g. Hennig 1965, 1966; Løvtrup 1977; Nelson 1978). Patterson (1981) points out that fossils have almost never been used to overturn phylogenies derived from Recent fauna, and that the widespread belief that fossils are the best way of determining relationships is a myth. Nevertheless, it remains true that a phylogeny that conflicted with the evidence from fossils could not seriously be upheld; so a knowledge of fossil remains is generally regarded as a necessary accompaniment to a sensible appreciation of relationships.

It is a widely repeated view that the invertebrate fossil record is so limited (because their soft bodies are poor material for preservation) that it can tell us nothing of invertebrate evolutionary history. Modern reviews make it very clear that the record available is by no means so inadequate, so we should expect to learn a good deal from an understanding of the preserved remains that are available. A recent quote (Stanley 1979) admirably reflects this more positive view: 'The role of palaeontology ... has been defined narrowly because of a false belief ... that the fossil record is woefully incomplete.' It may be true that some of the major events in the early divergence of invertebrate phyla are inevitably left open to conjecture, since they involved organisational changes in very small and entirely soft-bodied animals; but even there clues may be given by the modern interpretations of the Pre-Cambrian fossil beds, now amenable to new methodologies. Phylogenetic study should also be aided by an increasing appreciation of the extent and causes of the incompleteness of fossil records (Raup 1972, 1976*a,b*; Sepkoski 1978, 1979; Paul 1982; Holman 1985).

This chapter seeks evidence for early phylogenetic and organisational changes in animal design and complexity from the faunal patterns of fossil beds, especially in the earlier eras, and for subsequent radiations and extinctions that can reveal the important trends in invertebrate evolution. It is not intended as a review of the fossil histories of individual groups, as these are covered fully elsewhere (Moore & Teichert 1952-79; Clarkson 1979; Lehmann & Hillmer 1983; Conway Morris 1985) and will also be referred to where relevant in later chapters.

The first stages of evolution

Life on Earth seems to have begun relatively early during the planet's history, since the Earth is dated at around 4600 million years old and the first possibly biological remains (see Lehmann & Hillmer 1983) are dated at around 3800 million years old (fig 3.1), almost the same age as the earliest known rocks. From amongst the chemical diversity of this period, more complex self-replicating assemblages of molecules that could be dignified with the term 'living' arose. These were very simple procaryotic organisms, anaerobic heterotrophs utilising the simpler chemicals available in their aqueous surroundings, and little more than a membranous envelope enclosing a minimal amount of self-duplicating molecular machinery. These fore-runners of modern bacteria and blue-green algae then existed in a relatively unchanging state, in deeper waters where the harsh conditions of irradiation and electric

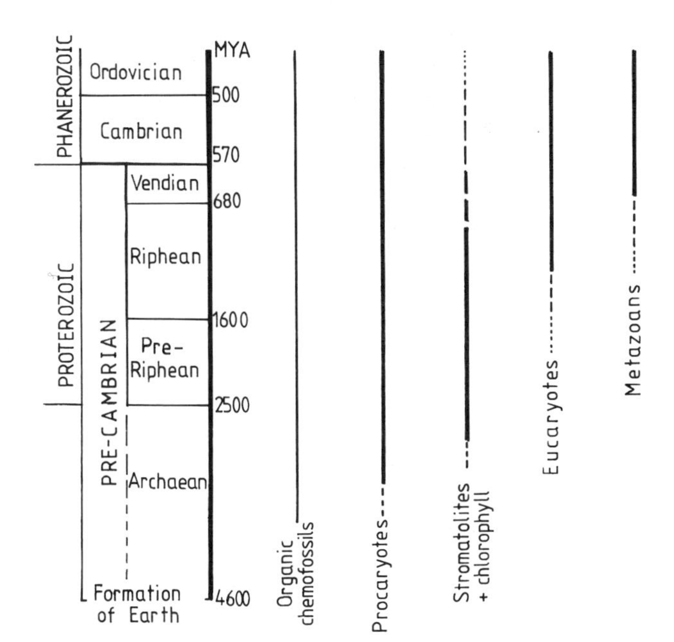

Fig 3.1. The time-scale and terminology of Pre-Cambrian geology and evolution, based in part on Lehmann & Hillmer (1983).

storms were ameliorated, for another 2000 million years (Sylvester-Bradley 1979). The major event that occurred some time during this period was the acquisition of photosynthetic abilities by some of the bacteria - traces of chlorophyll are known from rocks 3000 million years old, and blue-green algal reefs (stromatolites) occur from soon after this date. Thereafter evolution still proceeded only very slowly, but these organisms were gradually bringing about vital changes in their own habitat by oxygenating the atmosphere. When the oxygen levels rose to above 1% of the present atmospheric level (PAL) (fig 3.2), the 'Pasteur point' was reached and aerobic respiration became possible, yielding much greater energy from the glucose substrate than simple fermentation had allowed, and perhaps permitting a more complex cellular machinery. Some of these early cells may show the first traces of a eucaryotic condition as long as 1900 million years ago (MYA), though clear evidence does not occur until 1300-1400 MYA (Cloud 1976*a*; Ford 1979), and actual nuclei are not evident until remains from the Australian deposits at Bitter Springs, dated between 1400 and 1000 MYA (Schopf & Oehler 1976). Cloud (1976*a*) estimates that the eucaryotes arose at about 3% PAL (fig 3.2).

Subsequently, with the advent of sexual reproductive processes (evolution of which is itself a topic of fierce and unresolved controversy) the whole process of evolution by natural selection speeded up, with new possibilities for lifestyle arising. The acceleration of speciation processes brought about by the evolution of sex would eventually lead to adaptive radiation and changes at higher phyletic levels (Stanley 1976).

Estimates of the oxygen level at the start of the Cambrian have varied widely, from about 1% (the Berkner-Marshall theory, still much-quoted - see

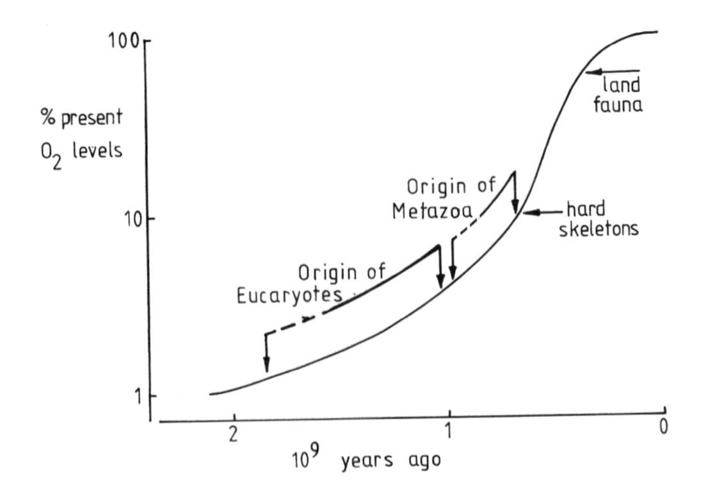

Fig 3.2. The composition of the Earth's atmosphere in terms of the percentage of oxygen relative to present atmospheric levels (% PAL); largely based on estimates by Cloud (1978).

for example Lehmann & Hillmer 1983) to nearly 100%. The best estimates may be those of Cloud (1976 *a,b*; 1978), the more recent paper putting the metazoan origins at about 7% PAL, the Cambrian boundary at about 9-10% PAL, and with a steady rise to around 20-25% PAL by the Cambrian/Ordovician boundary. Certainly towards the end of the Pre-Cambrian levels must have been high enough to allow an ozone layer to form, filtering out harmful ultra-violet radiation from the sun; and Cloud (1968) dates this effect at about 600-650 MYA (though a recent claim that ozone can be formed at only 0.1% PAL (Schopf 1980) might set it much earlier). Life in shallower seas became feasible with a protective ozone layer present, as did surface pelagic lifestyles; and eventually true animal life could evolve.

Presumably one or more groups of Pre-Cambrian eucaryotic algae lost their chlorophyll and began to feed on other organisms. Protozoans continued to do this within the confines of a single cell, and are not normally considered as part of the kingdom Animalia nowadays; but they are abundantly represented amongst the microfossils, in the form of foraminiferans, radiolarians and less familiar chitinozoans (possibly ciliates - Brasier 1979). True metazoan animals may have originated as clusters of these heterotrophic protists (see chapter 7), but we have no fossils indicative of these earliest stages. On present evidence, though, they may have arisen between 800 and 700 MYA, for the earliest incontrovertible animal remains are about 700 million years old (Glaessner 1984), in the late Pre-Cambrian period, when they are already diversified into several distinct groups (section 3.2). In the more oxygenated conditions complex tissue structures became possible, as did the manufacture of skeletal materials. In addition, evolutionary processes and rates may have been directly affected by the advent of these more advanced heterotrophs, 'cropping' the algal communities and producing a positive feedback on diversity levels (Stanley 1976). Thus some of the major faunal changes of the lower Cambrian may be partly attributable to biotic effects, as well as to physicochemical changes in the conditions of the planet's surface waters (see section 3.3). Many aspects of chemical evolution, and the earliest biological evolution, are reviewed by Sylvester-Bradley (1975), Cloud (1976*a*) and Glaessner (1984).

Fossilisation processes and patterns

From the Cambrian period, beginning 550-570 MYA, more complex mineral skeletons were secreted (see below), and fossilisation of animal remains becomes much more evident, particularly in shales, mudstones and limestones. Various different skeletal materials appear, though not necessarily at the same time or in the same sequence or proportions in different groups (Stanley 1975); their nature is therefore often of help in deciding a fossil's affinities (table 3.1). As a general rule, the tougher the skeleton the better fossil record a group leaves behind; molluscs and brachiopods have the most complete records, cnidarians the poorest, and echinoderms and arthropods are intermediate (Holman 1985).

Actual preservation mechanisms vary widely, as outlined in any palaeontology text. In only a few exceptional cases are animals without hard parts preserved, and only those hard parts that are buried quickly and in very special conditions after death are normally recognisably fossilised. Calcitic skeletons are most stable, but even if they are initially allowed to settle undisturbed they can be destroyed in acidic conditions or may be converted into other forms depending on the chemical equilibrium of the sediment, perhaps recrystallised with other minerals. If the original shell of the fossil is dissolved slowly enough, it may still leave behind recognisable internal or external moulds of itself, especially in very fine sediments. Occasionally the internal spaces of the animals may become filled with minerals such as pyrites or silica, and these can show very intricate details of the original anatomy. But the chances of any of these fossilisation processes happening are small and are selective between different types and sizes of animal, being dependent on local

Table 3.1 *Skeletal materials in invertebrates*

	Calcite	Aragonite	Silica	Apatite
Protistans	+	+	+	
Sponges	+	+	+	
Cnidarians				
Hydrozoans	+	+		
Anthozoans	+	+		
Molluscs				
Chitons		+		
Scaphopods		+		
Bivalves	+	+		
Gastropods	+	+		
Cephalopods	+	+		
Annelids	+			
Crustaceans				
Cirripedes	+			
Ostracods	+			
Bryozoans	+	+		
Brachiopods	+			+
Echinoderms	+			
Extinct groups:				
Archaeocyatha	+			
Conodonts	+			+

chemical conditions and sedimentation phenomena; and even those few organisms that are preserved initially are unlikely to survive later stratigraphic upheavals in a recognisable state. Just to make things worse, further bias in the fossil record is added by the palaeontologists who collect and study the specimens, both by their personal interests and theories and by the methodologies available to them.

Unquestionably, then, the fossil record is incomplete and gives only a few isolated windows on the pattern of ancient faunas. And these windows are inevitably less frequent the further back we go, both because mineralised parts were less frequent and because older rocks are bound to be more affected by distorting metamorphic and tectonic processes in the Earth's crust. However, such glimpses of the past history of animals can be immensely revealing, and many animals have been preserved in incredible detail, especially where they occur in limestone deposits and fine shales.

Given these problems, it is perhaps not surprising that the metazoan phyla show no consistent patterns in their relative species numbers currently as compared to the fossil record (fig 3.3). These relationships are inevitably somewhat suspect, given the additional difficulties of defining a 'species' from fossil materials; but it is fairly clear that some groups are apparently much more speciose now (cnidarians, gastropods and of course the insects), others

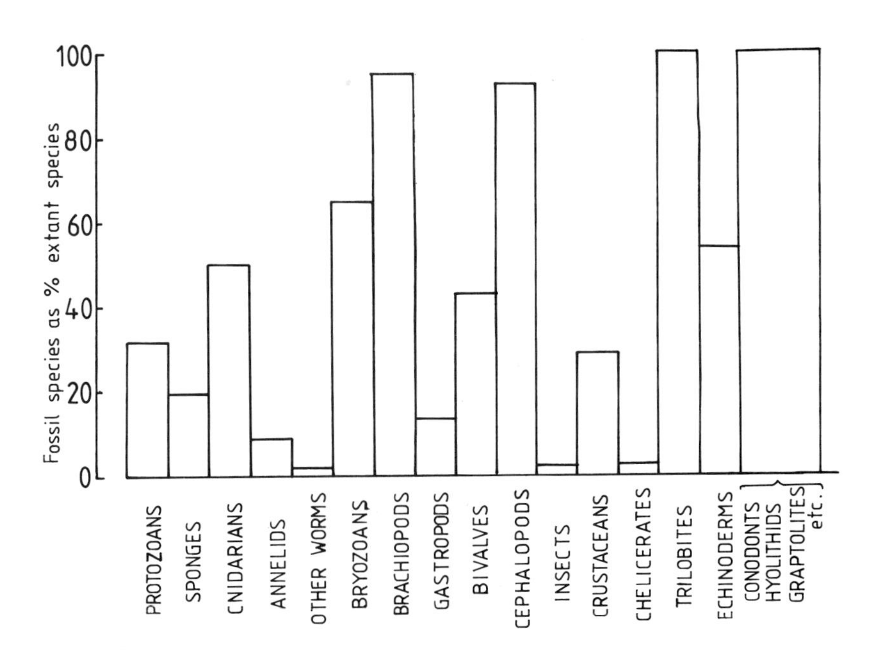

Fig 3.3. The abundance of different groups of animals in the fossil record, relative to their present diversity; based on data in Raup (1976*a*) and Lehmann & Hillmer (1983).

have far more fossil species than extant ones (brachiopods, echinoderms), and some are of course extinct, with *only* fossil species. The overall number of body plans has perhaps declined, as will be seen later; but there do not seem to be obvious patterns of abundance, diversifications or extinctions in relation to skeletal type or habitat or design (for example marine bivalved molluscs are now more speciose than in the past, while the exact opposite applies to the similarly constrained brachiopods). Therefore it is essential to look at specific groups and the biotic relationships between them, and specific conditions in each era, to understand what was happening to invertebrate animals and the ways in which they were evolving.

3.2 Pre-Cambrian fauna

The Pre-Cambrian period was dominated for most of its duration by autotrophic procaryotes, particularly the reef-forming bacteria and algae that are now termed stromatolites and that achieved reasonable diversity. These early organisms began to decline in abundance from about 800 MYA (fig 3.1), and thereafter they were largely confined to the subtidal and intertidal zones of osmotically more difficult habitats - both freshwater and hypersaline - where competition from metaphytes and metazoans was limited.

Brasier (1979) and Glaessner (1983, 1984) review some of the protistans that were next on the scene and have been found as traces in deposits from the late Pre-Cambrian. Both authors also refer to some indications of metazoan life between 1000 and 700 MYA - possible faecal pellets, burrows, feeding marks and other traces of activity. Indeed one set of trails from rocks 1400 million years old have been reported (Clemmey 1976). Trace fossils of this kind (see Seilacher 1977) continue to be more abundant then actual body fossils throughout the Pre-Cambrian, presumably bearing witness to an extensive soft-bodied fauna of worms; so there can be little doubt that metazoans existed for most if not all of this period. But it is not until the last part of the Pre-Cambrian, in the Vendian of 680-570 MYA (fig 3.1), that clear palaeontological evidence for metazoan life can be found, apparently post-dating a major glaciation episode. Indeed animals are relatively widespread in rocks of this age; they are known from Asia, Africa, Europe, and North America, and in particular from Australia where the important finds in the Ediacara Hills gave the entire faunal type its now familiar name.

The Ediacaran animals

These Ediacaran faunas, dated from around 680 MYA (see fig 3.1), have only become known in the last three decades (the site at Ediacara in South Australia was first discovered in 1946, though the similarly-dated Nama beds in Namibia were known earlier). However, despite many finds still awaiting full description, these Vendian beds have already revealed some surprising features. They generally show little similarity with the better-known Cambrian fauna that succeeded them about 100 million years later, and they show little or no mineralisation. They are remarkably alike in all areas where they occur, with relatively few of the known beds having species unique to themselves. All the Ediacaran beds are dominated by forms resembling cnidarians, annelids, and perhaps some arthropods; and they seem to have an abundance of filter feeders and detritivores, with few if any predators occupying higher trophic levels. The very diversity of the fauna makes it clear that these are not the earliest metazoans, though they are the first good view available of metazoan evolution. This vital gap between the protozoans and the earliest metazoan fossils can therefore only be filled by biological theory and speculations at present (chapter 7); here is one area where a direct appeal to palaeontology cannot help.

Nevertheless a quick survey of the animals to be found in this period (fig 3.4) will raise some interesting points. Of the common animals in the Ediacaran deposits, reviewed by Cloud & Glaessner (1982) and more fully by Glaessner (1984), those identified as Cnidaria seem to be most diverse. Traces of scyphozoan-like jellyfish are particularly abundant, and some sea-pen-like anthozoans and presumed floating chondrophoran hydrozoans are also present, indicating a high enough level of atmospheric oxygen to create and sustain an ozone layer screening ultra-violet radiation from surface waters. Many recent authors have continued to identify all these forms as cnidarians, though doubts are often expressed; the medusoids are little more than disc-like impressions with radial or concentric markings (Fedonkin 1982 suggests accommodating them all in a new high-rank taxon), and the frond-like 'pennatulids' could equally be unrelated to modern animal phyla (Scrutton 1979; Conway Morris 1985) and might even be plants (Pflug's 'Petalonamae', supposedly ancestral metazoans, discussed in chapter 7).

The 'annelids' from the Ediacaran have about six distinct genera, notably the polychaete-like *Spriggina* and *Dickinsonia*; and in the Nama beds a surprisingly modern-looking unsegmented echiuran-like animal has been described. Another group of Pre-Cambrian worms, the Sabelliditidae, are held by some to be ancestral to pogonophorans.

The 'arthropods' have at best only two genera, one of which (*Praecambridium*) is said to resemble trilobites and/or chelicerates, the other more crustacean-like. But neither of these animals has left clear traces of appendages that could show their real affinities (see chapter 11), and some authors, probably wisely, do not interpret them as arthropods at all. Further similar fossils from the Russian Vendian beds are also described by some as arthropods. But they might all be better included in the small group of unassigned specimens, along with the peculiar fossils *Parvancorina*, possibly a larval trilobite, and *Tribrachidium* which may conceivably be a proto-echinoderm.

The implication from these different phyletic diversities, if their usual tentative ascriptions are correct and the fossils are taken at face value, may be that cnidarians had a significantly earlier origin than the other groups represented, whilst the first attempts at arthropodisation had, perhaps, only recently begun, probably from a preceding annelid-like stock (chapter 11). Given the extremely patchy occurrence of soft-bodied cnidarians even in later Phanerozoic deposits, it is quite likely that cnidarians had indeed existed for some long time previously but had not been preserved. Since all three of the extant classes of cnidarian are perhaps present in the Ediacaran beds, there is no clarity about the direction of evolution within the phylum, or about the precedence of polyp and medusoid forms (see Scrutton 1979), issues which are relevant to the discussion of metazoan origins in chapter 7. Medusae are more abundant and more diverse than polyps, which might imply a radial hydrozoan/scyphozoan ancestry for the group (Scrutton 1979) as many have

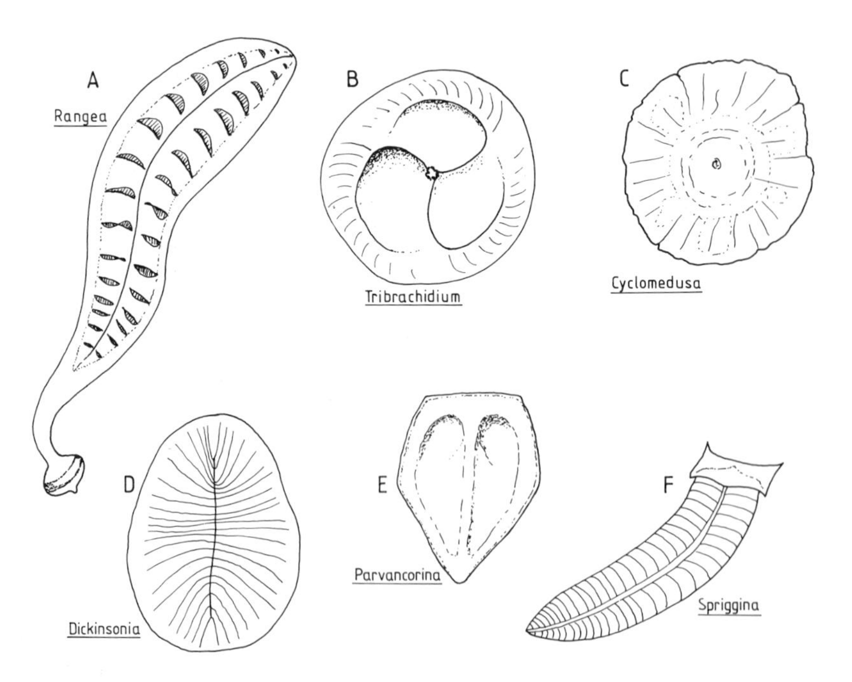

Fig 3.4. Representatives of the Pre-Cambrian Ediacaran fauna, redrawn from Glaessner (1984) and Seilacher (1984).

advocated; but medusae fossilise better and outnumber solitary polyps in virtually *all* fossil faunas, so this is hardly conclusive. There may be some further clues from the fact that advanced hydrozoans (Chondrophora) and primitive scyphozoans coexist at this time, suggesting that the hydrozoan radiation must have preceded that of scyphozoans (Glaessner 1984). Similarly, octocorallian anthozoans may be the most primitive of their class given the abundance of pennatulid-like forms, and again early radiation is implied; thus anthozoans and hydrozoans may be about equally close to the cnidarian stem form, with scyphozoans developing later (cf. Bischoff, in Glaessner 1984). But it should be stressed that the fossils being considered are of such uncertain affinities in themselves that drawing phylogenetic conclusions relevant to cnidarian evolution and metazoan origins is really pushing at the limits of credulity (Scrutton 1979).

Moving on to 'higher' issues, there are no obvious links between the numerous Pre-Cambrian cnidarians and the other groups present in the same fossil beds, which are all (apparently) segmented coelomates. This raises the possibility that the coelomates are rather direct descendants of cnidarian-like animals, with acoelomates as later developments, a view that would also be suggested by some variations of gastraea/enterocoel theories discussed elsewhere in this book. However, such a conclusion from the fossil evidence is by no means convincing, for there is in fact virtually *no* fossil record of the acoelomate and pseudocoelomate animals in any era; they are the worst possible material for fossilisation, being small and entirely lacking in hard parts. Hence their absence as fossils certainly cannot be taken as evidence for their non-existence in the Vendian period, or as support for 'archecoelomate' metazoan origins (chapter 7). Rather, the abundance of trace fossils attributable to worms even back in the Riphean period (fig 3.1) suggests that worms more probably preceded cnidarians, and that some of these worms were large enough to have had body cavities, whether coelomate or pseudocoelomate.

Since the 'deuterostome' groups are conspicuously absent from Pre-Cambrian beds, theories of enterocoelic coelomic origins (see chapter 2) that would require the 'deuterostomes' to be more primitive than schizocoelic 'protostomes' look somewhat implausible in any case. In fact, the postulated abundance of quite sophisticated coelomates in the Vendian, together with the earlier trace fossil evidence for extensive large infaunal life that has already been mentioned, probably indicate just how much metaozan evolution had already passed unrecorded. Perhaps 'protostome' worms in particular were the first major descendants of the earlier smaller worms, with the oligomerous lophophorates (readily preserved bryozoans and brachiopods especially) and then the 'deuterostomes' following as later Cambrian developments (see Clark 1979; Brasier 1979).

The Vendian Ediacaran faunas, then, suggest an origin for the Metazoa at least some 700-800 MYA, and a fairly rapid diversification into quite complex body plans, often apparently similar to those of modern phyla. They may bear witness to a particularly early origin and diversification for the cnidarians, though it is unlikely that these predominantly carnivorous 'diploblasts' pre-dated all the more complex 'triploblastic' animals. They give no direct evidence for an early appearance of other 'lower' groups, such as platyhelminths, ctenophores or sponges (even though the latter are endowed with spicules that should be readily preserved). Presumably whatever other simple metazoans were present lacked the toughened, perhaps chitinous, cuticles or mesogloea of those groups that are found as fossils in the Ediacaran beds. It is certainly not possible to establish any sequences for the appearance of the first metazoan phyla using the known Ediacaran fossils. Though they give intriguing glimpses of Pre-Cambrian life, these are surely rather selective, and it would be dangerous to try and read too much into them.

As a final warning on this point, it is noteworthy that Seilacher (1984) has recently challenged every previous interpretation of the Ediacaran fauna, and believes that none of its constituent organisms bear any affinity to modern animals; instead he sees them as an entirely separate set of experiments in metazoan architecture, subsequently wiped out by a mass extinction event (see also Gould 1985, chapter 15). For example, the supposed Ediacaran 'jellyfish' could not have functioned like modern medusae, since they had concentric structures (presumed to be muscle traces) centrally rather than peripherally, so could not have pulsed their bells. Similarly the so-called 'sea-pens' were much more close-set, with little evidence of branches on which individual polyps could function. Seilacher thus regards the Ediacaran fauna as quite different from cnidarians or any other extant fauna, and sees them as various manifestations of a two-dimensional architecture, (ribbons, pancakes, balloons - lacking in internal complexity but built more as expanded surfaces). If these observations prove correct, Seilacher's radical view removes any necessity to try and trace relationships across the Pre-Cambrian/Cambrian divide, and all the above speculations on the significance of the Ediacaran fauna for metazoan phylogeny would become irrelevant.

3.3 Cambrian fauna and the Cambrian radiation

Just beyond the Pre-Cambrian/Cambrian transition, about 550-570 MYA, a number of large-bodied animals with hardened skeletons (especially phosphatic in nature) are found in abundance in fossiliferous rocks, all fully distinct and complex on their first appearance (fig 3.5) and quite different from the Pre-Cambrian fauna already described (Durham 1978; Brasier 1979). The most familiar of these in the earliest Cambrian rocks (the Tommotian stage) are certain primitive gastropod and monoplacophoran-type molluscs, mollusc-like hyolithids, brachiopods, the first echinoderms, numerous sponge spicules, and a now extinct (perhaps parazoan) group called the archaeocyathids. There are also many tubular and cone-shaped shells of unknown affinities, and shelled protists such as foraminiferans. Particularly abundant are the conodonts, tooth-like phosphatic (apatite) fossils formerly of uncertain status but now linked to lophophorates and/or chordates (see below). In the next stage of rocks all these animals are joined by abundant trilobites; some tracks in Tommotian rocks suggest that trilobites were indeed present then too, but perhaps were too small or too little mineralised to be preserved.

Even where preservation of soft-bodied remains has occurred, as in the Lower Cambrian sandstones of Sweden, the faunas are very unlike the soft-

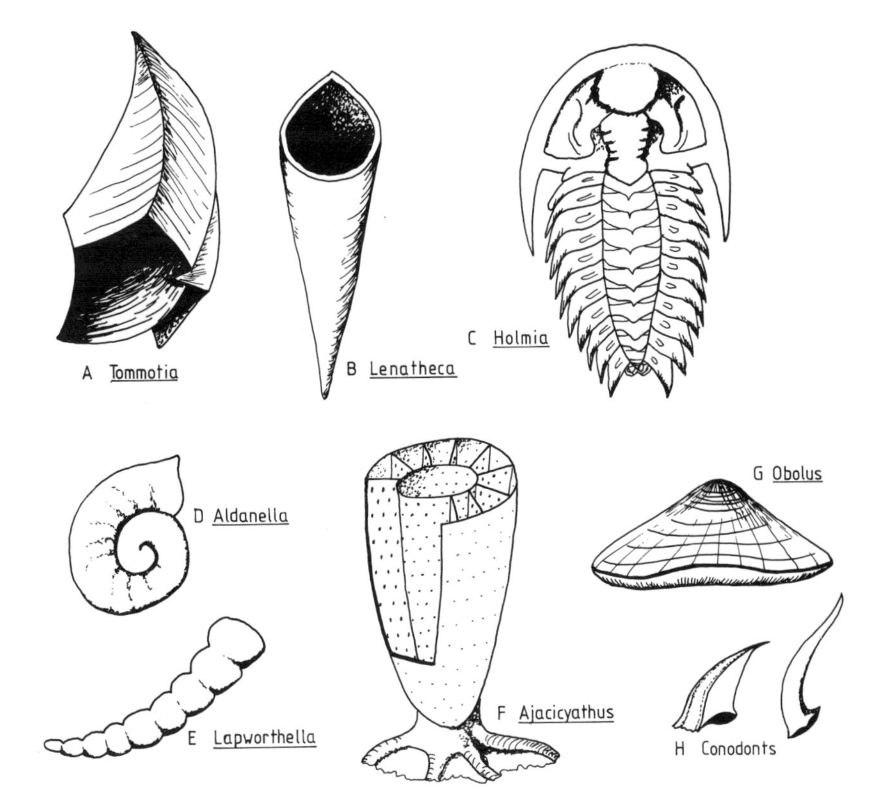

Fig 3.5. The earliest Cambrian faunas, including shelly fossils *(A, B, D, E)* some of which are probably molluscan, trilobites *(C)*; archaeocyathans *(F)*, now extinct but perhaps sponge-like; phosphatic brachiopods *(G)*; and conodonts *(H)*; (redrawn from various sources).

bodied Pre-Cambrian animals at Ediacara, nearly all of which (whatever their affinities) appear to have gone extinct. Only a few possible chondrophorans persisted, and the annelid-like *Dickinsonia* and Sabelliditidae may have left descendants as the living groups Spintheridae (Annelida) and Pogonophora respectively. This loss of the Pre-Cambrian fauna may have been because of an increasing diversity of predators and scavengers, before adequately protective chitinous or mineral skeletons were developed. However, the reason for the worldwide fossil discontinuity at the beginning of the Cambrian is still unclear; a lengthy discussion by Glaessner (1984) reaches no particular conclusion, and several authors continue to hold conflicting views.

The problems of skeletons and mineralisation

Certainly the most obvious aspect of the changed fauna of the early Cambrian is the sudden appearance of mineralised shells, yet it would be simple-minded to suggest that the soft faunas of the Vendian were replaced by more competitive armed and/or armoured shelled animals, not least because so many soft-bodied forms existed very successfully in the Cambrian as well, as indicated by the remarkable Middle Cambrian Burgess Shale fauna (see below). Why then did mineralised skeletons 'suddenly' appear so abundantly in so many different groups?

Lowenstam & Margulis (1980) and Glaessner (1984) point out that this effect may be somewhat mythical, since chitinous structures were probably present in Ediacaran faunas (in 'chondrophoran' bells, and in the supposed arthropods), and calcareous spicules could well have been there too but would not have been preserved in a recognisable form (any more than they are preserved in more recent rocks). Stanley (1975) also points out that skeletonisation was not simultaneous in the major taxa. However, more extensive biomineralisation clearly *did* occur in the Cambrian. It has already been suggested that rising levels of atmospheric oxygen may have been an important factor, and Towe (1981) particularly points to the necessity of oxygen for the effective production of collagen, a constituent of the connective tissues that must presumably precede mineralisation and allow muscles to be linked to shells and hard cuticles. Yet there was no apparent 'critical point' at the start of the Cambrian if Cloud's values of 9-10% PAL are to be accepted (see section 3.1). Another possibility is that the oceans became slightly more alkaline after the Pre-Cambrian era, allowing better precipitation of carbonates within proteinaceous matrices (Degens *et al.* 1967). In fact, though, phosphatic deposition seems to have slightly preceded the widespread use of calcium carbonate; and there were almost simultaneous increases in the use of silica, organic materials including chitin, and agglutinated deposits such as sand grains to make skeletons (Brasier 1979, 1986). Theories based on changing temperature or salinity of the oceans therefore run into similar difficulties, and are not as yet borne out by chemical analysis of the sediments concerned, though general rises in sea level (perhaps giving rise to many briny, anoxic isolated basins) do seem to have accompanied some major biomineralisation events (Brasier 1986).

It seems, though, that environmental biochemistry does not provide the whole answer to biomineralisation. Biotic considerations must also be invoked. Protection is an obvious advantage of a shell, and a positive coevolutionary feedback within an ecosystem can readily be envisaged whereby the acquisition of hard parts by a predator favours the evolution of shells by its prey and the effect proliferates through other groups. Yet protection may not be the fundamental 'reason' for the development of shells. Vogel & Gutmann (1981) argue that biomechanical advantages may initially be more important, the deposition of calcium salts giving greater rigidity to a shell and allowing better muscle attachments, greater mechanical efficiency, and hence energetic economy. Other possible reasons for biomineralisation are discussed by Brasier (1986); he stresses that interactions and chain reactions seem increasingly likely. It seems probable that several abiotic factors and ecological factors contributed to the selective favouring of shells at this period, and different groups of animals may have mobilised the inherent abilities of their constituent eucaryotic cells to achieve biomineralisation independently and for varying suites of reasons, in turn increasing the selective pressure on each other to become similarly shelled and/or effectively infaunal.

Soft-bodied fauna in the Cambrian

The Burgess Shale fauna of British Columbia, dating from the Middle Cambrian period (530 MYA), is perhaps the best known of all fossil finds now, and gives important clues about the wider spectrum of Cambrian life beyond the familiar mineralised remains. It was first discovered in 1909 by Walcott, who excavated many thousands of fossils from the site, and many thousands more were found in the 1960s by Whittington. Since then it has been extensively reported in papers by Whittington, Briggs and Conway Morris, and excellent reviews are available (Conway Morris 1979; Whittington 1980, 1985). It gives a remarkable insight into a relatively deep water ecosystem (perhaps about 160 m depth), with unparalleled preservation of the soft bodies of many of the benthic animals inhabiting a mudbank area.

It includes well over 100 genera, some of which are shown in fig 3.6, and of which many are the more typical armoured Cambrian trilobite types. But about 80% of the fauna is relatively soft-bodied (Conway Morris 1979), much of it being less sclerotised arthropod forms (some of them considered in chapter 11); arthropods together form about 35-40% of the total fauna. There is also an array of molluscs (perhaps *Helcionella* and *Scenella* - see chapter 10), echinoderms (including the earliest known crinoids and perhaps some ancestors of holothurians - see chapter 12), and brachiopods (chapter 13); while sponges become abundant and dominate the suspension feeders. Cnidarians are very rare, apparently in direct contrast to the Pre-Cambrian beds; perhaps this indicates that the Ediacaran 'medusoids' are not in fact cnidarians in any real sense, but do indeed represent an earlier and extinct design of radially organised animals, as Seilacher would maintain. One Burgess Shale form, *Mackenzia,* is the first known soft-bodied cnidarian; and another, *Fasciculus*, has been reinterpreted by Conway Morris as a ctenophore, greatly pre-dating any other known member of this phylum.

But in addition there are the much more peculiar worms and other oddities that have recently excited much attention, some of which are shown in fig 3.7. They represent about the best preserved fossil worms that we have from any era. Some appear to be polychaetes, such as *Canadia* and *Eldonia*, whilst others are clearly priapulids, up to 20 cm long and unusually abundant and diverse, outnumbering the polychaetes; perhaps the dominance of polychaetes in all subsequent faunas was due to the development of jaws by some eunicid-like forms in the early Ordovician (Conway Morris 1979). There are also

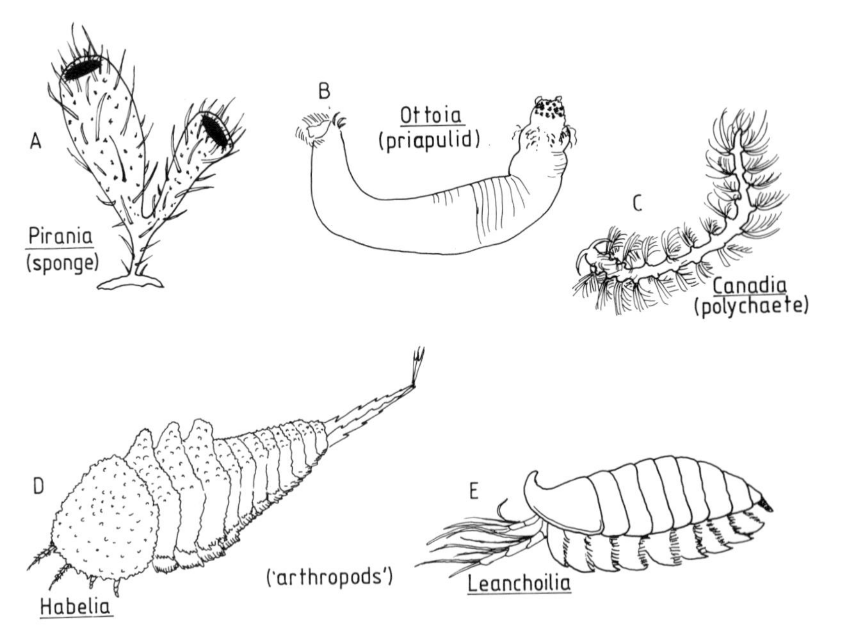

Fig 3.6. Representatives of the soft-bodied Burgess Shale fauna that can be assigned to known taxa, including sponges, worms and arthropods; redrawn from Conway Morris (1979) and Whittington (1985).

apparent hemichordate worms, graptolites, and a worm-like form with a clear notochord (*Pikaia*) that must presumably be an early chordate (see chapter 12).

Amongst the other curiosities are *Aysheaia* and *Opabinia,* both illustrated and considered further in chapter 11, as they are (or have in the past been) held to be members of arthropod groups. But *Aysheaia* may be closest to tardigrades, and *Opabinia* with its long mobile snout bearing terminal jaws has no obvious kin. Similarly, *Amiskwia* was once thought to be a chaetognath, or more recently a nemertean, but now is usually allowed to be a member of an otherwise unknown phylum.

Odontogriphus, though originally described as a brachiopod, has been reinterpreted (Conway Morris 1979) as having an annulated and flattened body 5-6cm long, with a distinct head, bearing both lateral palps and a curious central set of teeth arranged as a pair of loops. These teeth bear a marked resemblance to the conodonts, with which they are contemporary; and they have been interpreted by Conway Morris as supports for a food-gathering lophophore. This view raises interesting possibilities; the abundant fossil conodonts were sometimes in the past interpreted as chaetognath spines, or as the supports of lophophores, but the story of their interpretation and more recent identification is given by Gould (1985, chapter 16). From one Carboniferous fossil some conodonts have been clearly demonstrated as the teeth of a specific 'worm', *Clydagnathus* (Briggs *et al.* 1983). At first this was

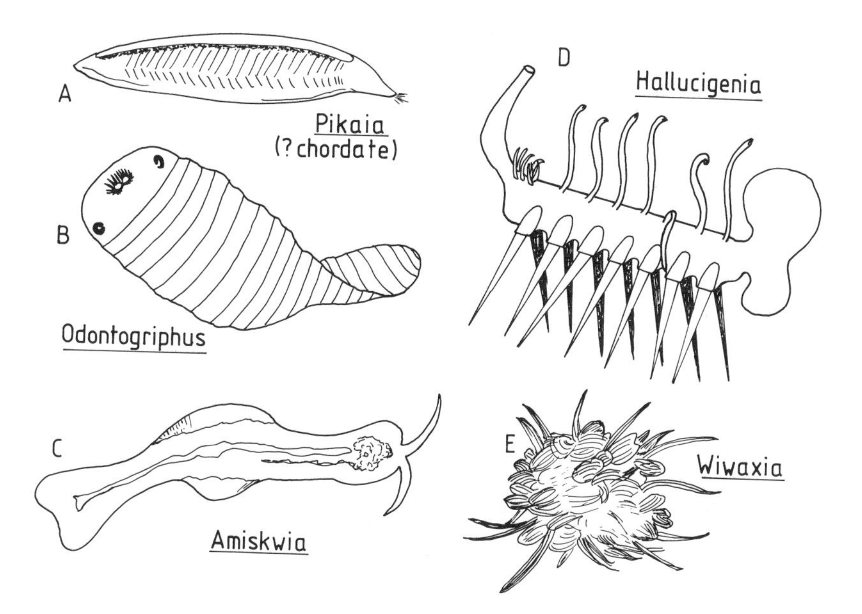

Fig 3.7. Further examples of the Burgess Shale animals, these being of less certain status or included in a broad category 'Problematica'; sources as in fig 3.6.

thought to be possibly related to chordates or chaetognaths, but more probably in a phylum of its own, the Conodonta (Gould 1985). But even more recent reports based on three newly found fossils from the same beds (Aldridge *et al.* 1986) suggest that these conodont-bearing 'worms' were also possessed of myotomes, giving stronger support to a chordate/craniate affinity and probably removing the necessity for a phylum Conodonta altogether (see Benton 1987). These findings therefore not only suggest a link between the Cambrian *Odontogriphus*, the Carboniferous *Clydagnathus*, and certain modern lophophorate worms, but also may make the phosphatic/apatite conodonts helpful in cementing together the lophophorates, deuterostomes and chordates (chapters 12 & 13).

Another peculiar Burgess Shale creature is *Wiwaxia*, bearing overlapping scales and lateral spines dorsally, somewhat like the elytra of modern scaleworms, and a feeding apparatus of rasping teeth reminiscent of a molluscan radula. Opinion still differs as to whether it is related to molluscs or is yet another variety of unclassifiable worm (Whittington 1985).

Even more curious than any of these is the delightful fossil named *Hallucigenia* (fig 3.7*D*), whose elongate trunk rests on seven pairs of long upright spines, and bears seven tentacles with snapper-like tips; how such a creature fed or moved remains totally obscure.

Many of these animals, and others present in the Burgess shales, are now accepted as being so peculiar that they are given unique status, on the assumption that earlier ascriptions were incorrect (Conway Morris 1979); in total, 16 % of the genera are said to belong to no known phylum. But this fauna cannot be dismissed as unusual or aberrant - it is only their preservation that is abnormal. Exploration of comparable mid-Cambrian rocks in North America is beginning to reveal similar though much less beautifully preserved animals; they appear to have been characteristic of the period over quite a large geographical area. Whatever other messages they carry, it is quite clear that they testify to the great radiation of both hard and soft-bodied fauna that had occurred in the preceding few tens of millions of years.

Cambrian radiation

Many authors have spoken of the 'Cambrian radiation event' or even 'Cambrian explosion' as a distinctive and special phenomenon (Brasier 1979). It occurred at the end of the Vendian period, leading through the Tommotian (Lower Cambrian), ending the period of early living forms (the Proterozoic aeon) and initiating that of obvious visible life (the Phanerozoic). A summary of the known fossil remains occurring on either side of the Pre-Cambrian (Vendian) and Cambrian eras (fig 3.8) clearly supports this idea. Most of the

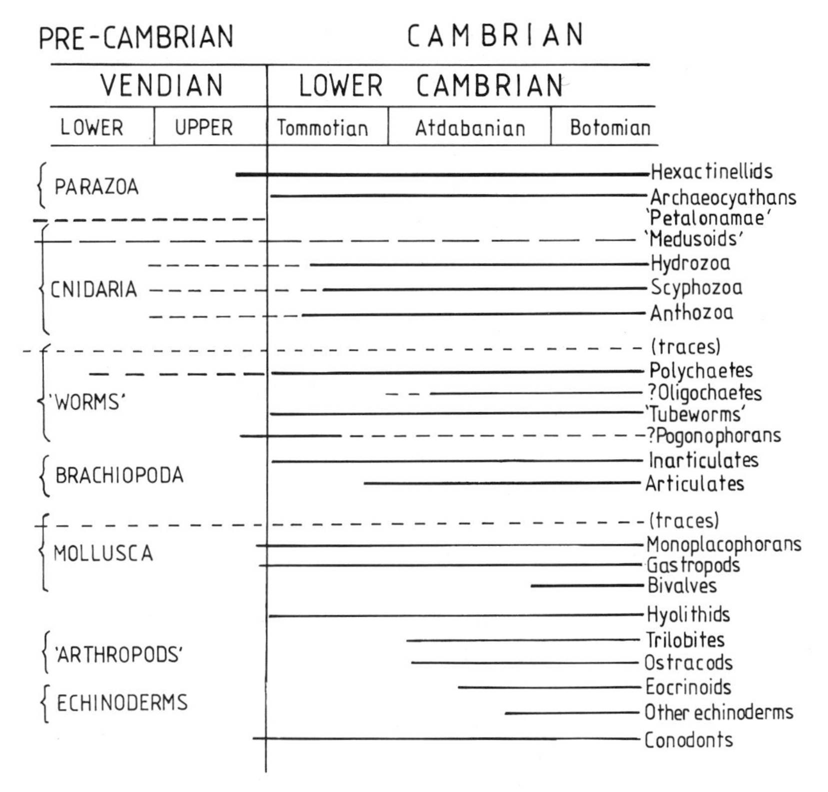

Fig 3.8. Events at the Pre-Cambrian/Cambrian boundary, when many of the existing phyla had their origins in the fossil record; modified from Brasier (1979).

familiar animal phyla of today date their appearance from about that point; of the 27 palaeontologically important classes of animals, the mean geological age is 533 ± 51 million years (see Raup 1983), reflecting the general phenomenon that the higher taxa radiate very early within groups, with the radiation of orders and families occurring later and at less predictable rates. Not only the invertebrates were diversifying at this time however. Over a period of only about 20 million years there were apparently similar surges in the diversity of protists, red and green algae and even procaryotes. Brasier suggests that these effects were allied to changes in food supply and trophic structures, with suspension feeding becoming dominant and microplankton developing organic skeletons. Soft-bodied Vendian faunas largely died out, while new biomineralised forms arose; but the soft-bodied fauna also diversified, as indicated by the Burgess Shale findings. There was apparently also a general increase in body size within groups. Some or all of these effects may have been linked to flooding of continental platforms due to a global rise in sea level.

Estimates of species diversity through time are notoriously hazardous (Valentine 1973*b*; Raup 1976*a,b*; Bambach 1975, 1977; Signor 1978). Some authors concluded that diversity rose rapidly to a Palaeozoic peak, declining in the Permian and Triassic before rising again to a much higher Recent peak. Raup argued that preservation and sampling bias must be taken into greater account, and that apparent species number can be grossly affected by the volume of sedimentary rock sampled from different eras. When this is taken into account (Raup 1976*b*) a trend of gradually increasing species diversity throughout the early Phanerozoic is revealed, and numbers then settle to a steady lower equilibrium value. Bambach (1977) argued for a middle line, with species number continuing to rise beyond the Mid-Palaeozoic but perhaps only four-fold rather than by Valentine's order of magnitude. All these authors agree that the initial rise in species number was a Cambrian phenomenon, though they continue to dispute the later patterns.

However Sepkoski, in two important papers, has analysed the recorded changes of diversity statistically, at the order (1978) and family (1979) levels, documenting more carefully the reality of an exponential surge in fossil diversity during the Cambrian era (fig 3.9). But an equally striking three-fold rise in faunal diversity occurred in the Ordovician, after a Late Cambrian plateau; and only after this did the Palaeozoic fauna increase more slowly and steadily, until it was roughly at equilibrium in terms of diversity. In fact much of the Cambrian fauna died out, as is shown in fig 3.9. And whereas the surge in the Ordovician was generally of rather 'familiar' types that can be readily

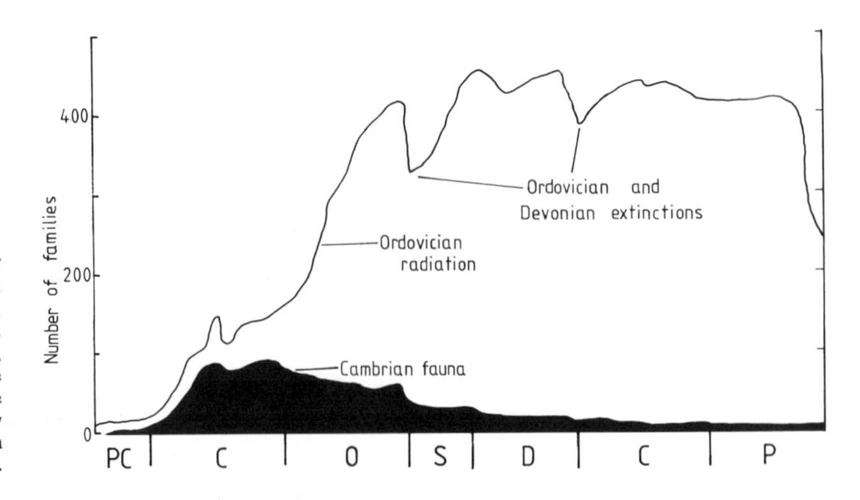

Fig 3.9. The pattern of animal radiation in the Phanerozoic, showing marked increases in species number in the Cambrian, and again in the Ordovician, with the Cambrian fauna largely going extinct; based on Sepkoski (1979).

assigned to an order (articulate brachiopods, molluscs, ostracods, bryozoans, crinoids), much of the Cambrian fauna is peculiar and cannot be related to the architecture of modern animals - this is especially obvious for the soft-bodied faunas described above, but also applies to hyolithids, archaeocyathids, and perhaps trilobites (chapter 11).

Sepkoski interprets his data as a three-stage process of metazoan diversification. Firstly there was prolonged slow evolution through the Pre-Cambrian era; then a diversification of rather generalist forms in the early Cambrian, having 'broad habitat and trophic requirements', an idea that would certainly fit with our inability to decide how many of these odd animals made a living beyond describing them as detritus feeders; and then a growth of the Palaeozoic fauna with 'more specialised and narrower ecological requirements', gradually ousting the Cambrian generalists and building up complex communities such as coral reefs. A recent overview of the problem of species diversity through time, assembled by all the recent protagonists on this issue, (Sepkoski, Bambach, Raup & Valentine 1981) comes to rather similar conclusions.

Thus the Cambrian radiation event seems to be a reality, though slower and less spectacular than earlier authors might have believed; but the reasons why it occurred when it did are still speculative and obscure, and it in any case may have relatively little to do with subsequent invertebrate diversity. Most of the phyla we now know were established by the time it happened, but many of them almost 'started again' in terms of design at the level of families and orders from the end of the Cambrian. It therefore seems rather likely that large parts of the Cambrian faunas in themselves may be unhelpful for studies of phylogeny.

However, the Pre-Cambrian and Cambrian faunas do bring us two very important messages. Firstly they indicate how many experiments in animal design have been attempted, and how commonly new designs converge on a rather limited number of successful options at particular times. And secondly, they stretch our imaginations and credulities to the point where a belief in phylogenies that show branching diversity through time, all branches converging back on a single point of metazoan origins, seems almost insupportable; Whittington (1985), reviewing these faunas, looks 'sceptically' upon such ideas, feeling that 'metazoan animals may have originated more than once, in different places and at different times'.

3.4 Subsequent radiations and extinctions

Phanerozoic evolution has now been proceeding for 570 million years, and in that period the invertebrate fauna has changed fairly considerably, both at the specific or generic level and in terms of gross faunal patterns. It has also changed in overall diversity, though the precise nature of these changes are controversial. An approximate indication of the abundance of different metazoan groups through time is given in fig 3.10, showing the extreme variability of the patterns; almost every phylum or class has a different sequence of abundance peaks and troughs, even for those continuously evolving in the relatively stable conditions of marine habitats. There are nevertheless some punctuations in abundance when major extinctions occurred; these usually terminate each of the main geological periods, which is hardly surprising as those periods are largely defined on faunal differences. The most important extinction episodes were at the end of the Cambrian, Ordovician, Devonian, Permian (the most profound of all), Triassic and Cretaceous eras; though the Devonian and Triassic events may not be statistically detectable beyond the 'noise' of variable fossil diversity.

Throughout the Cambrian, the trilobites, archaeocyathids and thin-shelled inarticulate phosphatic brachiopods were the most abundant animals to be preserved in fossil strata. But at the end of this era the fauna changed dramatically within a few million years as discussed above; within the trilobites many genera went extinct and new ones appeared (Whittington 1966), and some of the more peculiar Cambrian groups began to decline. Archaeocyathans, with their curious double-walled calcareous skeletons interpreted by some as intermediate between sponges and cnidarians, died out almost completely. Then in the Ordovician period there was a considerable radiation of animals from many phyla having calcareous skeletons: brachiopods, molluscs, corals, echinoderms, and new trilobite forms, together

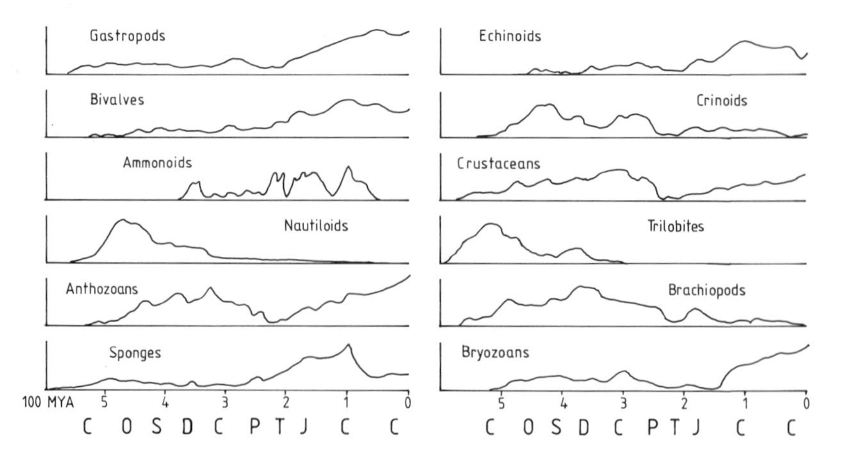

Fig 3.10. Patterns of radiation and extinction in different taxa, modified from data in Raup (1976*a*) and Lehmann & Hillmer (1983).

with early bryozoans and the graptolites that are generally assumed to be forerunners of hemichordate deuterostomes (see chapter 12). This radiation established the marine faunal pattern for much of the succeeding Devonian and Carboniferous eras as well. Apart from the graptolites no major extinctions occurred, and diversity increased rather slowly except in special cases such as ammonoid molluscs which became very abundant and diverse in the Devonian. There were no other major changes until the late Permian when the last trilobites vanished, together with many brachiopods, corals, molluscs and echinoids; perhaps 50% of all marine families died out at this time. Thereafter almost entirely new forms of corals and echinoderms emerged, and the bivalve molluscs largely replaced the brachiopods.

By the Triassic era the ammonoid molluscs were dominant and at their peak, but these were rendered nearly extinct at the end of that period, only one small group persisting and diversifying again in the Jurassic and Cretaceous before declining as the Cretaceous era ended. At this same time the molluscan belemnites, the conodonts, the remaining brachiopods, and many other groups were retrenching, with a 15% reduction in the families of marine invertebrates overall, whilst on land the final demise of the dinosaurs (part of the 'Cretaceous extinction', whatever its causes) was in progress. Thus moving into the Tertiary period the marine invertebrate fauna was much like that of today, just as on land the familiar mammalian types were becoming established in place of reptiles. Overall trends in the relative diversity of different phyla through time are shown in fig 3.11, based on the estimates of Raup (1976*a*).

One other point should be made in relation to extinctions, radiations and faunal patterns through time. Possibly the most interesting message the fossil record holds is the trend of decrease in diversity of organic design through most of the Phanerozoic, despite a general increase in species number. In other words, the present fauna is dominated by just half-a-dozen or so very abundant phyla, and though there may now be thirty five or so phyla in all, very many of them are insignificant in terms of species number, and the overall pattern of animal life is much less diverse than it has been in the past. The same trend is seen at the level of classes - for example there are now just five or six basic designs of echinoderm, but there have been about twenty in all, and some fifteen during the Ordovician era alone. All of this is to be expected given the early radiation at the phyletic and class level described earlier, since diversity at this level then has nowhere to go but 'downwards'; the phenomenon is discussed in more detail by Raup (1983).

Fossils then indicate very clearly that many other animal designs have been 'tried out' and have subsequently died out. In extreme cases we are left only with the curious but often quite distinctive remains now so often dismissed as 'Problematica' (see Gould 1985, chapter 16); such forms as the conodont animal *Clydagnathus*, the Burgess Shale *Hallucigenia* or *Opabinia*, even perhaps the entire Ediacaran fauna. Many weird and unclassifiable creatures clearly *did* exist in the Palaeozoic, and it is only the conservatism of palaeontologists that prevents each being given the phyletic status it probably deserves.

3.5 Conclusions

Fossil evidence leaves a state of tantalising uncertainty as to the first stages in metazoan evolution; and it seems likely that this will always be the case, since the earliest small metazoans, whatever their origin, would be unlikely candidates for preservation.

Given the evidence of trace fossils, there must have been metazoans around in the seas for hundreds of millions of years before their bodies were first preserved in the late Pre-Cambrian, and it is reasonably certain that these were both surface-dwellers and infaunal burrowers, though inevitably there can be

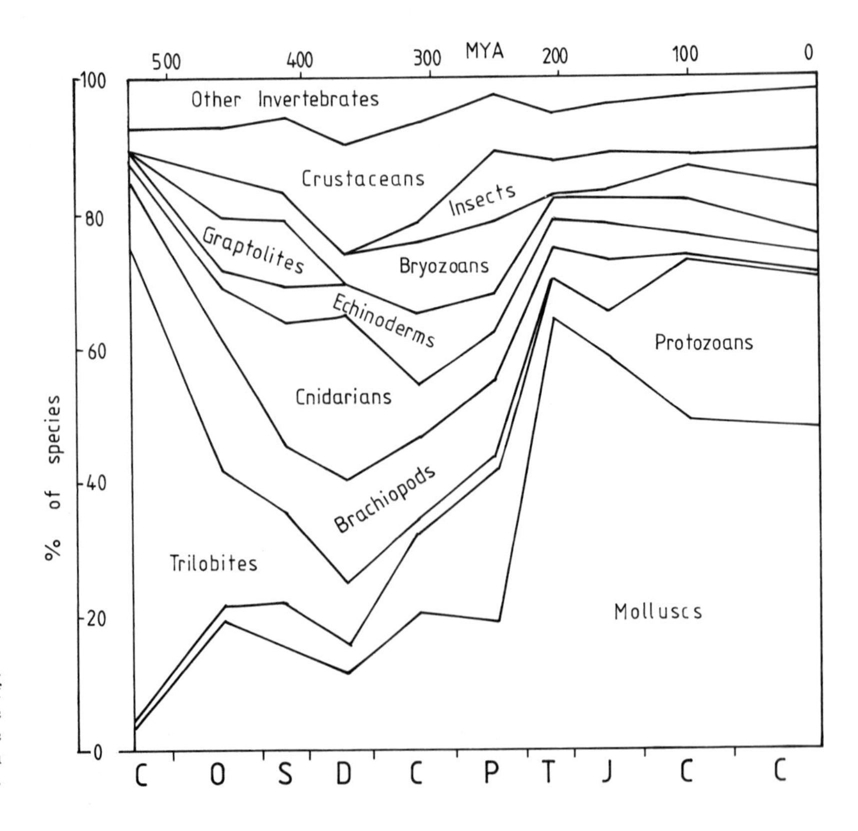

Fig 3.11. Proportional contributions of invertebrate taxa to the fossil record through the Phanerozoic, modified from Raup (1976*a*).

no certainty that pelagic forms existed. Many of these early animals, living up to 1000 MYA, were worm-like, and some were large enough to have had body cavities. From the evidence of the early Cambrian fossil beds (and perhaps also the Ediacaran fauna), they were possibly part of what is now called the protostome assemblage, resembling annelids, echiurans and priapulids. At some time during the Pre-Cambrian a range of radially organised and probably diploblastic animals had also evolved, debatably related to modern cnidarians. Some of these had tough enough coverings to be preserved by the last part of the Pre-Cambrian, but they left no earlier remains that could indicate their time of origin in relation to the worms. It seems likely that they were a separate evolutionary line, and they may have no direct descendants. They do not offer much assistance in arbitrating between opposing views of cnidarian phylogeny, nor can their relation with Pre-Cambrian worms be properly assessed to help with questions of metazoan origins; though polyphyletic views are supported by some experts on these early fossils.

At about the Cambrian boundary, diversity of both mineralised and soft-bodied faunas increased, probably for a variety of reasons that had more to do with biotic interactions than environmental chemistry or geology. The faunas preserved bear little relation to those they succeeded, although most of the modern phyla date from this period, with lophophorates and deuterostomes apparently post-dating the protostomes. In addition there were many novel animal designs (both hard- and soft-bodied) often referred to as 'experiments', which largely died out towards the end of the Cambrian leaving no descendants. A number of shelled groups seem to have occurred, of which only molluscs and brachiopods persisted beyond the Cambrian; and an even greater number of arthropodised groups are represented, only a few of which can be assigned to the extant crustaceans and chelicerates. Most of these groups can best be seen as separate developments from soft-bodied worm-like ancestors; again experimentation, convergence and polyphyletic origins of various designs seem to be indicated.

After some of these 'experiments' had been weeded out by a series of extinctions at the end of the Cambrian, the modern fauna was established in its gross patterns by the start of the Ordovician era, though the classes of animals present were often unfamiliar. A series of further extinctions and partial replacements followed within many phyla through the Palaeozoic, until the present-day balance was achieved by about the end of the Cretaceous.

Despite our best efforts, then, we have perhaps not achieved very much understanding of invertebrate relationships by taking a look at the fossil record. Some trends of diversifications of species, and loss of phyla and classes, have been revealed, and plausible reasons for them can be discerned. Some suggestions of multiple experimentations with body plans, and probable polyphyletic origins of certain groups, have been mentioned. Possible indications of the sequence of appearance of the 'super-phyla' have emerged, and certainly sequences of class and order evolution within the relatively well-preserved phyla such as molluscs, brachiopods, echinoderms and crustaceans can be established, as will be seen in more detail in later chapters. But the sequences of appearance of body plans and of phyla remain obscure, and the relations between them are thus still a mystery.

This is not just because the record is inadequate, implying that all that is needed is to split some more rocks, but perhaps because, as Patterson (1981) and others have recently claimed, fossils and their stratigraphic relations to each other are not after all the panacea for phylogeneticists. Extreme cladists would now deny that fossils can ever be useful in such issues. This seems excessively dismissive of what has undoubtedly been a major testing ground for the lesser aspects of phylogeny, within classes, orders or families; for a great deal is known about the time-scale of faunal changes, and the chequered history of groups such as cephalopods, brachiopods and echinoderms, that would be entirely lost without the fossil record. But unfortunately the anti-fossil lobby may well be right in its judgements for the problems of higher taxonomic weight considered here (Fortey & Jefferies 1982). It actually seems unlikely, from the literature, that any one author's views of metazoan phylogeny has ever been substantially formed, or substantially altered after formation, by reference to the palaeontological record.

4 Evidence from chemistry and genetics

4.1 Introduction

In accordance with reductionist principles, molecular distributions and structures should be the most fundamental and ultimately satisfactory level at which evidence for phylogenetic relationships can be sought. The paradigm of modern biology requires that the base sequence of DNA determines the ultimate form and properties of the organism, as shown on page 51, even if it is not yet clear how this works at a developmental level. In an ideal world, then, a final taxonomy of the animal kingdom should be in terms of similarity and dissimilarity of this nucleic acid sequence, and it is the evidence available from genes, proteins and other chemical products that is considered in this chapter, leaving the processes of development for chapter 5.

In practice the field of molecular phylogeny is necessarily rather new, being largely dependent on techniques developed in the last forty years and vastly accelerated by those of the last decade. Indeed, this is an area where a genuine revolution has occurred; whereas the early and pioneering attempts to trace molecular history depended almost entirely on presence and absence criteria, there is now a new scientific discipline devoted to deciphering the three-dimensional structures of proteins, the sequences of proteins and genes, and of the intervening DNA regions. Thus practical comparisons are now possible at the level of the essential make-up of the genetic material, and the rearrangements, deletions, insertions, and duplications of portions of nucleic acids as they have occurred through evolutionary history; and this is surely the most promising area for new insights into phylogeny in the near future.

However, most current workers in the field have approached their material very much from backgrounds as geneticists or as structural molecular biologists, and as yet have (perhaps understandably) failed to explore the realms of invertebrate relationships, preferring to test their skills and techniques on smaller issues of specific or generic status in well-known groups, or taking a broad sweeping approach at the whole living world. Thus the material available is at best thin, and cannot lead to any very firm conclusions at present. Where comparisons have been drawn they are frequently confined to primates, or perhaps extended to mammals and birds, or - if we are really lucky - may involve a handful of vertebrates, a solitary insect, nematode or sea urchin, and a protist or yeast.

Nevertheless, the field of molecular phylogeny is likely to be so important in the next few years that it cannot be left out here. Instead, this chapter considers the early comparative work, at a relatively gross level, and then summarises what little is so far available at the structural/sequencing level. On these latter issues this chapter will undoubtedly be out of date even as it is read; brief discussions of the principles and problems are included, in the hope of providing an appropriate framework for future results.

4.2 The traditional approach - 'diagnostic' molecules

Skeletal molecules

A number of different examples of the classic molecular constituents of skeletal elements and connective tissues have been invoked from time to time as useful indicators of phylogeny. These include the ubiquitous proteins (notably collagen), and a number of polysaccharide derivatives (cellulose, chitin); the former occur as protein frameworks modified by polysaccharide matrices, the latter as polysaccharide fibrous frameworks with a protein matrix. In addition there are specific tissue types, such as cartilage, assembled from several of these precursors in characteristic associations with cells.

Collagens may be recognised fairly easily by their amino-acid composition (about one third glycine, with high percentages of proline and hydroxyproline), their wide-angle X-ray diffraction pattern, and their characteristic 60-70nm banding when analysed by electron-microscopy. But collagen is perhaps the least useful skeletal molecule for present purposes, as it occurs in every metazoan group that has been tested, (and in some animals is by far the bulk of the total protein content). It is normally produced mesodermally, but may be ectodermal in the cuticles of worms such as nematodes and annelids. Simple presence/absence criteria therefore are of no help in phylogenetic analyses. The fact that collagen is found in all the Metazoa and is entirely absent in protists (though related and possibly precursor molecules do occur) could be taken to support theories of a single origin for metazoans from amongst the unicells (see chapter 7). However, the existence of related molecular sequences in protists, and the facts that collagen is such a necessary accompaniment to effective multicellularity and can only be made once sufficiently oxygenating conditions are present (see chapter 4), suggest that it could readily have been synthesised independently in various groups of incipient metazoans in the Pre-Cambrian.

A variety of fibrous proteins of non-collagenous nature occur throughout the animal kingdom, and their classification remains uncertain. In chordates the epidermal fibrous proteins are termed keratins, and these molecules are supposed to be restricted to terrestrial vertebrates, but several authors have proposed the wider term 'epidermin' to include all epidermal structural proteins, given that the criteria which distinguish supposed keratins are ambiguous. However, if the definition is spread too wide, then all phylogenetic significance at a presence/absence level is likely to be lost; all groups that have relatively tough cuticles are included, and those with simpler epidermal coverings are excluded (see section 6.3).

A further major group of structural proteins, the elastins, appear to present similar problems of diagnosis and characterisation, and are equally subject to convergent and divergent occurrence arising from ecological and mechanical requirements (see Mathews 1967).

Non-protein structural molecules are generally easier to define, and polymerised hexoses or pentoses are particularly common as skeletal elements, especially as surface sheets rather than internal permeating connective tissues. Cellulose is the best known of all, but though dominant in plants it is represented in animals only in the test of ascidian tunicates, in a few worms, and in a modified form in mammals. By contrast, chitin (a homo-polysaccharide of repeating N-acetyl- D-glucosamine units) has long been a classic case quoted in support of molecular phylogenies (see Rudall & Kenchington 1973 for a review of pertinent early literature). Three types are known, with differing chain orientations, though their distribution is sporadic and they may occur in different parts of the same animal (Jeuniaux 1971, 1982). Chitin occurs in most phyla of animals, and while it is particularly evident in the arthropods it is also found in all the major traditional 'spiralian/protostome' groups (annelids have chitinous chaetae and crop linings, molluscs have a chitinous radula and there is usually some chitin in the shell). The lophophorate phyla use it in their cuticles, as do priapulids, some hydrozoan cnidarians, and some protozoans. Nematodes and sponges use it to protect their resistant eggs or gemmules. But it is conspicuously absent from both echinoderms and vertebrates, supporting the alliance of these phyla in the 'deuterostomes'. Jeuniaux (1963) propounded a well-known phylogeny based on chitin distribution (fig 4.1), largely supporting accepted patterns.

There are a few problems with this view however (see Løvtrup 1977). Given its presence in fungi and many protists, chitin synthesis must be a very fundamental property of cells, and it seems clear that this synthetic pathway has merely been lost in a few groups; a shared *lost* character is unlikely to have much value in phylogeny. Thus while it remains true that chitin has not been demonstrated in echinoderms, hemichordates or vertebrates, it is also apparently absent in flatworms (but not gnathostomulids), in nemerteans, and in sipunculans, all good 'protostome'-affiliated phyla. In contrast, pogonophorans and chaetognaths possess it, though formerly both were assigned to the deuterostomes. This is not too much of a problem, as both groups are sufficiently enigmatic to be set aside from this major dichotomy (see chapter 12). But more seriously it has been claimed that both the gill bars of the cephalochordate *Branchiostoma* and the 'peritrophic membrane' of tunicates are chitinous, rather undermining the idea that the deuterostomes are a natural assemblage in which some stem group lost the ability to synthesise chitin. These findings have, however, been challenged (Rudall & Kenchington 1973) on the grounds that the tissues concerned do not show the requisite X-ray diffraction patterns to be chitin. Florkin (1966) and Rudall & Kenchington (1973) maintain that the hypothesis of a loss of chitin synthetase in primitive deuterostomes is fully supported, and that this enzyme has also been secondarily lost in certain protostome groups (fig 4.1). However, there appears to be some element of interpreting the data to fit preconceptions here,

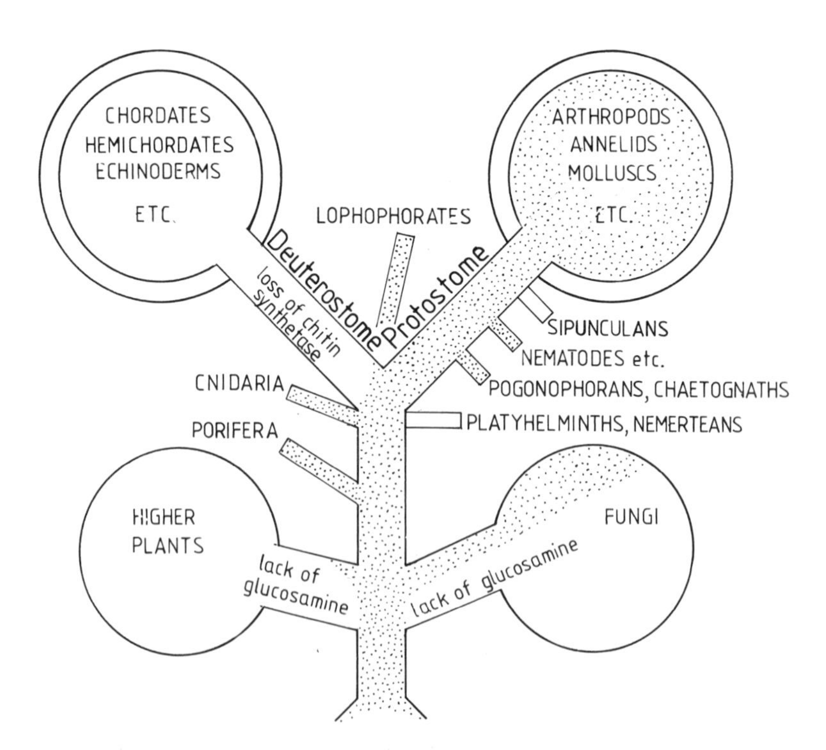

Fig 4.1. The distribution of chitin in invertebrates, based on Jeuniaux (1963); stippled areas indicate the presence of this molecule.

rather than using it in its own right; many other phylogenies would be at least equally plausible. A summary of the known occurrence of chitin in the Metazoa is given in table 4.1.

Other polysaccharide molecules contribute to skeletal materials, particularly the mucopolysaccharides where the regular *B* 1-4 linkages of chitin are replaced by alternating *B*1-4 and *B*1-3 links (the latter characterising glucans). Hyaluronic acid, chondroitin derivatives and keratosulphates are included here. However, the relationships between these various molecules are a little obscure, and their occurrence only partially determined; a brief review is given by Løvtrup (1977), but leads to no helpful conclusion. Sialic acids, frequent components of glycoproteins and important for the functional and structural contact between cells, were used phylogenetically to support the protostome/deuterostome distinction by Warren (1963), who maintained that they only

Table 4.1 *Occurrence of chitin in invertebrates*

Present	Absent	Unknown
Some Protozoans (Thecate amoebae, and ciliate cysts)	Most Protozoans	Placozoans
Sponges	Echinoderms	Mesozoans
Cnidarians	Hemichordates	
Nematodes	Tunicates ?	
Nematomorphs??	Cephalochordates ?	
Acanthocephalans??	Vertebrates	
Priapulids	Ctenophores	
Entoprocts	Platyhelminths	
Rotifers	Nemerteans	
Gnathostomulids	Sipunculans	
Gastrotrichs		
Annelids		
Echiurans		
Molluscs		
Arthropods (Insects Crustaceans Chelicerates)		
Onychophorans		
Tardigrades		
Pogonophorans		
Phoronids		
Bryozoans		
Brachiopods		
Chaetognaths		

occur in the latter super-phylum. In fact he identified these compounds in all vertebrates, cephalochordates, hemichordates and echinoderms that were tested, though they seemed surprisingly absent in the tunicates (conventionally a sub-phylum of the chordates). A few instances of sialic acid presence in molluscs, crustaceans and acoel platyhelminths were also described (see also Løvtrup 1977). But a general absence in protostomes, lophophorates, and diploblasts still seems to hold; in fact according to a more recent review (Segler *et al.* 1978) only exogenous sialic acids ever occur in all the protostomes tested (usually in the digestive glands), whereas there is clear evidence for the endogenous production of these compounds in all deuterostomes including some tunicates. These authors suggest that the presence of sialic acids may contribute to the embryological differences found in deuterostomes (chapter 5), where cell contact phenomena are often involved.

Finally, cartilage is a complex tissue assembled from collagen, acid polysaccharides (particularly hyaluronate and chondroitin sulphate), and polymorphic cells, forming a gristle-like material. Once regarded as a typically vertebrate phenomenon, it is now clear that precisely similar tissues occur in several invertebrate groups (Person & Philpott 1969). They have been reported in cnidarians and annelids, and are particularly characteristic as the molluscan odontophore and cephalopod cranium, (though the latter additionally contains chitin). The chelicerate *Limulus* also has cartilage, as do some tunicates. Thus this complex material appears to occur in some representatives of all the classic 'super-phyla'; either it was invented rather early in the metazoans and has been lost by many groups, or it has been assembled uniquely by a number of different phyla at different times, and is of little assistance to phylogeneticists.

Physiological molecules

Haemoglobins and other oxygen-carrying pigments are a favourite source of phylogenetic speculations, as are a number of other protein-based moieties that function as carriers or transducers in animals. In almost every case, these proteins are either almost ubiquitous (absence suggesting secondary loss) or occur so sporadically that independent evolution from rather fundamental sub-units seems certain. In either of these situations simply scoring presence or absence is unlikely to be very helpful. So this section passes rather quickly over such molecules.

The distribution of respiratory pigments certainly makes little sense in relation to any current classification (table 4.2). Haemoglobins and

Table 4.2 *Invertebrate respiratory pigments and molecular weights*

Haemoglobin - Iron-porphyrin protein

Either in cells (usually low MW) or in solution (usually large polymers)

Group	Genus	MW
Vertebrates		68 000
Mollusc		
	Planorbis	1.6 million
	Arca	33 600
Insect (dipterans, hemipterans)		
	Gasterophilus	34 000
	Chironomus	31 400
Crustacean (branchiopods)		
	Artemia	250 000
	Daphnia	400-700 000
Echinoderm (holothurians, plus one ophiuroid)		
	Thyone	23 600
Annelid		
	Lumbricus	
	Arenicola	2.6-3.6 million
	Serpula	
	Notomastus	36 000
Nematode		
	Ascaris	330 000
Platyhelminth (trematodes, turbellarians)		
	Dicrocoelium	22 000

Also some echiurans, phoronids, nemerteans, protozoans; and a few plants.

{**Myoglobin**

Muscles - most - 17 000}

Chlorocruorin - Iron-porphyrin protein (always in solution)

Group	Genus	MW
Annelid (four families of polychaetes)		
	Spirographis	3.4 million

Haemerythrin - Protein linked with iron (always in cells)

Group	Genus	MW
Sipunculan		
	Sipunculus	66 000
	Phascolosoma	120 000
Annelid		
	Magelona	
Priapulid		
	Priapulus	
Brachiopod		
	Lingula	

Haemocyanin - Protein linked with copper (always in solution)

Group	Genus	MW
Mollusc (all except bivalves)		
	Helix	6.7 million
	Octopus	2.8 million
Chelicerate		
	Limulus	1.3 million
Crustacean		
	Palinurus	447 000
	Homarus	803 000

myoglobins, based on the same sub-units, are the commonest proteins used, though they vary markedly in quaternary structure and molecular weight (about 15 000 up to 3 million). They are composed of varying haemoglobin polypeptide chains; in general the vertebrate varieties are tetrameric, the protochordate and invertebrate forms monomeric or dimeric (though some molluscs and annelids have tetrameric haemoglobin, whilst *Ascaris* and *Glycera* haemoglobins may be octomeric). A survey of the occurrence of different structural forms is shown in fig 4.2, based on an excellent review by

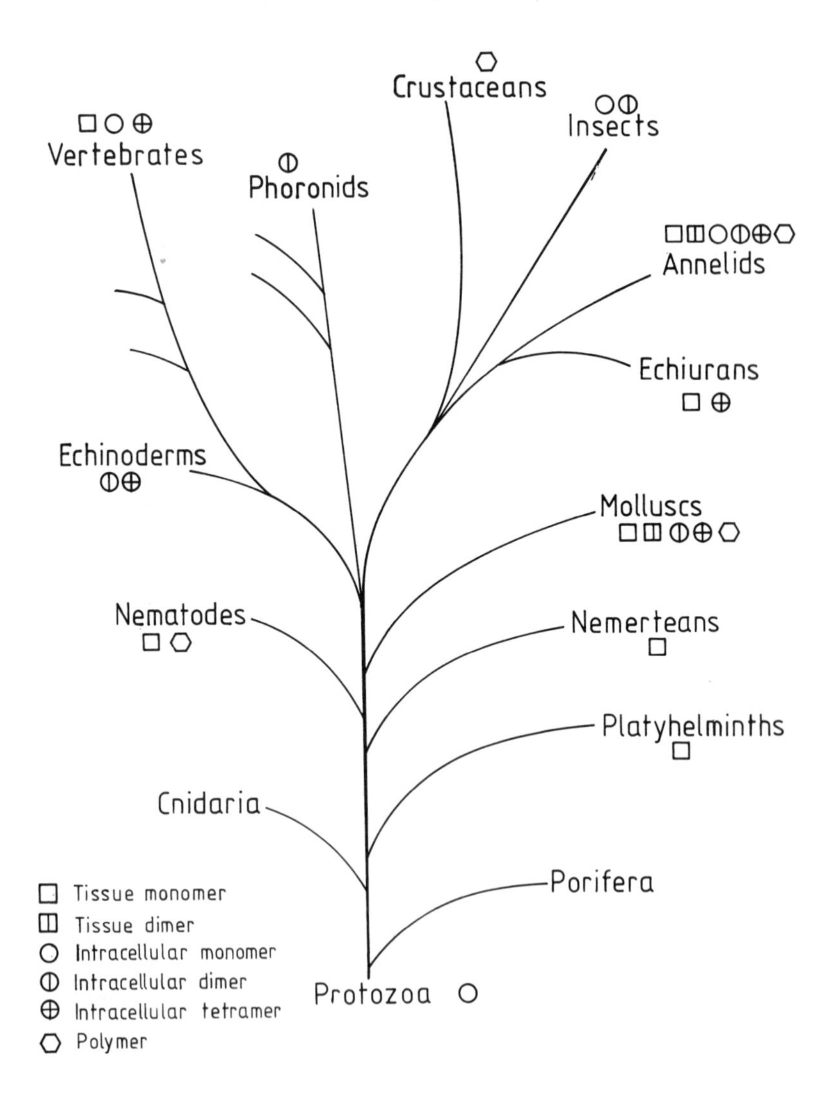

Fig 4.2. The structure of haemoglobin molecules in invertebrates, based on Terwilliger (1980).

Terwilliger (1980). Some interesting points emerge from this. It does appear that the conventional protostome groups have a greater diversity of haemoglobin forms; and there is some indication that only the 'higher' groups of Metazoa routinely build the monomeric subunits of globins up into larger molecules. The clear differences between crustacean and insect haemoglobins (the former being rather large polymers, the latter relatively small and with some clear resemblances to vertebrate forms), may be relevant to the issue of arthropod relationships discussed in chapter 11. And while it is clear that even dissimilar invertebrate haemoglobins are based on a constant oxygen-binding peptide unit of about 14-17 000 daltons, the major exception is *Ascaris* which seems to have a larger 40 000 subunit. Unfortunately no other 'pseudocoelomate' globins have been analysed to indicate whether this is of any phylogenetic significance.

The other respiratory pigments are infrequent, though haemocyanin is ubiquitous within the molluscs and fairly common in crustaceans (Linzen *et al.* 1985). To indicate the futility of phylogenetic inferences here, haemerythrin occurs in three groups that may be fairly closely related (annelids, sipunculans, priapulids) but also crops up in the conventionally distant brachiopods; while the single class Polychaeta exhibits within itself three of the four respiratory pigment molecules, in apparently random patterns amongst the families.

Another of the cornerstones of early comparative biochemistry was the use of phosphagens as diagnostic characters. Initially it was noted that vertebrates principally used creatine phosphate (CP), whilst invertebrates relied on arginine phosphate (AP). Subsequently creatine phosphate was identified in echinoid and ophiuroid echinoderms, and in hemichordates and cephalochordates; it thus seemed to be the 'deuterostome' phosphagen. However, Roche *et al.* (1957) gave particularly damning evidence of the ubiquity of AP and the presence of CP in a great many widely separated groups, and this issue has become clouded further by the realisation that phosphagens may not be endogenous but can be accumulated from the environment. Additional investigations also undermine the supposed dichotomy; for example Rockstein (1971) reports both AP and CP in all classes of echinoderms, while tunicates appear to have only AP. However, Watts (1975) has reinterpreted some of the data relating to deuterostomes (see chapter 12) and believes that CP, a biochemically more efficient molecule, only occurs there in advanced groups (the 'higher' echinoderms, for example) or in specialised sites (certain muscles, and highly motile spermatozoa). He suggests there is a phylogenetically significant pattern, and in particular that the absence of CP in most *adult* tunicates supports the idea of CP-containing vertebrates and cephalochordates as neotenic offshoots from earlier protochordate larvae. A summary of the present situation is given in table 4.3. Matters are clearly more complex and confused than the earlier literature allows, and no fundamental dichotomy for this character can be claimed. A good critique of the topic by Kerkut (1960), in which he points out many of the snags of this kind of analysis, amply underlines the dangers of simple presence/absence criteria in biochemical phylogenies.

Transmitters, hormones and pheromones provide another possible source of useful comparisons, though with the obvious proviso that presence of particular messengers is likely to be affected by functional need, and thus to

Table 4.3 *Distribution of phosphagens in invertebrates*

	Arginine	Creatine	Others
Protozoa	+		+
Porifera	+	+	+
Cnidaria	+	+	+
Nematoda	+		
Platyhelminthes	+		
Nemertea	+		
Mollusca	+	+	
Annelida	+	+	+
Sipuncula	+	+	
Crustacea	+		
Chelicerata	+		
Insecta	+		
Echinodermata	+	+	
Hemichordata	+	+	
Urochordata	+		
Cephalochordata	+	+	

vary with habitat and lifestyle. But since these molecules are often small and structurally simple, they are often chemically invariant between groups and an analysis of presence/absence is potentially useful. Neurotransmitters are perhaps the least promising, and seem to be conserved throughout the Metazoa and even inherited from the protists; classic examples are acetyl choline, gamma-amino-butyric-acid (GABA), glutamate, and the catecholamines. However, identification of peptide transmitters and ultimately of their amino-acid sequences (Barnstaple 1983; Iversen 1984) may eventually prove a useful phylogenetic tool. Another aspect of neural chemistry has also been discussed from a phylogenetic viewpoint recently (Hall & Ruttishauser 1985), demonstrating the presence of a similar neural cell adhesion molecule (NCAM) in all vertebrates and molluscs, but its absence in insects, crustaceans and nematodes.

Hormones (whether derived from the nervous system or of endocrine origin) may offer more scope for comparisons as they are commonly either small polypeptides or steroids. Ecdysone, for example, occurs with little modification in insects, crustaceans, chelicerates, nematodes, and some plants. Unfortunately this seems most unlikely to have phylogenetic significance, but rather to indicate that when an animal's design creates the need to moult, ecdysones are always enrolled to control the process. Other cases of sterols being used as taxonomic features within echinoderms and sponges are quoted by Kerkut (1960), though at a phyletic level they have proved diagnostically unsuitable.The polypeptides that in other groups frequently control growth and regeneration, or oogenesis and spawning, may have more taxonomic potential.

Pheromones are likely to be unhelpful, as the possible chemical structures that achieve the right balance of properties are rather constrained. As yet the data collected on terrestrial insect pheromones show no obvious taxonomic patterns, revealing a small range of volatile organic acids repeatedly occurring and with specificity improved by varying the mixture. There does not seem any reason to suppose that water-borne pheromones in other phyla will offer greater assistance in phylogeny. Once again, the prospects at this level are disappointing.

Simple approaches to molecular phylogenies thus turn out to be fraught with problems, inconsistencies, and errors. Scoring the presence of a molecule in a few samples of a few species from a phylum is never going to be good enough as grounds for comparisons between higher taxa, even without the problems of purification and uncertainties as to whether the molecules found are endogenously synthesised or derived from the food or environment. Some of the classic success stories of early comparative biochemistry appear to be untenable, whilst others retain a degree of plausibility in separating echinoderms and vertebrates from other invertebrates but tend to leave question marks over other supposed deuterostomes. Clearly there is a need to move on rapidly to a more sophisticated consideration of molecular history and interrelationships.

4.3 Molecular and genome evolution

In recent years, studies of molecular evolution have increasingly concentrated on the informational macromolecules - nucleic acids and proteins - that directly influence and reflect the evolutionary process. Ultimately, the raw materials for evolution are changes in the base sequence of nucleic acids, and this is manifested as changes in the amino-acid sequence of proteins and their three-dimensional structure (or perhaps equally commonly in their replication, if the DNA control regions are changed). These molecules therefore represent a fundamental library of evolutionary change (Zuckerkandl & Pauling 1965), and the extent of similarities or differences between homologous and isofunctional molecules in any two animals should indicate their degree of relatedness. If there is a constant mutation rate in the genetic material, as was early indicated by Zuckerkandl & Pauling (1962), then a numerical estimate of molecular difference will also actually indicate the time at which they diverged (the idea of a 'molecular clock'), and construction of strictly correct phylogenetic trees becomes a feasible goal. In effect, such a tree normally shows the branching order within protein families, leading to present-day molecular structures from a common ancestor coded by a 'primordial gene' (see Richardson 1981; Jacob 1983). The constancy of the 'clock' is debatable (e.g. Wilson *et al.* 1977; Goodman *et al.* 1982; Kimura 1983), since different genes do evolve at different rates (Dickerson 1971). But the principle holds good within gene/protein families, and a molecular evolutionary clock may also be useful on a wider scale if derived from average rates of evolution of a sufficiently large sample of proteins.

Ideally, then, molecular phylogeny could be the answer to all our problems. Furthermore there are now adequate techniques to analyse molecular structure. Literature in the field is expanding hugely, and new and somewhat unexpected properties of the genome are emerging. The presence of introns and exons, transposable elements, various types of repetitive DNA, and pseudogenes, together with the occurrence of multiple copies of genes in clusters (see Nei & Koehn 1983, for reviews), all contribute to a marked fluidity of the genome over evolutionary time. Notes of caution are thus appearing, and in particular, the old problems of detecting and taking due account of convergent and parallel changes is becoming ever more apparent. Molecules are subject to selective pressure just like all other traits; therefore they may end up with similar structures for purely functional reasons. This clearly occcurs with the mammalian trypsin and the bacterial subtilisin enzymes, which have very different tertiary structures but identical active sites where they split peptide bonds. Conversely, lysozyme and alpha-lactalbumin have about 60% sequence homology and very similar tertiary structures, yet very little functional homology. At the level of sequence these problems might be somewhat lessened (but not abolished), since the genetic code is degenerate and apparently non-adaptive changes can occur (see Kimura 1983) where convergent similarity is unlikely; for example several base codons may give the same amino-acid ('silent substitutions'), or similar amino-acids at relatively unconstrained sites in the molecule may allow normal functioning. But the realisation that most organisms contain an apparent vast excess of DNA, much of which is highly repetitive, rather mobile within the genome, and not translated, has both complicated and enlivened the whole issue of sequence interpretation in recent years. It has also led to the view that tertiary and quatenary protein structure may be the best hope for elucidating relationships, as these levels of organisation at least change less rapidly than the relatively unconstrained primary structure.

Proteins, as the expressions of gene action, have been analysed in a number of ways. Early studies concentrated on comparing molecular weights, amino-acid ratios, or chemical and electric charge properties by electrophoresis. X-ray crystallography of three-dimensional structure supplemented these approaches in the early 1960s, and may still provide vital insights (see Phillips *et al.* 1983) particularly in combination with nuclear magnetic resonance analysis. The ultimate chemical deciphering of linear sequence (from which secondary and tertiary structures can increasingly readily be predicted, now that the analyses are computerised), is clearly complementary to these physical techniques.

Nucleic acid studies have also been progressing. Crystallographic studies of nuclear material were of relatively early vintage, but are perhaps of less use now that the main problem of molecular structure of DNA has been settled. Ratios of purines to pyrimidines, and ratios of actual bases, were also used as evidence in elucidating the essential structure of nucleic acids in the 1950s, but may still provide some interesting comparisons at a gross level. An example is shown in table 4.4, comparing ratios of G+C to the total base complement in mitochondrial DNA from a variety of phyla (Brown 1983), and suggesting relatively low ratios in invertebrates. The ability of the DNA strands to separate shows an even more striking vertebrate/invertebrate distinction.

Analysis is concentrated nowadays on structure and actual sequencing, aided by the use of recombinant DNA and restriction enzymes. Schopf (1983) describes nucleic acid structure and composition as the new fourth 'level' in

Table 4.4 *Gross DNA composition and properties in animals*

	% Guanine + cytosine	Ease of strand separation (as mg/ml CsCl gradient)
Vertebrates		
mammals/birds	37-50	31-43
fish/amphibians	39-41	13-23
Echinoderms		
Strongylocentrotus	39	< 5
Molluscs		
Mytilus	43	< 10
Insects		
Drosophila	21	< 5
Nematode		
Ascaris	31	-
Platyhelminth		
Hymenolepis	32	-

approaching comparative biology. Russell & Subak-Sharpe (1977) analysed the DNA composition of a number of deuterostomes when seeking evidence for chordate ancestry, giving the cluster analysis shown in fig 4.3; craniates proved surprisingly distinct from all invertebrates and supposed protochordates tested, while echinoderms, brachiopods and tunicates seemed remarkably mixed up. These authors indicate that other invertebrates are even more different from craniates than are the deuterostomes, perhaps supporting traditional dichotomies in the invertebrates though hardly giving strong backing to any particular view of chordate affinities. A further example of nucleic acid studies is available in the works of Ishikawa on ribosomal RNA structure, summarised in a review (1977) that indicates a fundamental split between protostome and pseudocoelomate RNA units on the one hand (having a 'hidden break' in the 28s rRNA), and deuterostome units on the other, lacking this break. Interestingly, the lophophorate groups (or Tentaculata in the author's terminology) had RNA much more like that of protostomes. The significance of these results is still uncertain though, given that in a few cases (nematodes and aphids) aberrant RNAs without the break occurred, within otherwise cohesive taxa.

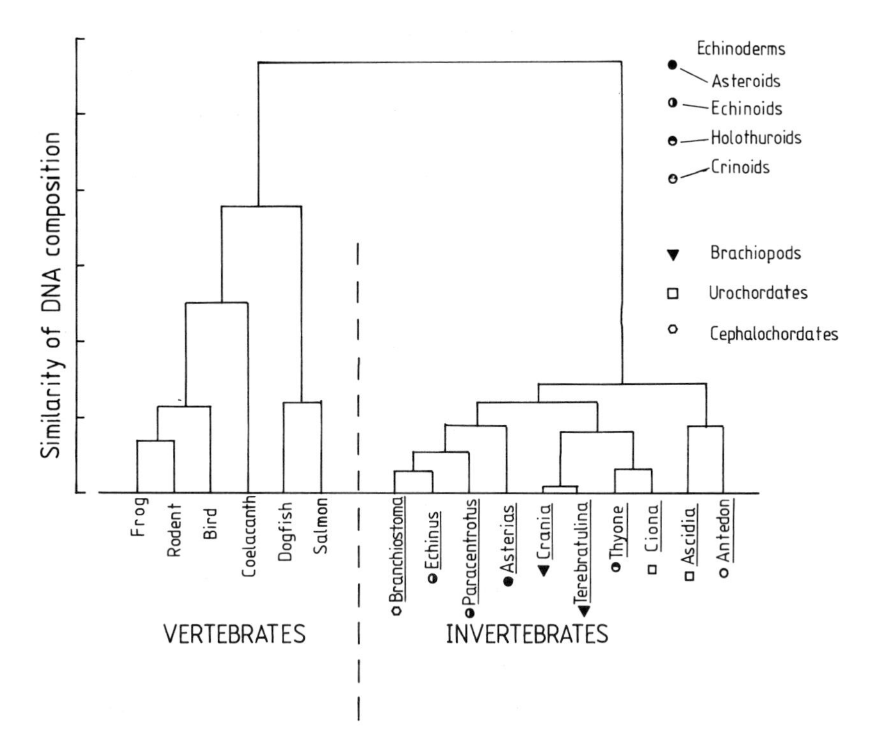

Fig 4.3. Gross similarity of DNA composition in a range of vertebrates, deuterostomes, and lophophorates; from Russell & Subak-Sharpe (1977).

What then has been the outcome of this concentrated technological activity? As far as invertebrates go, really not very much can yet be said with certainty. The best approach we can adopt here is to review the literature on just a few of the 'classic' protein and gene families, for though quite a number of proteins have been used, most are as yet of no use at all to invertebrate phylogeny studies (for example the immunoglobins, insulins, and albumins). But a few examples should serve to show the possibilities for molecular analysis, and indicate if any new phylogenetic patterns are yet emerging - or if traditional ones are being confirmed.

Collagens

The protein collagen is possibly the most widely studied of all molecules; some analysis has been made of it in every metazoan phylum, and useful reviews are given by Adams (1978), Solomon & Cheah (1981), and Runnegar (1985). Vertebrate collagen is inevitably best known. Individual polypeptides form alpha chains (procollagen) about 1000 amino-acids in length, with the classic repeat sequence -glycine-X-Y-, where X is frequently proline and Y is frequently hydroxyproline. Three of these polypeptides assembled together in a helical conformation make a single collagen molecule, forming a semi-rigid microfibril with a molecular weight around 300 000; and about a million such molecules together make up the characteristic banded collagen fibril detectable with the electron microscope and forming the basis of most connective tissues. Most vertebrate collagens are made of three identical alpha chains.

In vertebrates as a whole, though, there are at least nine different possible alpha chains, and some parts of unrelated molecules (for example acetylcholinesterase) also contain helical collagenous sequences; hence there must be at least 15 separate genes in the collagen family of vertebrates. These presumably descended from a primordial gene formed by repeated duplications of the gly-X-Y tripeptide genetic sequence. A common repeat of 54 base pairs (18 amino-acids) is particularly evident, as is the larger 'D unit' of 234 amino-acids (78 gly-X-Y sequences). Collagen genes are extensively interrupted by untranslated sequences (introns), so that simple base-pair counting has its hazards; though the positioning of the introns has helped to confirm the primordial 18 amino-acid unit, which is rarely interrupted by insertions.

Turning to invertebrates, the collagens of at least some phyla bear a striking resemblance to those of vertebrates. The cnidarian *Actinia,* the acanthocephalan *Macracanthorhynchus*, and the mollusc *Haliotis* all have tripeptide molecules with a molecular weight of about 300 000. In all these cases the three alpha chains are identical. However, in some cephalopods, crustaceans and echinoderms a mixture of two chain types occurs in one molecule. Even more unusual are the collagens of the nematode *Ascaris*, having a molecular weight of 900 000, with disulphide cross-links; and of the annelids *Lumbricus, Nereis* and *Pheretima,* all with molecular weights of about 1.8 million and apparently assembled from chains up to six times the length of those found in vertebrates. Nematode and annelid collagens do appear to have the same fundamental sub-units as vertebrates (especially the 54 base-pair repeat), but they also differ strikingly from 'normal' in their complements of proline and hydroxyproline; in the former group the ratio is 20 proline to one hydroxyproline, while in the latter the same ratio is 0.05. At a further extreme, the tough silk collagen of the sawfly *Nematus* has no hydroxyproline at all. This last point underlines an important feature of collagen variability, as it relates composition to function. Indeed it is now known that there are at least three quite different collagen types in *Ascaris* (cuticle, body wall, and basement membrane), with a greater variation in properties than occurs in the whole of the vertebrates.

At the sequencing level best estimates from a variety of collagens now indicate that the basic molecule evolved around 800-900 MYA (Runnegar

1985). This date conveniently corresponds quite well with the point when synthesis of collagen would first become possible in relation to molecular oxygen supplies (Towe 1981), and to the probable time of the appearance of fossilised metazoan groups.

Although collagens are relatively well studied, and provide some interesting insights into molecular evolution, it cannot be said that they have helped much as yet in invertebrate phylogeny. We have little comparative data on the occurrence of different alpha chains in different phyla, or of the deletions, insertions or replications in those chains, though these would be fruitful areas for study. Nematodes and annelids seem more aberrant (in the sense of being more unlike vertebrates) than do some supposedly lower groups (cnidarians, flatworms), but it seems most likely that this reflects functional constraints as both nematodes and annelids use collagen as a major component of a thick, mechanically important, cuticle. There are virtually no good data for many of the key invertebrate phyla on which phylogenetic debate centres (molluscs, echinoderms/protochordates, lophophorates), so we are inevitably left at something of a loose end for the present.

Globins

Whilst the patterns of occurrence of haemoglobin have proved of little use phylogenetically (section 4.2), the family of globin genes has been a favoured tool for molecular phylogeneticists, as reviewed by Jeffreys *et al.* (1983). Globins occur throughout much of the animal kingdom, and even in some plants as 'leghaemoglobins', (though these *may* be a case of gene transfer between species, by retroviruses or other agents). Detailed sequence comparisons have allowed the construction of phylogenetic trees for the globin gene family, which is exemplified by a constancy of divergence rate. The tree for the globin genes of humans is particularly well known; we have eight globin genes in two sub-families (alpha and beta) contributing to haemoglobin tetramers, differentially expressed in embryo, foetus and adult.

Runnegar (1984) suggests that globin genes are clearly monophyletic but share a common ancestry with type *b* cytochromes, from which two of their primitive three exons were derived. The most ancient change in the primordial globin gene was then a duplication about 1100 MYA, giving rise to the ancestors of modern haemoglobin and myoglobin genes; actual myoglobin was established as a separate entity perhaps 500-800 MYA. The split between alpha and beta families occurred about 500 MYA, probably soon after the origin of vertebrates. The most recent major changes to give the full set of eight genes is dated about 40 MYA. Unfortunately there is still a large gap in our knowledge between 1100 and 500 MYA, when invertebrate globins must have been evolving; nearly all the recent studies have concentrated on differences within primates, sometimes including other vertebrates, but rarely straying into an analysis of soft-bodied forms. Jeffreys *et al.* (1983) remark that 'the entire invertebrate kingdom of haemoglobins is completely virgin ground for DNA analysis'.

There are just a few known sequences of invertebrate globins though. Several molluscan varieties have been analysed, and show great variability; *Aplysia*, for example, seems to have haemoglobin more like that of a sperm whale than like another mollusc, *Busycon* (though the latter does seem to be particularly abnormal amongst the molluscs). A partial analysis of phoronid haemoglobin puts it about equally distant from the (protostome) mollusc and the (deuterostome) whale. Some details of annelid haemoglobins, and of that of a holothurian, have also been published, but no obvious patterns could be said to be emerging as yet. There can be no doubt that useful comparisons will soon be available though.

Cytochrome c

As a protein found wherever electron transfer chains occur, whether in photosynthetic or respiratory processes, this molecule is particularly attractive as a key to relationships, and has been used to unravel the origins of basic metabolic pathways in the different kingdoms of living material (Dickerson 1980). The sequencing of cytochrome *c* was an early showpiece of molecular phylogeny within the animals, too (Dayhoff & Park 1969; McLaughlin & Dayhoff 1973). The eucaryotes apparently gave rise to zooflagellates and later to green plants, whilst both fungi and metazoans developed later still. Only four invertebrates were tested by McLaughlin & Dayhoff, but data from these suggested (inconclusively) that molluscs were closer to chordates than were insects, all being derived from flatworm-like ancestors.

Actins and tubulins

For these two proteins, involved in motility systems, the situation improves a little as there have been some concerted attempts to consider invertebrate issues. The actin genes have been considered in some detail by Davidson *et al.* (1982); a nematode, an insect and a sea urchin are already well studied, all with multiple genes, and a yeast with just a single actin gene has also been characterised. Unlike the globin genes, the actins have proved very variable, especially in the siting and length of intervening sequences. Davidson *et al.* neverthless identify common insertions at three sites in the genes of an echinoderm and of vertebrates, absent in *Drosophila*; and they cite this in support of the deuterostome assemblage. Their attempt to use absolute sequence data in evaluating the same dichotomy seems less convincing, as it implausibly relies on designating the nematode *Caenorhabditis* as a protostome (see chapters 8 & 9). However, it does highlight the point that insect actins, though used in striated muscle, are more like the cytoplasmic actins than the skeletal muscle actins of mammals. New sub-families of actins may thus have arisen in the deuterostomes from which the vertebrate skeletal actins evolved.

Clearly these data are scanty and any postulates based on them are likely to be undermined by new findings. However, they do begin to indicate the possibilities of such comparative work.

Tubulins, in contrast to actins, may be among the more highly conserved proteins (reviewed by Roberts & Hyams 1979), though heterogeneities of alpha and beta tubulins are beginning to emerge within species and across distantly related species (Cowan & Dudley 1983; Adouette *et al.* 1985). In particular, it seems that in the Metazoa there may be more difference between the flagellar and cytoplasmic tubulin within one organism than between the highly conserved flagellar tubulins of different species. In protists the same tubulins perform both roles. The implication is that duplication and evolution of tubulin genes really only began within the early metazoans, the primitive type being retained in the flagellar axoneme while new cytoplasmic tubulins diversified. This view was confirmed with antibody studies by Adouette *et al.* (1985), who showed that *Paramecium* tubulin antibodies reacted with all ciliate tubulins, and with most ciliary/flagellar tubulins of a range of invertebrates and vertebrates, but never with the cytoplasmic tubulins of the latter groups. This again suggests a rather fundamental split between protists and metazoans. It may also suggest that cytoplasmic tubulins are particularly good candidates for future sequence comparisons.

Histones

Two recent reviews of histone genes (Hentschel & Birnstiel 1981; Maxson *et al.* 1983) have stressed their highly conserved nature, which should make these genes good candidates for phylogenetic comparisons between higher taxa. Histones form complexes in fixed mass ratios with DNA, and occur in at least five varieties (H1, H2A, H2B, H3, H4), of which the last two are particularly invariant. All the coding genes seem to undergo exceptionally slow evolutionary change, and are characterised by a lack of introns. Whereas the 1981 review suggested that invertebrate histones (echinoderms, nematodes) had a clear tandem repeat structure that was lacking in vertebrates, more recent work discussed by Maxson *et al.* indicates an almost complete lack of phylogenetic pattern in the histones; closely related species may have random or repetitive sequences, and similar repetitions may occur in conventionally distant groups. The histones have thus proved disappointing so far from the point of view of invertebrate relationships.

Homeobox gene sequences

There has recently been some excitement over the nature and occurrence of a particular DNA sequence that has become known as the 'homeobox', and which after the work of McGinnis *et al.* (1984) was thought to be associated with metamerism in metazoans. The DNA sequence involved is highly conserved, about 180 base pairs in length, and was originally identified in some of the homeotic and segmentation genes of the fruitfly *Drosophila*. Its protein product appears to mediate its functions by DNA binding, and seems to be confined to the nucleus.

Data in the original paper recorded multiple copies of the sequence in annelids, insects, and chordates, and suggested its absence in nematodes, and echinoderms. This cut across the conventional protostome/ deuterostome divisions, and led McGinnis to suggest the link with trunk segmentation. Subsequently (McGinnis 1985) this postulate was somewhat modified when the sequence was also found in echinoderms and urochordates, whilst platyhelminths and cnidarians were added to the list of those phyla lacking the sequence. McGinnis thus suggested a general correlation with metamerism (though echinoderms are not metameric, and urochordates cannot really be so designated), and he associated the gene sequence with 'higher Metazoa'.

A further paper by Holland & Hogan (1986) has clarified the issue somewhat by identifying homeobox sequences in further groups (table 4.5), including molluscs, brachiopods, and (probably) nemerteans and tapeworms. Sipunculans are also thought to have this type of homeotic gene (Holland, pers. comm.). It thus appears that this particular gene sequence occurs throughout many metazoan phyla, irrespective of their conventional affinities, but that it may still have a certain association with metameric groups (for example, it may occur in the cestodes, with bodies assembled from repeating units, but not in the closely related flukes). It seems to be absent in the lower radiate groups and in worm-like pseudocoelomates; but interestingly it has been found in all coelomate groups tested. Since such a complex DNA sequence of around 180 base pairs is most unlikely to have arisen convergently, phylogenetic schemes that derive different coelomate groups independently from a protistan stage (e.g. those of Nursall or Inglis, see chapter 1) look very suspect, though coelomates could of course still be polyphyletic from a flatworm stage. The development of more precise probes is likely to give greater clarity to this issue in the very near future. It may well

Table 4.5 *Homeobox genetic sequences in invertebrates*

Present	Absent
Vertebrates (several)	
Cephalochordates	
Branchiostoma	
Urochordates	
Ascidia	
Phallusia	
Echinoderms	Nematodes
Paracentrotus	*Caenorhabditis*
Sphaerechinus	*Trichinella*
Brachiopods	*Nippostrongylus*
Terebratulina	*Strongylus*
Insects	*Ascaris*
Drosophila	
Tenebrio	
Crustaceans	Cnidarians
Leander	*Aequorea*
Sipunculans	*Sarsia*
Golfingia ?	
Annelids	
Lumbricus	
Herpobdella	
Molluscs	Molluscs
Aplysia	*Patella ?*
Biomphalaria	*Loligo ?*
Nemerteans	
Lineus ?	
Platyhelminths	Platyhelminths
Echinococcus ?	*Hymenolepis ?*
(tapeworm)	(tapeworm)
	Schistosoma
	(fluke)
	Yeasts, slime moulds
	& bacteria

turn out that the homeobox sequence codes for a specific protein concerned with binding the nucleic acid in many cases amongst 'higher' metazoans where genes are required to interact closely during development.

4.4 Conclusions

Molecular taxonomy will, within the very near future, make significant contributions to our understanding of invertebrate phylogenetics - this can hardly be doubted. Unfortunately it cannot really be said to have done so as yet. Nevertheless we do have a few positive indications to offer encouragement, and it is probably easiest to summarise these in terms of the traditional super-phyletic divisions of the animal kingdom.

(1) 'Metazoa' are probably distinct from protists in their diversity of cytoplasmic tubulins, and in their possession of collagen; but these molecules are clearly related to existing protistan precursors, and do not greatly help in deciding whether metazoans as a whole are monophyletic.

(2) 'Radiata' are not usefully united by chemical features. Although ctenophores have rarely been included in comparative studies, we can be reasonably confident of one major difference here in that many cnidarians possess chitin, but ctenophores do not.

(3) 'Pseudocoelomates' are not chemically cohesive in any respect, and the implausibility of this assemblage is clearly confirmed. Some groups share certain features with the protostome assemblage, and others with deuterostomes or acoelomates. Nematodes are particularly aberrant and cannot readily be allied with any other phylum; they have peculiar collagen, peculiar rRNA, and unique haemoglobin. Given the diverse chemical attributes of pseudocoelomates, it is clearly not possible to decide whether the more controversial groups (entoprocts and priapulids, for example - see chapter 9) belong here.

(4) 'Protostomes' (at least the classic annelid and various arthropod groups), can usually be scored together in chemical terms. But there are no molecular data to give unequivocal support to the monophyletic reality of the protostome assemblage (see chapter 8). Molluscs, platyhelminths, nemerteans and sipunculans each differ from the 'core' protostomes in at least some features; in particular, chitin is absent from the last three of these groups. Molluscs seem to share some features with the vertebrate line (for example the neural cell adhesion molecule), but of course do synthesise chitin. Sipunculans and molluscs share with annelids the presence of creatine phosphate, but arthropods and the acoelomate protostome worms lack this molecule.

(5) 'Arthropods' resemble each other closely in some respects, though they differ in both the occurrence and macromolecular structure of their respiratory pigments. Insects and crustaceans may both possess haemoglobin, but in the former it always seems to be a cellular monomer or dimer whereas in crustaceans a polymer of much higher molecular weight occurs. Chelicerates do not possess haemoglobin; indeed arachnids generally have no respiratory pigments, though xiphosurans possess haemocyanin. The arthropods, though clearly united (and constrained) by their chitinous cuticles, could thus represent several separate evolutionary lines as discussed in chapter 11.

(6) 'Lophophorates' do not have sialic acids, and do have chitin; in these respects, and in their rRNA, they are quite distinct from the deuterostomes. However their DNA composition is more like deuterostomes than protostomes, and the haemoglobin of phoronids seems about equidistant from the known equivalents in those two super-phyla. Chemically the lophophorates could therefore be tentatively described as a third and independent coelomate assemblage of equivalent rank to these two, or as an intermediate or transitional form between them (see chapter 13).
(7) 'Deuterostomes' (if we exclude the highly suspect pogonophorans and chaetognaths) are united by the presence of sialic acids in all groups except tunicates, by an absence of chitin (though with a query over some protochordates), and by the presence of creatine phosphate. None of these are exclusive features, but the group does seem reasonably chemically coherent. In addition, vertebrates and echinoderms share some common inserted sequences in their actin genes. However, within the deuterostome group vertebrates are rather distinct, particularly in the general properties and composition of their DNA, and it is impossible at present to see clear chemical evidence for their evolutionary route from invertebrate ancestors.

Overall, it would be fitting to leave this field on a few cautionary notes. The difficulties of detecting secondary losses of molecules, and the possiblities of inter-species gene transfer by viral agents, both of which may obscure phylogenetic relations, have already been mentioned; but above all the spectre of convergence needs to be reinvoked. This is no easier to ignore in molecular evidence than elsewhere in phylogenetic studies, and several of the similarities summarised above may be no indication at all of common ancestry but merely of convergent solutions to similar requirements for molecular functioning. Perhaps the elements of redundancy in the genetic code, and hence less-constrained features of molecular organisation, will make detection of convergence somewhat easier for molecular phylogeneticists than for those pursuing other lines of enquiry, particularly when highly conserved molecules with extremely long evolutionary histories are analysed; but at present there are very few invertebrate studies that have reached this degree of sophistication.

5 Evidence from embryology and larvae

5.1 Introduction

The way in which genes and proteins evolve and can reveal phylogenies has been dealt with, as far as possible; this chapter moves on to the later stages of the processes and interrelationships shown on page 51. Development and morphogenesis require a complex series of cellular behaviours and interactions, all ultimately dependent on cellular proteins. These include cell division, cell determination, cell shape changes, and cell sorting and movement (Gerhart *et al.* 1982; Wessells 1982; Maynard Smith 1983). Increase in morphological complexity thus arises by epigenetic effects, in an apparently hierarchical fashion. Though biologists are still far from understanding the relations between genetics and developmental morphology, an enormous investment of research effort is being expended in these fields and good reviews of the underlying principles and problems, set in an evolutionary context, are fortunately available (see Bonner 1982; Goodwin *et al.* 1983; Raff & Kaufmann 1983), though again much of the work is purely vertebrate- and insect-orientated. Nevertheless from such work it is increasingly apparent that the developing organism is itself a 'target' for evolutionary change and that developmental processes may directly affect the expression of mutational genomic changes (Lewin 1981). This is clearly not the place to consider such issues in any detail, but it should be borne in mind that new insights in genetics are beginning to have profound effects on our appreciation of the role of embryology in evolution and hence in phylogeny, and that further shifts of view are very likely over the next few years.

From a historical viewpoint, the role of embryological studies in establishing the relationships of animals has been paramount for at least a hundred years now, stemming largely from the profoundly elegant analyses of Haeckel in the late nineteenth century. He was particularly instrumental in promoting the idea that each change during embryology reflects changes undergone through the evolutionary history of a group, from single cells through early multicellularity, to hollow ball stages and intucked gastrulas, worm-like embryos and even tadpole-like juvenile vertebrates. His views, published in 1874 and 1875, were based on very much earlier observations of the resemblance of embryonic higher animals to adult lower animals, particularly stressed by von Baer and Müller; characters were said to be visible during embryology that were diagnostic of related lower taxa but were lost in the adult, a classic example being the oft-quoted resemblance of mammalian embryos to their fish-like, gill-slit-possessing ancestors. Haeckel also noted the similarity of embryology in groups where adult form was dissimilar. If these ideas are really true, and the embryonic stages of a higher animal do resemble the adult stages of its ancestors, then clearly embryology must be the fundamental tool of phylogeneticists, representing a library of evolutionary changes, and its evidence must be heavily favoured.

Ernst Haeckel was thus the instigator of what became known as the 'biogenetic law', embodying the principle of embryological events as recapitulations of evolutionary changes - hence the much-repeated adage that 'ontogeny recapitulates phylogeny'. This phrase was a particular target for Walter Garstang (1922, 1928*a*) who was a key figure in pointing out all the problems that recapitulation theory creates or leaves unsolved, notably in his wickedly apt verses (published in 1951), and most memorably in the equally quoted retort 'ontogeny does not recapitulate phylogeny - it creates it'. In this phrase he encapsulates the contrary idea that embryological and larval forms are themselves so subject to selective forces that they are highly modified structures in their own right, a constrained base on which subsequent adult morphology must be built, and indeed that early morphology may be a starting point itself for major evolutionary innovation by the process of neoteny or 'paedomorphosis'. According to Garstang and his successors (notably de Beer, in 1958), evolution could no longer be thought of as a modification of adult forms, and development was not a speeded-up trip through past ancestral forms. Embryonic stages did not resemble adult ancestors, but only embryonic stages of the ancestors up to whatever point it was in ontogeny where divergence occurred - and this point could be very early on indeed in certain conditions. Looking to embryos and larvae for evidence of relationships was therefore seen as simplistic and even dangerous.

The complex issues of the relationship between ontogeny and phylogeny, and the history of recapitulation, have been taken up and examined in a more theoretical and modern framework by Gould (1977), Løvtrup (1978) and Patterson (1983). Riedl (1979) has also used recapitulation as a central theme in a novel work on the ordering systems of the natural world, attempting to recast the concept in a more acceptable light. All these works provide useful insights on the relationships between development, evolution and phylogeny and could be taken as a useful background for this chapter; they reveal how, in a sense, Haeckelian views are making a comeback.

Despite a good many clear and logical critiques of 'traditional' evolutionary recapitulation in the early part of this century, much of the influence of Haeckel's doctrine has always remained, almost undiminished, in invertebrate phylogeny. After the 1880s most of the major divisions of the animal kingdom that are still accepted or debated today were established, with their roots very firmly in Haeckel's comparative embryology. From this period come the distinctions between diploblastic (supposedly gastrula-like) and triploblastic phyla, the separation of acoelomates, pseudocoelomates and coelomates, and most importantly the division of the major coelomate groups into protostomes and deuterostomes, this being almost entirely based on the embryological differences of cleavage, blastopore fate, mesoderm and coelom formation, and larval form.

Therefore the modern perspective on Haeckelian precepts, arising from work over the last fifty years, creates two problems. Firstly there has been some tendency to be unduly critical and dismissive of views that actually do make useful contributions, simply because they are labelled as Haeckelian and recapitulation is unfashionable. Recapitulation *per se* is certainly a somewhat oversimplified view of embryology, but even a cursory understanding of embryology will reveal that it may nevertheless be true that early stages are less modified than later ones. Hence a good comparative embryologist may be able to discriminate between genuinely primitive traits and specialised highly adapted embryonic features. It is unhelpful to dismiss the biogenetic law out of hand.

But on the other side of the coin it is still possible that the hidden effects of recapitulationist ideas may go undetected in areas where they are less obviously relevant but have nevertheless exercised an insidious influence. For example, most students are taught that the Haeckelian blastaea-gastraea view of the origin of metazoans (discussed in chapter 7) is outdated and recapitulationist, but relatively few have appreciated how much the whole framework and terminology of classical embryology is based upon these ideas. Many textbook views of evolutionary sequences among lower animals still depend on Haeckelian logic, having protozoans giving rise to diploblastic cnidarians and these in turn to triploblastic acoelomate and coelomate worms.

It is therefore very difficult to achieve a balanced view of the diverse processes of animal embryology, and nearly impossible not to approach the subject by considering the invaluable Haeckelian synthesis first and then looking at the issues it raises. So that course is followed in this chapter, taking a deeper look at the last century of embryological work, in an attempt to put old insights into a more modern framework and to arrive at an overview of the relation between embryological events and evolutionary events.

5.2 Embryological processes; theme and variations

The classic course of animal embryology, as described and interpreted by Haeckel, is shown in fig 5.1. After the fertilised egg has undergone a few cleavages the embryo takes the form of a hollow blastula, with a single layer of cells surrounding a central blastocoel (as we saw in chapter 2). At some point on the surface of the blastula a dimple appears, and this gradually tucks inwards in the process called gastrulation, to form a double-walled embryo, thus initiating germ-layer formation. The dimple has thus become the blastopore, and is destined in due course to be the mouth and/or anus of the animal, with the invaginated pocket becoming the gut. Simultaneously, or shortly thereafter, the formation of the third germ layer (mesoderm) occurs, and organogenesis is initiated. At about this stage the embryo is commonly transformed into a larva, being freed from any protective coverings and having some mechanism of motility to ensure dispersion. Only later does the final metamorphosis into the adult form occur.

Obviously this pattern is very much modified in different groups of animals, particularly where the lifestyle does not involve a dispersive larval stage but relies on direct development, when large quantities of yolk tend to occur in the embryo and can obscure this classic series of changes. But the important question is how far these events *are* really fundamental to (or homologous in) all animals, and whether there are useful dichotomies in the manner of their occurrence or in the resultant morphology. Are we really looking at a 'theme and variations', or at numerous quite separate scores? To decide on this, it will

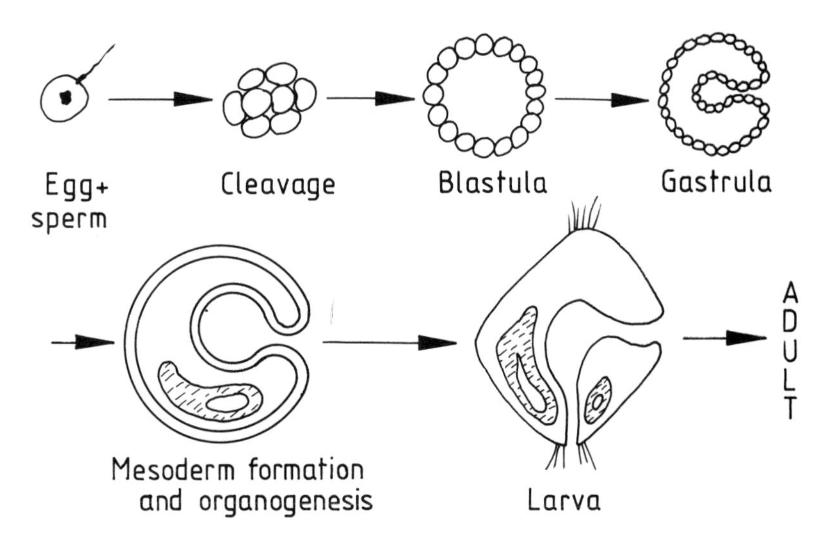

Fig 5.1. The basic sequence of processes in classical animal embryology.

probably be most helpful to look at each stage in turn, concentrating particularly on the earliest stages of embryology which should be most relevant to phylogeny if there is any truth in the idea of recapitulation.

Cleavage

Cleavage represents the partitioning of an embryo by successive mitotic divisions into an organised multicellular form. Cleavage divisions separate daughter cells along specific cleavage planes, and result in specific positional patterns of the resultant blastomeres, which may be further distinguished by varying sizes and varying investment with yolk materials.

To achieve an orderly progression from a one-cell stage to a multicellular embryo, there are relatively few ways in which patterned cell cleavage can occur. Up to the four-cell stage there is little variation, as the first two cleavage planes are vertical from pole to pole, splitting the egg vertically into four equal blastomeres. Thereafter, it has been traditional to subdivide embryonic cleavage into two types, shown in fig 5.2, these only becoming distinguishable from the four-cell stage. In **radial** cleavage, the third cleavage plane is equatorial, and the second quartet of cells lie directly over the first quartet, with no rotation of the cleavage axis. In **spiral** cleavage the axes are rotated, the cleavage spindles forming oblique to the polar axis of the egg, so

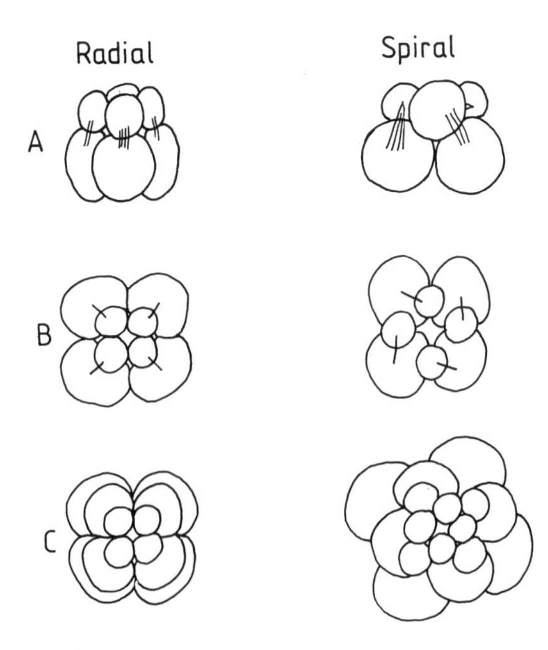

Fig 5.2. Radial and spiral cleavage patterns through the 8-cell and 16-cell stages.
A: side view.
B, C: top views.

that the second set of four cells lie somewhat offset to the first set, effectively lying in the depressions between the original four cells. This latter type of cleavage would actually be better described by the term 'oblique', since each successive cleavage in fact occurs in alternating directions; if the first split was to the right (dextrotropic) the next will be to the left (laevotropic) and so on, so there is no actual continuous spiralling in the embryo. In theory, radial cleavage is diagnostic of deuterostome animals (echinoderms, hemichordates and chordates in particular), while spiral cleavage occurs in the protostomes (annelids, molluscs, sipunculans and arthropods), which together with nemerteans and platyhelminths are often given the alternative name Spiralia (see chapter 1).

A second possibility is to characterise cleavage by the fate of the blastomeres it creates. In some animals, separated blastomeres will only develop into whatever part of the animal they would have formed if left in situ; their fate is already determined very early in development, so this is termed **determinate** cleavage. In other cases each blastomere on its own can form an entire animal; this is **indeterminate** cleavage, and animals that show it can produce identical twins when embryos split accidentally. Again the patterns of occurrence follow the classic dichotomy; determinate cleavage is held to be a protostome feature, with indeterminate cleavage being found in the deuterostome groups (hence of course the possibility of identical twins in chordate humans).

Thirdly, cleavage may be characterised by the amount of yolk present in the egg and the effect this has on divisions (fig 5.3). Where there is a massive yolk supply, organised cleavage patterns may be entirely obscured and little phylogenetic inference can be drawn; in many ways, the amount of yolk in a particular species' development is the single most important determinant of its embryological patterns.

These various characterisations of cleavage types are not entirely independent of each other. Both radial and spiral forms can really only be seen where cleavage is holoblastic, their patterns being lost when there is too much yolk. Further, it is frequently stated that spiral cleavage is normally determinate, and radial cleavage indeterminate; therefore these combined features become one of the major dichotomies between protostomes and deuterostomes.

Spiral, determinate, holoblastic cleavage has proved particularly useful in comparative studies because the fate of each cell can be traced. A common system of naming the blastomeres, whose history is reviewed by Costello & Henley (1976) is widely used as shown in fig 5.4. At the eight-cell stage, the four upper blastomeres (usually smaller, as any yolk present sinks to the

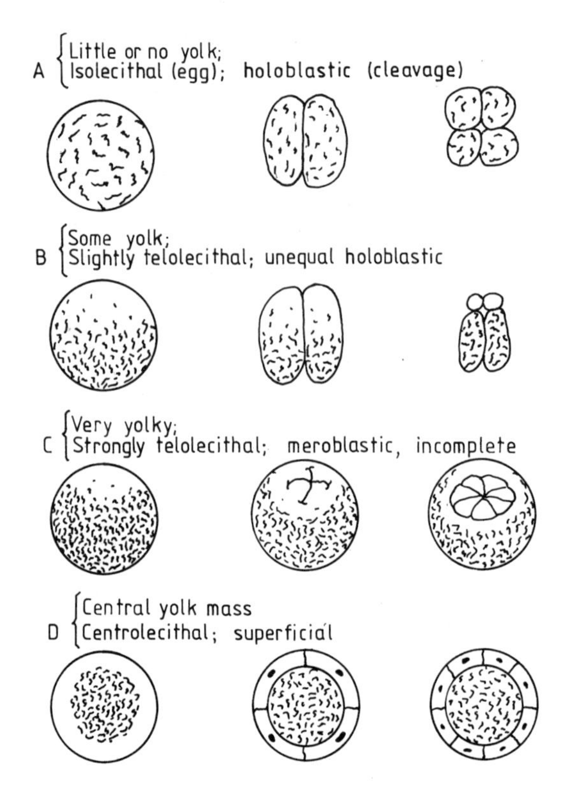

Fig 5.3. Terminology of embryos and cleavage patterns, showing the effects of varying amounts of yolk.

vegetal pole, and so called micromeres) are designated 1a, 1b, 1c and 1d, while the lower macromeres become 1A-D. The macromeres split transversely again to give micromeres 2a-d, and are thereafter called 2A-D themselves. When micromeres divide, their products are distinguished by exponents (e.g. $1a^1$, $1a^2$, then $1a^{11}$ and $1a^{12}$), whilst new numbered sets of micromeres are derived from macromeres. At the 32-cell stage, the macromeres (now 3A-D) are already fated to form the whole of the endoderm, and they now give rise to a fourth set of micromeres, 4a-d, of which the first three also become endoderm. The fourth of these cells, 4d, is destined to be the origin of all the animal's mesoderm; it may also be called the mesentoblast or M cell. A further important group of cells is the quartet $1a^2$-$1d^2$, which form the prototroch ciliary band in the larva (see below). In all typical spiralian animals, then, an exact correspondence between embryonic cells can be seen (particularly clearly set out by Siewing 1969, 1979, 1980), and there is hope of detecting degrees of relationship by examining the similarity of behaviour and position of these cells (Anderson 1982).

For radial cleavage, also best seen in holoblastic embryos, the fates of the various cells generally remain plastic at least until the start of gastrulation. It is in any case much less easy to see correspondences between cells in these embryos, largely because the radial pattern of cleavage, where cells perch on top of each other, forms a less compact and stable ball of cells, and the pattern soon breaks down to give a rather irregular cell mass.

Cleavage patterns therefore seem to give a number of possible clues to phylogenetic relationships, whether at a very specific level in the spiralian groups or more generally by the absence of spirality and the occurrence of a more disorganised radial condition. Perhaps this less precise radiality, now seen in deuterostomes, could be seen as arising from the even less organised condition in cnidarians, and as being the precursor of spiral cleavage; this would accord well with enterocoelic/archecoelomate views of metazoan relationships. Maybe then all the groups with determinate spiral cleavage should be seen as a very tightly-knit derived clade, more coherent than any other super-phyletic grouping; this has certainly been the implication of a good deal of recent European literature. Or, as a very different viewpoint, perhaps the spiralian condition should be taken as a non-specialised starting point, with radially cleaving deuterostomes as derivatives of it, as Salvini-Plawen (1980*a*, 1985) has advocated.

But before taking any of these views, as always we need to inject a few notes of caution. Firstly, and inevitably, there are exceptions to all the generalisations we have made. Many animals turn out to have a form of cleavage best described as bilateral, not readily fitted in to any neat phylogenetic scheme, and leading almost to a continuum of types rather than a neat dichotomy. A few groups have instances of unique cleavage patterns; a

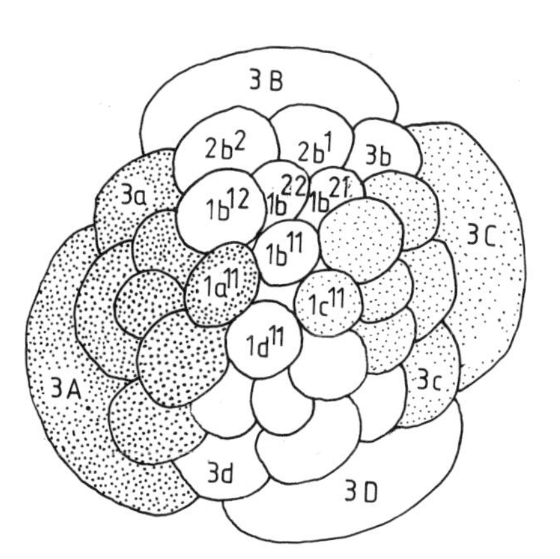

Fig 5.4. The numbering system used in describing spiral determinate embryos where each cell has a precise fate.

classic case is the nematodes, some of which are odd as early as the four-cell stage when the cells are arranged in a T-shape rather than as the usual regular quadrant. Even amongst the classic spiralians there are problematic cases; within the platyhelminths only polyclad turbellarians really have spiral cleavage, and amongst molluscs cephalopods have no trace of it. Very few arthropods show it at all, though they are usually assumed to be close relatives of the classically spiralian annelids; the nearest thing to spiral cleavage is found in cirripede crustaceans. And a few slightly unexpected groups, not normally included in the protostomes, do show spiral cleavage - the pseudocoelomate rotifers are a good example. Some of these exceptions do not matter too much, as they clearly can be explained by the presence of disruptive yolk (notably in cephalopods and most arthropods), correlated with direct development or much-reduced larval stages; but some are more worrying. As we noted earlier, there really are not very many ways in which organised cleavage can occur, and it is quite possible that unrelated groups have convergently adopted the same basic patterns.

Related to this first problem is a second one, for as was realised even within the nineteenth century by Lillie, and as Costello & Henley (1976) have recently pointed out, the cleavage patterns of each species have to be adapted to the needs of the future larva. That is to say, adaptation is working even at this very early stage. Spirality itself could be seen as the obvious way of creating an organism - the trochophore larva - effectively built as a 'spinning top', with the use of alternating oblique cleavages giving the extra advantage that for post-larval bilaterality the alternation could be stopped in the middle like a pendulum. Variations in spiral cleavage found in rotifers, cirripedes and advanced molluscs (like the bivalves with glochidia larva) could then be interpreted as ways of achieving more definitely bilateral larvae for specific purposes; and where there is direct development, as in cephalopods, cleavage is bilateral so that the embryo can achieve the necessary adult bilaterality as efficiently as possible. In support of these views, in the asymmetric adult structures of gastropods and some bivalves the direction of asymmetry is causally linked with the direction of initial spirality in the embryo - there is a direct link between larval cleavage patterns and adult form. (However, since in both of the classic super-phyletic groupings adult form is satisfactorily achieved even in meroblastic eggs where all traces of organised cleavage are lost, it would be inappropriate to give much weight to this link).

A third possible problem in relation to cleavage patterns was raised by Løvtrup (1977) when he suggested that this feature is not an independent character but is determined by the presence or absence of egg membranes.

Where there is no continuous outer surface around an egg, spiral cleavage inevitably occurs because it represents the most dynamically stable condition, one set of blastomeres automatically sliding into position in the furrows created by the other set. Where an outer membrane is present, radial cleavage is feasible because the blastomeres are supported in less stable positions by their mechanical interaction with this surface; and it is possible that sialic acids (apparently absent in protostomes - see chapter 4) may be involved in these cell-surface interactions. This point may also be relevant to other aspects of the embryological evidence for super-phyletic assemblages, as we shall see; and the whole issue of controlled cell movements as cell masses and cell sheets is considered further by Wessells (1982). If spiral cleavage is indeed the normal and 'easy' way to organise early embryonic growth, then Salvini-Plawen's (1980*a*, 1985) idea that radially cleaving deuterostomes are a regressed development from it (possible only when egg membranes were developed) perhaps becomes more attractive than the opposing view that deuterostomes are closer to a primitive archecoelomate with spiralian protostomes as a derived offshoot.

Finally, Løvtrup also points out that the issue of determinate and indeterminate cleavage is somewhat unsatisfactory. Urochordates have proved to be determinate in their cleavage even at the two-cell stage, despite their normal status as deuterostomes. Perhaps more importantly, determinacy is logically a continuum and cannot really be treated as a dichotomy; it depends on the relative rates of cell determination and of cell division. If the former proceeds relatively slowly then the embryo appears indeterminate. In protostome groups cell determination always appears to set in very early, and it is a reasonable generalisation that spiral cleavage is always determinate; but radially cleaving groups show the full range of possible degrees of determinacy.

On balance, then, cleavage patterns may not be as useful as at first appears, and may reflect the constraints on the embryo itself and on the future larval and adult forms as much as they reflect phylogenetic constraints. Whilst radially or bilaterally or indeterminately cleaving forms may be merely reflections of other embryonic factors, or may be essentially negative in character, it nevertheless seems likely that there are also clear positive indications here. One clearcut statement that can be made is that spiral cleavage does not occur in embryos of echinoderms, hemichordates and chordates, the classic deuterostomes, even where these are holoblastic and spirality would be entirely feasible; some degree of relatedness may therefore be implied (chapter 12). And it also seems most likely that the cleavage sequences matching precisely, cell for cell, in various spiralian groups are an indication of common ancestry (chapter 8).

Blastula formation

After a certain number of cleavage divisions (a number which may be widely different according to species) the animal embryo becomes organised into a blastula stage, conventionally a hollow ball of small cells arranged around a central space known as the blastocoel (see chapter 2). From this point cell division does continue but is no longer termed cleavage; the divisions are less regular, and development largely proceeds by organised cell movements.

The blastula varies considerably in different kinds of animal, largely according to the amount of yolk present. Where a spacious blastocoel is present the term coeloblastula (or sometimes archeblastula) is used; and the blastocoel may be central if the egg was isolecithal or strongly displaced to the upper animal pole if it was telolecithal (fig 5.3 clarifies the terminology). Alternatively the blastula stage may be a solid stereoblastula, with no blastocoel. Again this can be of two types. If cleavage was meroblastic in a telolecithal egg (with only the upper part of the zygote dividing) then the blastula consists merely of a disc of cells perched on top of a mass of undivided yolk, and is known as a discoblastula. If the egg was centrolecithal a periblastula is produced, with a layer of cells surrounding the yolk mass. All these alternatives are shown in fig 5.5.

Gastrulation and the fate of the blastopore

After the blastula has been formed, the next step is always an organised sequence of cell movements to produce the classic germ layers, discussed in

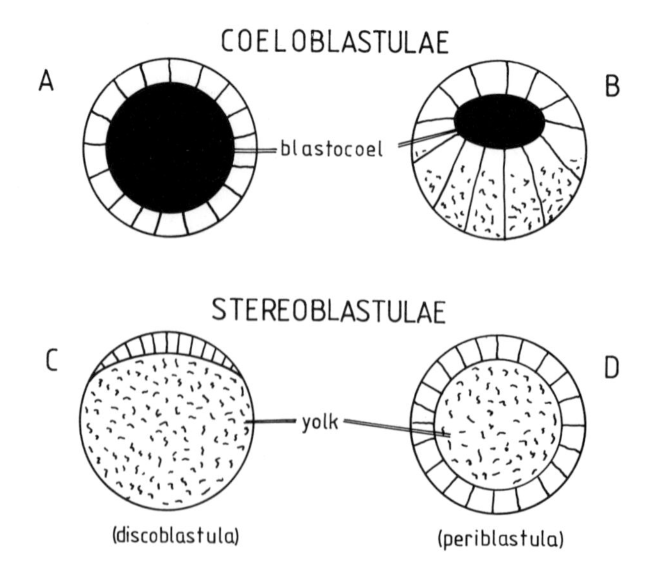

Fig 5.5. Different types of blastula found in invertebrates, largely dependent on the amount of yolk present.

chapter 2. Although the processes by which this occurs are very variable, they are all termed gastrulation, and the next stage of the embryo's development is termed a gastrula; this is clearly a throwback to the Haeckelian synthesis, and it is by no means certain nowadays that all 'gastrulas' are homologous or merit this constancy of terminology. However, although there is no a priori reason why an embryo should not develop its three layers in situ, no actual cases of this are known: all metazoans do undergo some process of three-dimensional rearrangement to achieve their three germ layers. Hence it may in fact be reasonable to regard 'gastrulation' as a universal event with a degree of common ancestry (see Wessells 1982).

In theory, the 'gastrula' is a simple two-layered structure with the outer cell layer forming ectoderm and the inner layer endoderm. In practice, mesoderm is commonly present from a very early stage, so that the gastrula is not in fact two-layered; and the manner in which the endoderm actually forms is extremely variable (fig 5.6), this being highly dependent on the preceding cleavage patterns and the form of the blastula.

The most straightforward method of gastrulation is the classic method always given in introductory texts - the process of invagination. This is particularly characteristic of coeloblastulae. It involves the vegetal part of the ball of cells moving smoothly inwards at a specific point called the blastopore, visible as a dimple on the surface of a late blastula. The cells continue tucking into the ball to form an inner sac (fig 5.6*A*), known as the archenteron (or gastrocoel, or gut - see chapter 2). Sometimes these vegetal cells invaginate until the blastocoel is entirely obliterated, but more normally invertebrates retain a space between the outer ectoderm and the invaginated endoderm. Haeckel regarded this as the primitive mode of gastrula formation (from which all others were secondary derivations), and he therefore termed the result an archegastrula.

Where cleavage has been unequal and holoblastic, giving a yolk-laden vegetal area and a small blastocoel near the animal pole, the vegetal cells seem to be too inert to invaginate and gastrulation occurs instead by epiboly. This involves the small animal cells growing down over the vegetal area, which thus becomes enclosed as endoderm (fig 5.6*B*). At the point where the converging animal cells meet a small invaginated area does commonly occur, and the entrance to it is again termed the blastopore. Thus it seems reasonable to regard epiboly as a modified form of invagination.

In meroblastic eggs where the blastula is of the discoblastula type, something rather similar again happens, a few cells tucking in under the disc of ectoderm to become the endoderm, all still lying on top of the mass of yolk

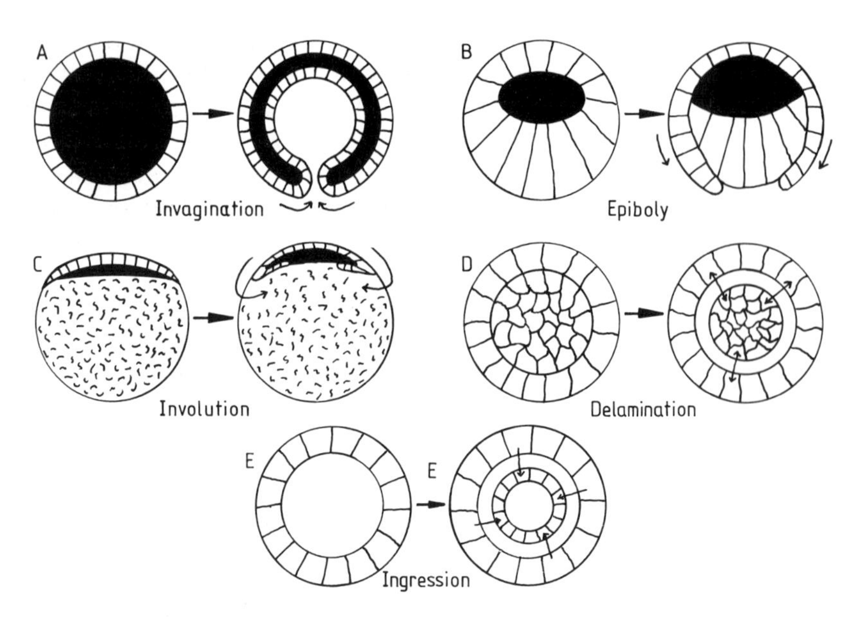

Fig 5.6. The five main types of gastrulation. Further comparison is given in the text.

(fig 5.6*C*). This is termed involution, and the end-product is a discogastrula.

The remaining two types of gastrulation are fundamentally different from those so far described. Delamination occurs where cleavage has led to a stereoblastula, and involves the separation of the outer layer of cells from the inner cell mass (fig 5.6*D*). Ingression is characteristic of a few groups such as sponges and cnidarians, possessing hollow blastulae but with no trace of invagination processes; instead the outer cells simply bud off to the interior to form a second cell layer (fig 5.6*E*), sometimes from the vegetal area (polar ingression), but often more or less randomly from all over the blastula wall. In neither of these cases is there any blastopore formation, and an archenteron is not produced. The rather solid result is therefore called a stereogastrula.

There are two very important issues to follow up in relation to gastrulation. Firstly, the Haeckelian view that invagination is primitive (and hence that a gastraea-type organism was ancestral to all metazoans, as will be discussed in chapter 7) is clearly running into difficulties. For there are several quite different types of gastrula and different ways of making the endoderm; in some animals the gastrula is hollow and the endoderm is a polarised,determined phenomenon, derived from specific cells, whereas in others the gastrula is solid and any blastula cell may contribute to the endoderm. This makes it difficult to believe that the endodermal germ layer is homologous throughout the Metazoa, and the germ layer theory (chapter 2) looks a little shaky. More seriously, if recapitulation 'works' then the more primitive animals should conform more closely to the original embryological types, yet the groups which Haeckel and most subsequent zoologists have been happy to regard as the 'lowest' of the metazoans - the Porifera and Cnidaria - are (as Haeckel himself recognised) the very groups most prone to having stereogastrulae and delaminated or ingressed endoderm. More modern treatments of the whole problem have tended to follow Hyman (1940, 1959) in regarding ingression as the original method of making endoderm, and invagination as a derived and shortened route for achieving the two-layered condition. It seems fair to conclude that there are two basic and alternative processes for achieving increased structural complexity in animals, which can be seen as 'infolding' and 'inwandering', and both were probably present in Metazoa very early on. Either is used according to existing circumstances and preceding constraints, and in relation specifically to gastrulation either may occur in quite closely related species. Løvtrup (1977) and others would again maintain that the major determinant is the presence or absence of coating surface membranes; where these are present, cells can move over them in organised sheets (infolding - invagination, epiboly), but where absent cells move more independently and randomly (inwandering - delamination and ingression). The mode of gastrulation may therefore have little to do with phylogeny.

However the gastrula itself remains diagnostically important because it is supposed to be another point at which the major super-phyla Protostomia and Deuterostomia can be distinguished. Here we meet the second major problem to arise, which concerns the blastopore and its fate at the end of gastrulation. In theory, as Grobben described in 1908, the blastopore as a point of access to the archenteron will become the mouth in protostomes (i.e. the mouth is primary, as the name implies), whereas a mouth forms secondarily, at some distance from the blastopore, in the deuterostomes, the blastopore either closing up or becoming the anus. In some cases, though, gastrulation clearly does not involve a blastopore at all; and where one is identified, it is not at all obvious that the innovative intucking dimple of invaginating embryos is homologous with the retrospective ectodermal meeting point of epibolic embryos. The homologies of blastopores and mouths in evolutionary terms, and the whole issue of their apparent relationships in embryology, are strongly criticised by Williams (1960); Nielsen (1985) and Ax (1987) also give useful commentaries.

Even if distinctive blastopore/mouth/anus relationships do exist, the apparent 'end result' may actually depend very largely on the mechanism of gastrulation (see Løvtrup 1977), in which case it is an unhappy character on which to put so much weight. Where invagination occurs (at its most classic in the echinoderm embryo), it always begins at the vegetal pole and thus ends up near the animal pole, where the tip of the archenteron fuses with the ectoderm to form an opening. Since the animal pole generally produces the first nerve cells, thus defining the head, it follows that this secondary opening will be near the head and will be called the mouth; and the original blastopore opening becomes the anus. Invaginating gastrulas ought therefore to produce deuterostomes. By contrast, where epiboly occurs the lips of the 'blastopore' are formed by the advancing ectoderm, usually at the vegetal end, and the blastopore tends to be slit-like and to get gradually smaller as the ectodermal sheets close up together (see fig 5.7). It may close completely, or close in the middle and leave a hole at either end, or of course remain open only at one end - so any epibolic embryo can become any variety of protostome or deuterostome (or an unclassifiable intermediate) almost at random. All these alternatives do seem to occur in spirally-cleaving epibolic annelids and molluscs. Finally, in large and yolk-laden eggs with stereogastrulae, which can occur in both super-phyla, it is illogical in any case to try and analyse the fate of a non-existent blastopore.

So yet another of the embryological features on which the major phylogenetic dichotomy of protostomes and deuterostomes rests appears to have its problems. It is clearly not independent of earlier embryological events, and it may not be independent of mechanical factors such as the presence of egg membranes. And nothing has been said of the exceptions as yet, which are regrettably abundant as could be predicted from the analysis given above: in almost every phyla there are animals that form their blastopores into anuses when they should be forming mouths from them, or form both mouth and anus quite independently of a blastopore.

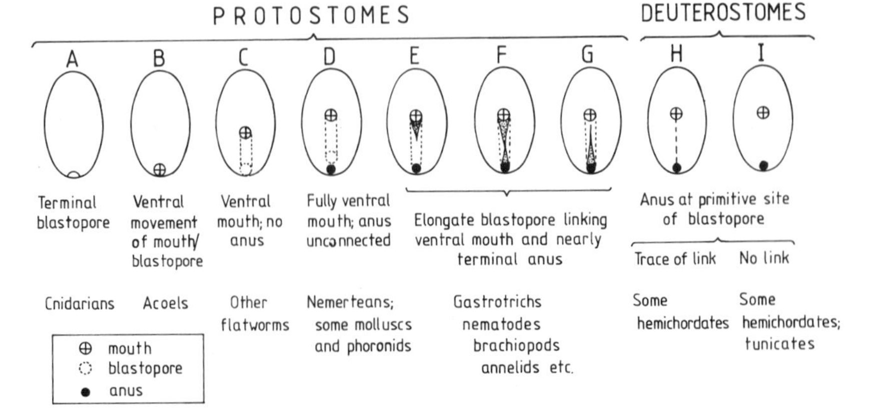

Fig 5.7. The relationships between blastopore, mouth and anus in different groups of animals, according to Salvini-Plawen (1980*a*); the deuterostomes are seen as derived from the protostomes, by an abbreviating of the usual developmental processes.

This is one generalisation that perhaps does not stand up very well in the light of modern embryological knowledge. However the analysis of the problem by Salvini-Plawen (1980*a*) may indicate why matters have become somewhat confused. He believes that the relation of blastopore to mouth and anus can be seen as a continuum, with the deuterostome condition derivable from the spiralian system as in fig 5.7. The original blastopore lengthens in the early spiralian groups, becoming less terminal and increasingly ventral, with the mouth opening gradually moving further to the end of it. The mouth thus becomes markedly ventral, and the blastopore begins to close in the middle; but in some cases it closes more rapidly ventrally and remains as a long slit at the terminal end, so that only the terminal anus is seen as forming from the blastopore and the embryo is designated a deuterostome. If this view is correct (and many examples are given in support of it) it would not be surprising that anomalies can occur in particular groups where progress along the continuum it represents has been arrested or even reversed. And at the same time if most representatives of most groups do conform to these patterns, then perhaps the fate of the blastopore does have something important to tell us about major trends in phylogeny.

The mesoderm and organogenesis

After the process of gastrulation, by whatever means, the embryo has established two cell layers and it rapidly proceeds to form a third, the mesoderm, from which the bulk of its adult body and most of the familiar tissues and organs will arise. The later stages of organ formation are evidently fairly specific to particular animal habits and habitats, and are unlikely to be helpful in phylogeny; therefore only the early parts of these processes need be of concern.

The process of mesoderm formation was considered in chapter 2, as it is so closely related to the origin of the coelom in many phyla. Here it should be particularly recalled that mesoderm can arise in several different ways, somewhat different in protostomes and deuterostomes. In the former, mesoderm arises by proliferation within the blastocoel from the invaginated 4d cell (see fig 2.6*A*), which in many cases is also the cell that initiates the process of invagination and thus determines the site of the blastopore. In deuterostomes, the cells may not be determined prior to gastrulation as discussed earlier, and it is the site of invagination which determines the fate of all other areas; then from the walls of the invaginated archenteron pockets arise and push out into the blastocoel (fig 2.6*B*), their walls becoming the mesoderm.

It is worth noting that these two types of mesoderm formation are roughly equivalent to the two broad mechanisms of endoderm formation outlined above, infolding and inwandering. Thus, Løvtrup (1977) would say, they may once again relate simply to the presence of a covering surface, making virtually all the embryological features that are supposed to separate the major super-phyla reducible to this one underlying factor. If an embryo possesses a covering hyaline coat or other superficial membrane then it will be likely to have radial cleavage, an invaginating gastrula and secondary mouth, and enterocoelic mesoderm and coelom formation, all these features involving cell interactions with this coating; and it will be what is called a deuterostome. Without such a coating it is destined for the protostome assemblage. Unfortunately the correlation between the presence or absence of egg coatings (somewhat imprecisely defined by Løvtrup anyway) and the designation of protostomes and deuterostomes is by no means absolute. But at least Løvtrup's analysis underlines the degree of interdependence of the features commonly tabulated separately to support this long-standing dichotomy, and may give some idea of how various groups could have derived from others (in whichever direction, but perhaps most plausibly from protostome to deuterostome) in respect of a whole suite of superficially unrelated characters.

Embryological evidence thus underlies some vital issues in animal phylogeny, and its role in establishing these in the received wisdom of current teaching and textbooks can hardly be overstressed. But it is quite apparent that matters are not as simple as might at first be supposed, that there are manifold exceptions to all the rules, and that features are rarely simply dichotomous and even more rarely fully independent of preceding processes or current embryonic needs. Perhaps attention should therefore move on to the next stage, the evidence to be obtained from larval forms, with these lessons firmly in mind.

5.3 Larval forms

The larvae of invertebrates have perhaps been used even more than the preceding embryonic stages by phylogeneticists seeking support for their favourite theories. Whole edifices of invertebrate relationships have been erected on the basis of larval similarities, and complex transformation sequences between dissimilar larvae and between similar larval and adult forms have been devised. Inevitably these theories also involve particular views on the evolution of life histories, and there is a long-standing dispute between authors such as Jägersten (1972) and Nielsen (1979, 1985) on the one hand, who regard an alternating pelagic larva/benthic adult life-cycle as the primitive mode for marine invertebrates (thus requiring that larvae are necessarily a primitive phenomenon), and others who argue that larvae are a secondary and purely facultative invention, less likely to have overriding phylogenetic significance. All are agreed, though, that evolutionary studies should take account of all stages of an animal's life, and the concept of 'life-cycle phylogeny' has become increasingly popular (see Stearns 1980; Olive 1985; Nielsen 1985). Therefore both the types of larvae that occur, and the functional reasons for larval form and diversity, need to be discussed before the importance of invertebrate larvae to the phylogenetic debate can be judged.

Larval types and diversity

Following through with the protostome/deuterostome dichotomy that was a recurring theme in the embryological issues, it has long been held that there are two basic types of invertebrate larva corresponding to these two super-phyletic divisions. Thus a complete assemblage of dichotomous embryological features relating to this classification can be put together as in table 5.1, and is to be found in all elementary texts.

By far the best known of these larvae is the trochophore (sometimes called the trochosphere) that is deemed to be characteristic of the protostome/spiralian phyla. Found in its most generalised form in certain polychaete families, this larva (fig 5.8*A*) is approximately top-shaped or biconical. It has an outer single layer of ectoderm, and an equally simple complete endodermal digestive tract leading from the anterior mouth to a roughly ventral anus. The intervening blastocoel is partly filled with mesenchyme and muscle, and to either side of the gut is a band of mesoderm derived from the 4d cell. Few organs are present in the larva; there is one pair of protonephridia (cyrtocytes - see chapter 6) terminating internally in flame cells or solenocytes, and an 'apical organ' consisting of a dorsal tuft of cilia and usually a pair of ocelli, overlying a ganglionic nerve mass. The most useful diagnostic features of the trochophore are the external ciliated bands that provide both the motile system and a food-trapping device. The principal one is the prototroch, derived from the $1a^2$-$1d^2$ quartet in the embryo, and running as a pronounced ridge around the larval equator just above the mouth - it may involve one or more rows of ciliated cells. Another band, the metatroch, may also be found equatorially but below the mouth; and a third, the telotroch (or paratroch), may surround the

Table 5.1 *Embryological and larval features of protostomes and deuterostomes*

	Protostome	**Deuterostome**
Cleavage	Spiral	Radial
	Determinate	Indeterminate
Blastopore	Forms mouth, so mouth is primary	Does not form mouth, so mouth is secondary
Mesoderm	From 4d cell, proliferation	From gut wall, infolding
Coelom	Schizocoelic (split within mesodermal bands)	Enterocoelic (pouch off gut wall)
Larval type	Trochophore	'Dipleurula'

anus. This larval organisation, with varying degrees of modification, is described very commonly in lower invertebrates, including annelids, molluscs, flatworms, sipunculans, echiurans, and - much more dubiously - the lophophorate phyla. It is also somewhat similar to adult form in the ctenophores, and in particular in rotifers, both sometimes thought to be neotenous derivations from it. These facts led Hatschek as far back as the mid-nineteenth century to propose that it represented a remote common ancestor of all the Protostomia, including acoelomates and pseudocoelomate groups.

In contrast to the trochophore, the Deuterostomia are said to be possessed of a larval form based on the dipleurula (or sometimes the general term pluteus is used). It must be said at once that the dipleurula (fig 5.8) does not strictly exist as a larval form, and may never have done; it is merely an archetype, hypothesised as the simple 'lowest common denominator' of all the larvae currently exhibited by deuterostomes. It is described as a bilaterally symmetrical larva, again with simple ectoderm and complete digestive tract, with mouth and anus both somewhat anterior (or somewhat ventral, for some authors), but this time with the mesoderm developing from lateral coelomic pouches off the gut. The blastopore is posterior and protonephridia are absent, though there is again an apical organ with a tuft of cilia. One continuous band of enlarged and strongly ciliated cells (prototroch) passes above and below the mouth and laterally along the sides of the body. Strathmann (1978) regards the single-band system of ciliary feeding in these larvae as a vital link uniting deuterostomes and lophophorates (collectively the 'Oligomera') in both larval and adult forms; he contrasts it with the opposed band system of most spiralian larvae (and rotifers) where the prototroch and metatroch work together to trap food and transfer it to an intervening food groove.

In all existing echinoderms the hypothetical dipleurula larva is actually manifested in a more complex form, as for example the bipinnaria of starfish, the pluteus of brittle stars and urchins, or the auricularia of sea cucumbers. Hemichordates also have a larva that can be construed as a modified dipleurula, though in the chordate phylum no such larval form persists.

It would be very convenient if there were really just two basic types of invertebrate larva and all groups could be referred, using them, to the two classic super-phyla. Certainly some authors have come close to implying just this in the past. But as always matters are not so simple. Larvae are built from relatively few morphogenetic elements, and restrictions are inevitably imposed by the limited material; the larva must also be functionally viable and must allow the correct subsequent morphological changes. Both the larval forms under consideration are small, roughly spheroid, and with ciliated bands in varying positions providing locomotion. Most of the described larval characters are clearly just manifestations of the other distinctions between protostomes and deuterostomes - the site and mode of mesoderm and coelom

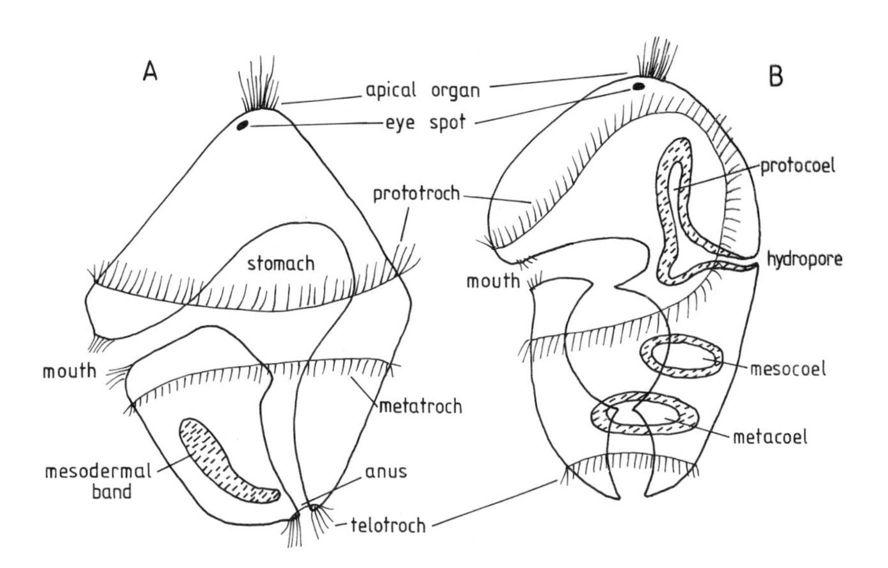

Fig 5.8. A comparison of the prototroch larva supposedly characteristic of protostomes *(A)*, and the hypothetical dipleurula larva *(B)* taken to be ancestral for deuterostomes.

formation, and the mouth/anus/blastopore relationships. (For example it seems a little unfair to say that because an animal has a coelom developing from lateral enterocoelic pouches it is a deuterostome, and because it is a deuterostome its larva must be a dipleurula larva, confirming that this is so by spotting the enterocoelic pouches in the larva!)

The ciliary bands thus provide the best, and perhaps only, independent means of distinguishing the two larval archetypes, but even then there are problems. In trochophores the principal prototroch band runs above the mouth; whereas in dipleurulas it should run in a ring around the mouth (shown diagrammatically in fig 5.9). The tendency then is to call any larva with the former arrangement a modified trochophore, and any with the latter pattern a modified dipleurula, thus making the a priori assumption that there *is* a simple and all-embracing dichotomy as no other arrangement is actually possible. As a result a great many larval forms have at various times been called 'modified trochophores' even though they bear little resemblance to the classic form shown in fig 5.8, and their possessors have become associated with the protostome assemblage; the lophophorates, discussed in chapter 13, are a classic case where this has confused matters, since most authors would now prefer to ally them with deuterostomes.

There is another problem linked with the larval ciliary bands, since this character too may not be entirely independent of earlier embryonic events. In a pre-gastrulation embryo the blastopore site is normally ventral/vegetal to the incipient ciliated prototroch. Thus, where the mouth is primary, it will inevitably be ventral to the prototroch, automatically giving a trochophore larva. But if the blastopore forms the anus and a secondary mouth opens, the point at which the invaginating archenteron merges with the ectoderm (or at

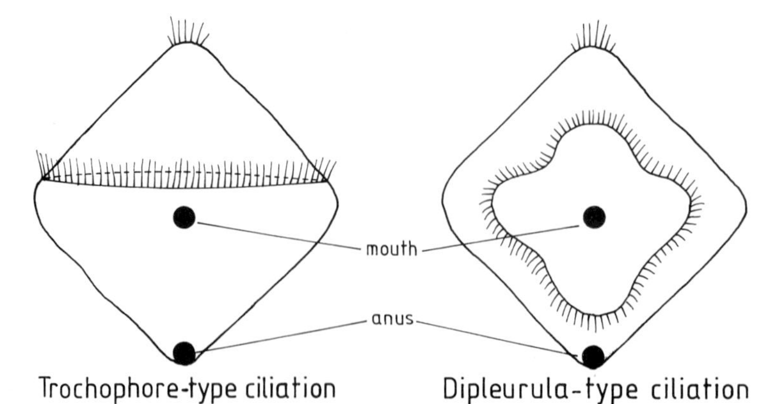

Fig 5.9. Diagrammatic view of the different disposition of the main ciliary band (prototroch) in the two main types of larva.

which the elongating blastopore ceases to elongate) will entirely determine mouth position and may be either above or below the prototroch cells. If above, the ciliated ring appears to go around the mouth and we have a dipleurula. But if the mouth appears below the ring we have a trochophore, even though the secondary nature of the mouth should characterise a deuterostome. It is probably true that the relation between ciliary bands and mouth is much more variable for the deuterostomes, even within one phylum such as the echinoderms. One is left with the unpleasant suspicion that an unbiassed observer might well have classified one or two of the larvae from this phylum as trochophores. And, indeed, that the theoretical distinction between trochophores and dipleurulas is not terribly clearcut.

Turning to practical difficulties, the situation has arisen where a great many phyla are said to produce a trochophore; and others, while not having actual dipleurulas, are said to have derivatives of a dipleurula ancestor. In the former group are the Müller's larva of certain polyclads, with multiple bands of cilia and eight posterior ciliated lobes; and the pilidium of nemerteans, helmet-shaped with ciliation around the rim (see fig 5.10 *A,B*). Neither form has an anus, and their likeness to the polychaete larva is limited. Molluscs also belong here, with an apparently much more 'standard' trochophore occurring in some members of all groups except the cephalopods; but this frequently lacks a metatroch and may not have a telotroch, and it is often of very brief duration before giving rise to a veliger (see fig 10.5) or other more specialised larval form. (It should be noted that some authors (Riedl 1960; Salvini-Plawen 1973, 1980*b*) therefore regard the molluscan trochophore as an independent form, merely convergent with the annelidan model; see below and chapter 10.) Arthropods of course *ought* to belong here, but actually never do have a trochophore. Most uniramians and chelicerates have centrolecithal eggs and more or less direct development, but even in crustacean groups that generally do produce pelagic larva there is no real trace of a trochophore. The simplest crustacean larva is a nauplius (fig 5.10*D*) with a cuticular covering and three pairs of appendages for swimming in place of cilia. Smaller phyla such as echiurans and sipunculans whose larvae are said to be trochophores should also be mentioned, as should the actinotroch and cyphonautes larvae of lophophorate phyla (chapter 13) that have been similarly designated on even more suspect grounds.

The point may have been reached where the trochophore seems to be almost totally devalued, and is a useless catch-all term. As a result, Salvini-Plawen (1973, 1980*b*) has offered a reappraisal of its meaning and believes it should henceforth only be applied to annelids and the closely related echiurans (see chapter 8). Flatworm, nemertean and entoproct larvae are, in his view, unrelated forms. The molluscan larva, primitively lacking a metatroch and without classic protonephridia, he calls a 'pseudotrochophore', deeming it probably of convergent origin with the annelidan larva. The larvae of sipunculans are even more dissimilar in his opinion, though perhaps all of these forms, and the phoronid larva, could be derived from an even more simple ancestral form that he terms a 'pericalymma'. Certainly an increased caution about the term 'modified trochophore' seems to be in order in phylogenetic studies; and attempts by some authors and texts to apply the term to the very different larvae of cnidarians (planulas and actinulas), ctenophores (cydippids), and even sponges (amphiblastulas), simply because it is felt that these groups 'ought' to be closer to protostomes, seem most unreasonable.

For deuterostome larvae there is even less cohesion, and variation can be extreme within one phylum, accentuating the issue of functional adaptation in pre-adult forms. It has already been shown that the hypothetical dipleurula is rather hard to define, except as the inevitable outcome of the other features that define a deuterostome; and while it is possible to derive many of the classic isolecithal echinoderm larvae (fig 5.11) from a dipleurula form, some ophiuroids, holothuroids and crinoids have a quite different larval form called a vitellaria (fig 5.11*F*) which bears greater resemblance to a trochophore and

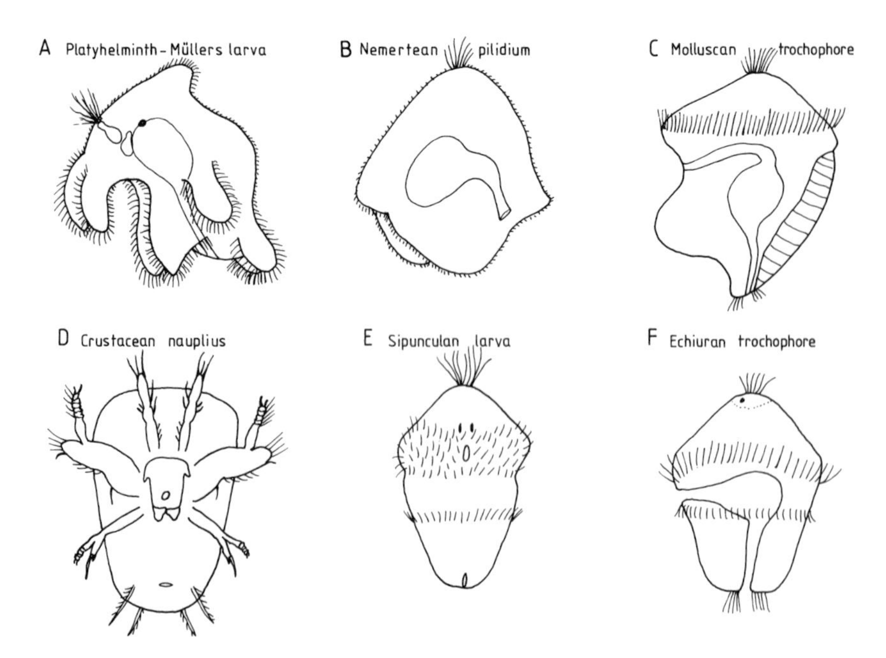

Fig 5.10. Some of the invertebrate larvae that have been described as 'modified trochophores' in the literature.

cannot be fitted into the scheme at all. Worse still, any classification of echinoderms based on larval form is totally discordant with adult morphology and with fossil evidence (Fell 1948; and see chapter 12), so that they are frequently cited as strong evidence against any phylogenetic use of embryology and larvae.

However, deuterostome larvae do yield one noted example to redress the balance on the value of early stages as a guide to phylogeny. This is the tornaria larva of the hemichordates, which very closely resembles an early auricularia echinoderm larva (fig 5.12). It must be admitted that this could also

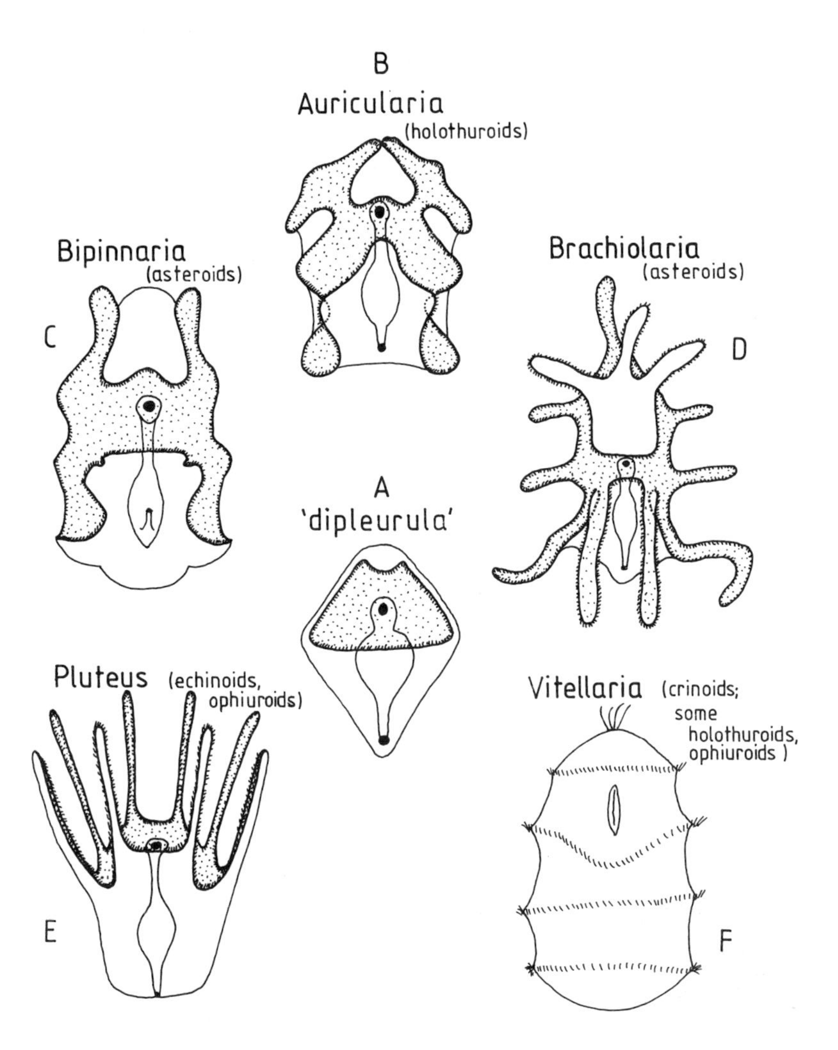

Fig 5.11. Varying larval forms in the echinoderms: types *B* - *E* can perhaps be derived from the ancestral dipleurula, but the vitellaria *(F)* is apparently unrelated and occurs in forms with more yolky eggs.

be due to convergence (chapter 12); but a clear link between echinoderms and the rest of the deuterostomes would be somewhat harder to make without this larva.

Finally we should continue this last theme by noting that larval types can sometimes be of unrivalled value in settling issues of identity and relationship at a lower taxonomic level. There are a number of well-known examples to back this up: the parasitic barnacles (Rhizocephala) that can only be identified as cirripedes by their larvae, or the parasitic mollusc *Entoconcha*, again only correctly classified by virtue of its veliger larva. Clearly there is a real role in taxonomy for the study of larvae, and larval form has on occasion been the only feature that has allowed the correct placement of an animal within its phylum or lesser taxon.

Functions and constraints

Larvae are clearly even more subject to modification for their own functional ends than are the earlier embryonic stage of an animal. They have very precise tasks to accomplish, and the selective pressure on them may be intense. If attention is restricted to marine pelagic larvae (Mileikovsky 1971; Crisp 1974; Horn *et al.* 1982), they must primarily achieve dispersal and habitat selection, establishing new populations where competition with the parents and/or with siblings is reduced, and outbreeding promoted. Planktotrophic feeding and

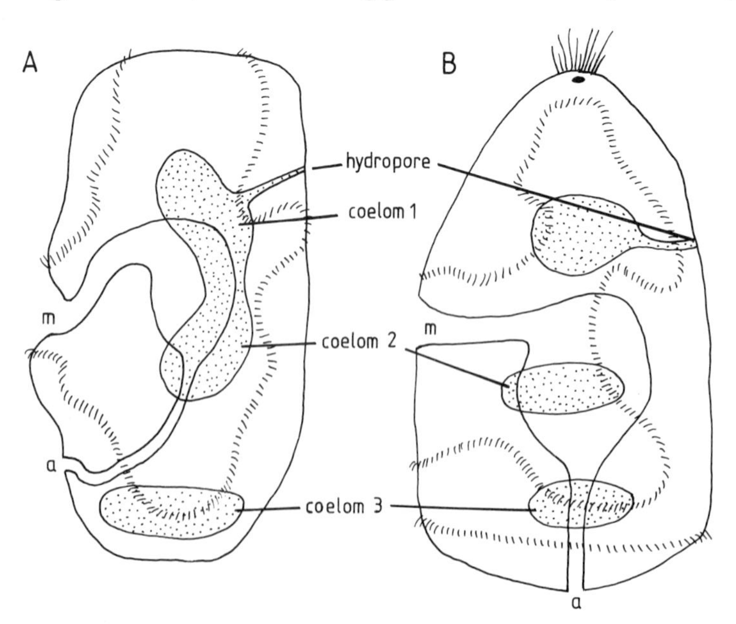

Fig 5.12. The similarity of the hemichordate 'tornaria' larva *(B)* with one type of echinoderm larva, the 'auricularia' *(A)*.

locomotory systems, and suitable sensory modalities, are therefore required, and the design constraints will be stringent. On the other hand, for many terrestrial invertebrates the adult achieves dispersal and site selection, and the larva is instead adapted as a rather stationary feeding and growth stage. In either case, larval function is very precise, the larval and adult stages clearly have quite different design constraints on them, and a drastic intervening metamorphosis may be needed.

For marine larvae, there is a profound morphological convergence in virtually all phyla in the early stages of larval life, since they are of similar size and perform such similar functions. The plankton abounds in microscopic larvae at certain times of year, and the only easy distinction that can be made by an amateur is between the crustacean forms with cuticle and articulated legs, and 'the rest' all having roughly spherical transparent bodies with an apical tuft and one or more flickering bands of propulsive cilia. Further identification relies on establishing the numbers and sites of these bands, and the lobes and protrusions on which they are borne. When the larvae are somewhat older they often become more distinctive in form, as they begin to acquire adult features (see Garstang 1928*a*; Jägersten 1972; Freeman 1982). The molluscan veliger begins to resemble the adult with shell, head and foot appearing; the annelid trochophore begins to add segments posteriorly; the hemichordate tornaria elongates and grows a tail. This 'adultation' may be an important source of potential evolutionary change, by varying the temporal relations of larval and adult features (Freeman 1982). However, more drastic and unexpected changes may also occur: the extension of the molluscan prototroch into a vast ciliated lobe propelling the creature, or the contortionist morphological changes of some of the echinoderm larvae. In other words, the larva does not simply unfold into a miniature adult, but continues to manifest adaptation of its own appropriate to changing size and habit.

Larval evolution

The next important issue is whether in fact the marked similarity of so many early larval forms is actually any indication of a primitive common ancestral larval type. Jägersten (1972) has strongly advocated this viewpoint, and much of the recent embryological work seems to follow his assumptions (e.g. Freeman 1982), regarding the larva as primitive and as a vital stage in the developmental programmes of metamorphosis, with direct development as a secondary phenomenon. Jägersten takes as his starting point a hypothetical 'bilaterogastraea' form, and a number of other European authors take a similar stand; their views will be seen again in chapter 7, as they bear heavily on questions of metazoan origins. The question for now is whether it is necessary to assume that a larval (pelagic) stage is primitively part of the metazoan life-cycle, or whether the marine invertebrate larva should be seen purely as an adaptation of the juvenile stages to pelagic life, occurring numerous times quite independently, from an initial state of direct development.

Jägersten argues for a primary alternation of pelagic larvae and benthic adults in the metazoan life-cycle, assuming that when the original blastaea-type ancestor (as hypothesised by Haeckel - see chapter 7) settled on the bottom as a bilaterogastraea, the juveniles remained in the pelagic zone. The two developmental stages thereafter diverged to different extents in different groups. Jägersten maintains that there is no evidence in favour of a primitively holobenthic life with direct development, and that where these occur nowadays they are clearly secondary. This is supported by the occurrence of the pelago-benthic lifestyle, together (he says) with primitive sperm, external fertilisation, coeloblastulas and gastrulation by invagination, throughout the Metazoa (fig 5.13). He also points out that where development occurs within protected egg membranes, the embryo still goes through stages that resemble the pelagic larva in related forms. Hence it follows that larvae are profoundly important in phylogeny, and do exhibit a degree of recapitulation of ancestral forms.

Extensions of Jägersten's theory lead to complex views of life-cycle evolution as recently proposed by Nielsen (1979; 1985) (see also Nielsen & Nørrevang 1985), where all larval forms are seen as interrelated, the primitive blastaea and gastraea being forerunners of the trochaea, which in turn gave rise to the true trochophore on the one hand and to a tornaea, ancestral to deuterostomes, on the other (see chapter 7, fig 7.4). Such theories seem attractive, but make a number of assumptions about rather complex morphological changes within the larvae, difficult to relate to functional need; they also of course rely on the underlying assumption that pelagic larvae *are* primitive.

The alternative view seems initially counter-intuitive as it has long been customary to think of direct development as a specific and secondary adaptation to difficult circumstances, such as fresh-water or terrestrial life. 'Primitive' animals are supposed to have a whole series of reproductive traits (Stearns 1980): simple spermatozoa (discussed in chapter 6), small oocytes, external fertilisation (see Franzén 1970), and planktotrophic larvae. Yet a review of the occurrence of these traits (Olive 1985) shows that they are not independent variables, but are associated particularly with larger-bodied organisms, accumulating gametocytes in a body cavity prior to synchronous release. Olive concludes that planktotrophy must have originated in relatively large invertebrates. So either large size was also primitive; or planktotrophic larvae are a secondary acquisition. The Jägersten theory strictly requires a rather small benthic adult ancestor, and the traits that he maintained were 'advanced' are actually only common among the smaller representatives of lower phyla. But all this could still be made compatible with Olive's observations, if the theories that early metazoans were relatively large and coelomate ('archecoelomate' theories, as discussed in chapter 2) are adopted; these theories invoke a big enough adult ancestor to show all the 'primitive' reproductive traits and can retain primitive pelago-benthic life-cycles. Alternatively, small ancestral invertebrates can be retained and the idea adopted that direct development is primary and all larvae are convergent secondary inventions. For example Ax (1987) rejects the idea of a primary pelago-benthic

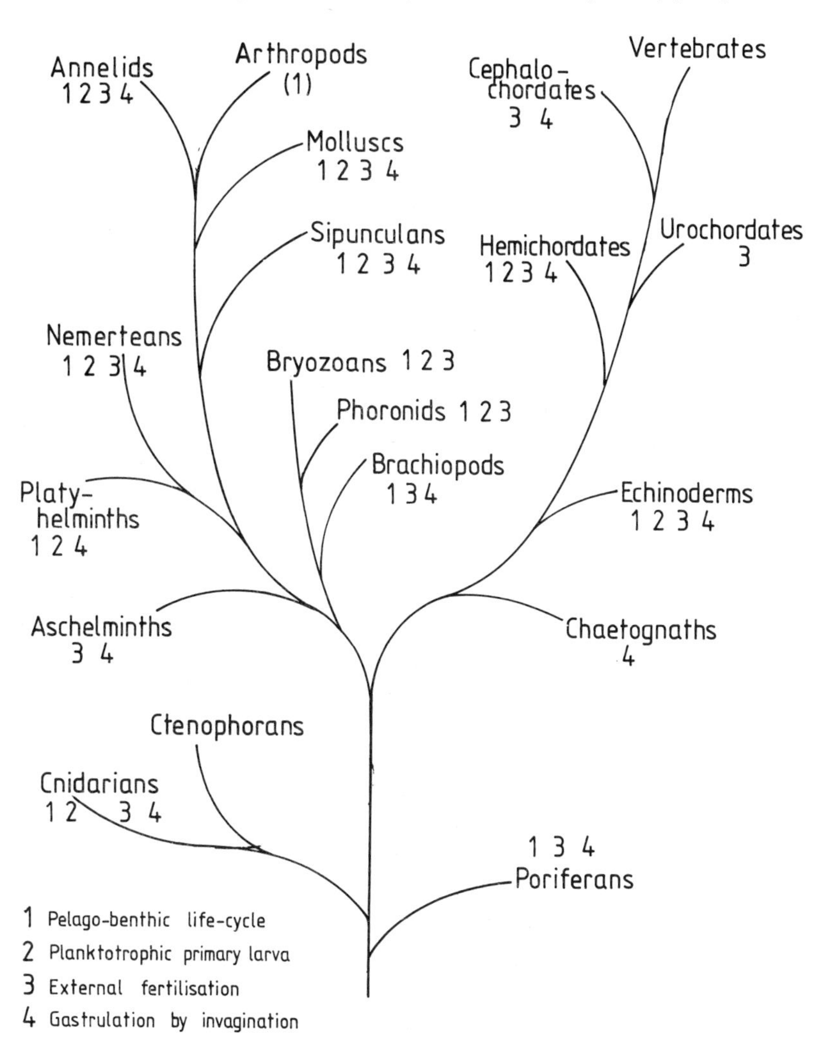

Fig 5.13. A review of the occurrence of supposedly primitive life-cycle characteristics in invertebrate groups, based on Jägersten (1972).

life cycle, arguing that a larva is lacking in the 'ground plan' of lower metazoans. Since it is still likely that in some cases a secondary direct development would arise later, this view is not inconsistent with the (few) cases that Jägersten quotes where directly developing larvae exhibit pelagic-larval features. All authors certainly agree that transitions between these modes of development have occurred, in either direction, numerous times (see also Strathmann 1978), so that such resemblances can never be conclusive evidence for the evolutionarily primitive mode.

It is perhaps only a matter of opinion as to which of these two alternatives, neither of them exactly reflecting conventional wisdom, involves the fewer unsupported assumptions and generalisations; but it is clear that Olive's work does make the idea of ciliated larva-like organisms as metazoan ancestors in themselves seem rather less plausible.

5.4 Conclusions

Embryological stages and larvae undoubtedly have an important role to play in phylogenetic studies. In interpreting their evidence, a central path between extreme adherence to the 'biogenetic law' and the alternative complete rejection of recapitulationist ideas must be found. Early stages in animal development are simpler and smaller, with fewer and more straightforward functions; for all these reasons they should be more conservative, reflecting relationships with other groups more clearly than highly specialised later stages. In modern terms, since development can be regarded as hierarchical it follows that evolutionarily viable changes would not so readily occur as additions or alterations at the end of sequences ('terminal additions' - Gould 1977). To that extent, a modified form of recapitulation makes perfect sense. But embryonic stages may clearly be adapted to their own ends (see also Horn *et al.* 1982), to the extent that ancestral conditions are obscured and convergent similarities with unrelated forms created. For larval stages, often larger, of longer duration, and with more diverse and precise functions, these problems are accentuated and it is impossible to be sure whether particular forms are indicative of common ancestry, or indeed whether the larva itself is a primitive phenomenon.

Within these reservations, a number of generalisations can be made. There appear, on first encounter, to be several neat dichotomies in the embryological development of metazoans, and these also appear to be reasonably aligned so that - for example - spiral cleavage, determinacy, proliferating 4d mesoderm and trochophore larvae all go together. However there are a great many exceptions both to the features themselves, so that few of them remain dichotomous, and to the correlating patterns. And at the same time there is some indication that they are not independent features, so that *some* association between those listed is only to be expected. Using them collectively as a basis for a fundamental dichotomous split in the metazoans does not seem advisable. This is partly because the tendency to include within the two major groups thus created a number of smaller phyla (sometimes *all* the smaller phyla), where the evidence from embryology is at best equivocal, becomes almost overriding and almost certainly counter-productive.

The best compromise then is to accept only the tighter relationships indicated from embryology, where functional reasons for convergence are not apparent. A degree of cohesiveness of the core group of spiralians then seems likely, including certain polyclads, nemerteans, annelids, uniramian arthropods, pogonophorans, echiurans, sipunculans and molluscs. But even here embryological and larval similarities are by no means as great as some texts suggest and it is probable that many of these groups are early offshoots within the Metazoa, perhaps independently derived from an ancestral flatworm stage with similar characters (see chapter 8). Anderson (1982) therefore unites only a sub-group of these phyla, those with clear 4d mesoderm, placing them within a broader spirally cleaving assemblage, the latter also including chelicerates, crustaceans, and some pseudocoelomate groups. So at best perhaps only annelids, echiurans, possibly pogonophorans, and probably uniramians truly belong together as a tight '4d-spiralian' assemblage, and the last two of these (chapters 8 and 11) certainly don't very clearly earn their place from their embryology!

Similarly the assemblage of phyla with radial cleavage and infolded enterocoelic mesoderm, often showing gastrulation by invagination, may be related (and by more than the constraints and convergences imposed by egg membranes), given the embryological and larval similarities of echinoderms and hemichordates. It is largely adult similarity that links chordates into this assemblage though, as we will see in chapter 12 - chordates being notorious for their aberrant schizocoelic coeloms, and having other aspects of their embryology generally distorted by excessive yolk.

The protostomes and deuterostomes are thus plausible as super-phyla, so long as they are not seen as the *only* two important branches of a dichotomous phylogenetic tree; at best, they might include three or four phyla on each side, leaving out more than three-quarters of all the known phyla. The features that are supposed to define the two super-phyla are frequently not dichotomous themselves, and there are no valid reasons to believe there should be only two possible permutations of these features; increasing knowledge of the invertebrates only too clearly reveals that many other permutations do exist.

If we *are* dealing, then, with a 'theme and variations' in considering animal development, we have seen that the unifying theme is over as early as the four-cell stage (and even then there are some exceptions such as the nematodes). Thereafter we may have two major themes that take most of the attention, but it is evident that there are a host of minor themes that are equally important to an understanding of the whole.

Thus, as will be seen in some chapters of part 3 of this book, there remain far too many cases where embryological patterns are merely confusing seen in isolation - within the phylum Echinodermata, or in the pseudocoelomate and lophophorate groups, for example - so that we can never have a great deal of confidence in any single case of embryological similarity. The message must be that embryology has got to be used with great care, with an awareness of possible functional convergences and divergences at least equal to that used in judging other evidence; and it must therefore be taken in balanced association with all other possible lines of comparative biology. It is not uniquely to be favoured as revealing the true course of evolution, as Haeckel would have averred; but nor is it to be viewed with total suspicion, as anti-recapitulationist dogma has sometimes implied, not least because of those cases regarding relationships within phyla where its testimony has proved quite invaluable.

6 Evidence from cell ultrastructure

6.1 Introduction

Any biological phenomenon must be dependent upon the fundamental activities of cells. Most obviously, cellular differentiation gives rise to the structural features of tissues and organs, and thus to the overall embryology and morphology of organisms that have been the classic materials for phylogenetic studies. In addition, short-term changes in cell structure and behaviour often initiate physiological or mechanical processes affecting the whole animal, influencing its relations with its environment and the selective pressures it experiences. Hence comparative studies of cells themselves should obviously have some role in an analysis of invertebrate relationships.

Some aspects of cell behaviour were considered in the previous two chapters. Virtually all the chemical constituents of the animal (chapter 4) are cellular products, either within the cell membrane or as secretions from it. And the processes of differentiation of embryos (chapter 5), from cell fusion at fertilisation through all stages of ontogenesis, are major cell behaviour phenomena. But now the concern is rather with cell structure and its bearing on phylogenetic matters. This has been an area of growing importance in the last fifteen years, with a literature largely derived from European sources in general and German authors in particular. An appreciation of some of the more important topics by English-speaking audiences has been lacking until very recently.

Clearly comparisons of cell structure rely on different techniques from earlier morphological studies, and have therefore only been possible since electron microscopes became available in the 1950s. Their increasing prevalence has led to the period since 1970 being termed the 'era of ultrastructure'. In principle there are no differences between ultrastructural details and larger-scale morphological features as phylogenetic indicators, so that the same strengths and weaknesses apply to both. Ultrastructure is in fact likely to be of greater use than conventional morphology in certain groups, where there is otherwise a paucity of useful systematic characters; and electron microscopy has certainly shown its worth (and has become a routine tool) in taxonomic studies of lower plants, protists and fungi. Thus it is not surprising that electron microscopy is most widely utilised amongst the lower Metazoa by invertebrate systematists. However its use in these areas remains somewhat controversial. Several authors have argued that ultrastructural features are so small that they are necessarily simple, and could thus have arisen convergently in widely separate evolutionary lines. It has also been said that variations in ultrastructure are particularly likely to be the result of functional requirements, and unrelated to evolutionary trends; and it is certainly true that ultrastructural study is far in advance of our understanding of cell and organelle function at a comparable level, and comparisons often proceed in a functional vacuum. Some of these problems are discussed by Storch (1979) and by Tyler (1979); and a more theoretical treatment of the problems of identifying homologies in ultrastructural studies is given by Rieger & Tyler (1979). There is little doubt that great care is needed in interpretation of ultrastructural comparisons, but clearly many of the key workers in this field are very aware of the hazards, perhaps more so than is the case for other possible approaches. This chapter is therefore devoted to a consideration of the evidence from recent work on cellular ultrastructure in the invertebrates, taking examples both from general cellular apparatus and from the morphology of specialised cells.

6.2 General cellular components

The ultrastructure of most cell organelles is sufficiently well known to need no recapitulation here, and has proved remarkably constant throughout the animal kingdom; it is doubtful if much can be gained from such material when seeking phylogenetically significant differences. Such comparative detail as is available has been summarised by Welsch & Storch (1976) and by Dillon (1981), though the phylogenetic remarks in these reviews commonly have to be limited to comparing numerous vertebrate tissues with a very few invertebrates and some protists, giving little scope for helpful conclusions. However there are a few special cases of cell features that vary in complexity or detail in an apparently consistent manner, and have been studied in a reasonable range of animals; these merit further attention.

Cilia and flagella

The distinction between cilia and flagella, and their distribution in different cell types and different phyla, have been long-standing subjects for debate, particularly in relation to the transition from a unicellular to a multicellular condition and to the status of lower metazoans (chapter 7). The planar or helical patterns of movement, length of shaft, and number of organelles per cell are the familiar features used to distinguish the two types (e.g. Sleigh 1974), and such a distinction has clear pragmatic value; but it is still not clear whether the two structures are really different or have separate origins. One commonly held view is that they represent a continuum rather than a dichotomy. Thus cells, especially within the Protista, may have more than one long, helically-beating 'flagellum', or just a few short 'cilia' that produce a sculling effect, contravening the usual distinctions. Or groups of 'cilia' may move in a far from two-dimensional manner with a path more nearly resembling a spiral. Even more confusingly, in just a few cases (e.g. the protist *Chlamydomonas*) inter-conversions occur between a classic flagellate state and a state where the action of the shortened organelles more resembles ciliary 'rowing'. Therefore many authors now use only the terms 'monociliate' (one organelle per cell, the traditional 'flagellate' condition) and 'multiciliate' (many organelles, traditionally 'ciliate') in describing cell types; whilst others, particularly protozoologists, have adopted the term flagellate in all cases. The former terms are probably more helpful in dealing with the Metazoa and will be used throughout this book.

Structure. The basic ultrastructure of these organelles is well known from a wide diversity of tissues and animals (see Afzelius 1969; Sanderson 1984), and has proved almost completely constant in the animal kingdom for all cases where the organelle retains a function in motility. A phylogeny of the axoneme system in its earlier stages, giving comparisons between all the kingdoms of living organisms, is given by Dillon (1981; page 197); but in metazoans the shaft of a motile cilium always has the familiar nine peripheral doublet microtubules bearing dynein arms and two core microtubules (fig 6.1*A*), all these being apparently constant in protein (tubulin) constituents (chapter 4) and in action. Only in the special cases of spermatozoa, receptor cells and some protonephridial excretory cells (all discussed in section 6.4) has variation in the shaft patterns of microtubules been detected. Clearly there cannot be much scope here for phylogenetic speculation.

However there are some features of the ultrastructure of these organelles that do show variation and have therefore been used as systematic characters. Some of these are relatively minor and have not yet been studied in a sufficient diversity of tissues for conclusions to be possible: surveys of such features can be found in Dillon (1981) or Sanderson (1984). There are some better-known cases, though. Firstly the basal apparatus, anchoring the cilium to its cell, is variable. There is always a centriolar structure (the basal body, or kinetosome), with triplet microtubules, at the base of the shaft, but the transition between this and the axoneme shaft is variable. Pitelka (1974) identified two forms: type 1, where the transition zone is flush with the cell surface, is found in some flagellates and all ciliates, and in mammals and birds; whilst type 2, with the transition zone set well above the cell surface and somewhat modified in form as shown in fig 6.1*B*, is found in some flagellates, virtually all invertebrates, fish, and amphibians. This seems to be of decidedly little help in deciphering invertebrate relations. But another feature of the basal apparatus may be more useful, for in many cases a second centriole-like body lies to one side of the first and at a varying angle to it, giving a 'diplosomal' basal body (fig 6.1*B*). This second centriole may be the vestige of a second flagellum according to some interpreters (Hibberd 1975; Salvini-Plawen & Mayr 1977). Its presence is virtually always correlated with a flagellate/monociliate condition (and hence it is absent where there is more than one cilium); and it is oriented in relation to the direction of the ciliary beat (Holley 1984). Possibly we have here a sensible means of distinguishing 'cilia' (monosomal) and 'flagella' (diplosomal), though it would cut across the traditional dichotomy.

In addition to the centrioles the basal apparatus often includes a basal foot and one or more striated rootlet structures (fig 6.1*B*), also known as

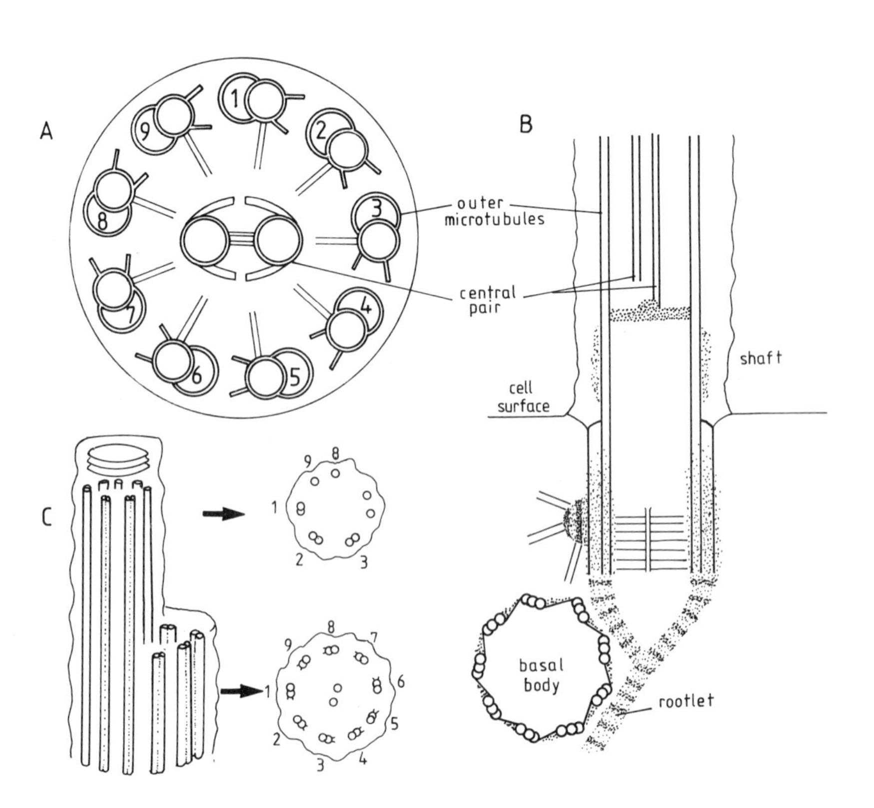

Fig 6.1. Structural aspects of cilia and flagella. *A*, the arrangement of microtubules in the axoneme, common to all cilia; *B*, the basal apparatus and associated structures, where variation does occur; *C*, the tip of a cilium from Acoela.

rhizoplasts, serving to anchor the organelle in its epithelium as the cell changes shape. Here again there is considerable variation in form, with one or two main rootlets of varying lengths and orientations in most cases, but exceptionally there may be no rootlets (as in many human cells) or as many as four of these structures. Attempts have been made to link this to phylogenetic schemes (e.g. Rieger 1976; Tyler 1979), with, for example, a constant arrangement in the main spiralian phyla. But Holley (1984, 1986) argues that rootlets are designed according to functional constraints, varying with the flexibility of the anchoring epithelium and the nature of the medium they propel (water or mucus), and may thus have little use as a systematic character.

Another feature of cilia is the pattern of termination of the microtubules at the apex of the shaft (Tyler 1979). In certain groups, rather than all the axoneme tubules terminating at the same point some are longer than others (fig 6.1*C*). In the acoel turbellarians the characteristic pattern is that four tubules end while five continue further up the shaft, giving a 'shelf' effect; this is found in all members of the group and in no other animals, and so appears to be a good taxonomic character (arguing against the primitive status of acoels - see chapters 7 and 8). But as yet this type of difference has not proved of much use except in rather special cases within phyla, and constancy remains the norm for most aspects of ciliary structure.

Finally, the 'ciliary necklace' (rows of particles on the membrane surface close to the base of a cilium) has been used as a phylogenetic tool by Bardele (1983), who reports that all invertebrates, including sponges, cnidarians and platyhelminths, have triple rows of particles forming the necklace when seen by freeze-fracture techniques (just occasionally increased to four or five rows). In this they resemble the flagellate protists but are quite distinct from the ciliates which have only a double-stranded necklace; such a finding has clear relevance to the problems of metazoan origins discussed in chapter 7.

While dealing with structure it may be appropriate to mention the findings of Stephens (1977) who suggested that while the outer sheaths of cilia and flagella may appear ultrastructurally similar they differ markedly in chemistry, the ciliary sheath containing a form of tubulin while the flagellar sheath is principally a glycoprotein. If this is substantiated for other tissues, and proves to correlate with other possible cilia/flagella differences, we may eventually be able to separate two distinct types of organelle with greater conviction, and their phylogenetic significance will perhaps become clearer.

Occurrence. Some animal phyla lack motile cilia entirely, notably the nematodes and the major arthropod groups, indicating a correlation with the presence of stiffened cuticles through which cilia could not function. All other phyla have at least some ciliated cells. Conventionally, multiciliate cells occur in most of the familiar 'protostome' invertebrate groups - platyhelminths, nemerteans, annelids, sipunculans and molluscs - and in some pseudocoelomate groups such as rotifers and entoprocts. Monociliate (flagellate) cells are known in sponges, cnidarians, lophophorates and all the classic 'deuterostome' groups, particularly in their larval forms (though regions of multiciliation may occur in some adults). Rieger (1976) and Rieger & Mainitz (1977) have also demonstrated monociliated cells in gnathostomulids, and in gastrotrichs where they co-occur with multiciliated cells. Rieger believes the monociliate condition to be more primitive in the gastrotrichs and in the Metazoa as a whole, and has argued for the monociliate 'collar cell' as an ancestral type (see below). As we will see, such a view has important implications for the origin of metazoans (chapter 7), and is also relevant to the protostome/deuterostome debate that recurs throughout this book.

But it is becoming apparent that functional considerations may again be obscuring matters. For example, the type of ciliation present may depend on the nature of the cuticle in a group; monociliation appears to be allied with primitive smooth cuticles, and multiciliation with more complex sculptured surface layers. Or it may relate to the habit of the animal, as epithelia used for creeping and gliding locomotion are commonly multiciliated whereas the feeding surfaces and tentacles of more sessile animals can be of either type. In fact both types of ciliation can occur not only in the same animal but in the same tissue, as shown for the tunicate endostyle (Holley 1986) where adjacent zones may have quite different functions and at least six distinct patterns of ciliation exist, some based on monociliated and some on multiciliated cells.

Since conversion between the two states has clearly been possible within the evolutionary history of one phylum and even one animal there may be limited significance in the precise occurrence of these organelles, and exceptions to any rule will always be found. It is unclear which of the two states came first in the Metazoa, and most author's views on this are coloured by their opinion on the origins of the multicellular forms from protists. But whatever the early sequence of evolutionary events, the generalisation that monociliation is very rarely found in the traditional Protostomia/Spiralia lines seems reasonably well established, whereas the Deuterostomia groups commonly do show this condition (and perhaps have retained it from an 'archecoelomate' stage according to some interpretations). A survey of the known occurrence of the two forms of ciliation, and their association with basal body types, is given in table 6.1 (based on Rieger 1976 and references therein; and M. Holley, pers. comm.). Possible phylogenetic relations suggested by these patterns are shown in fig 6.2.

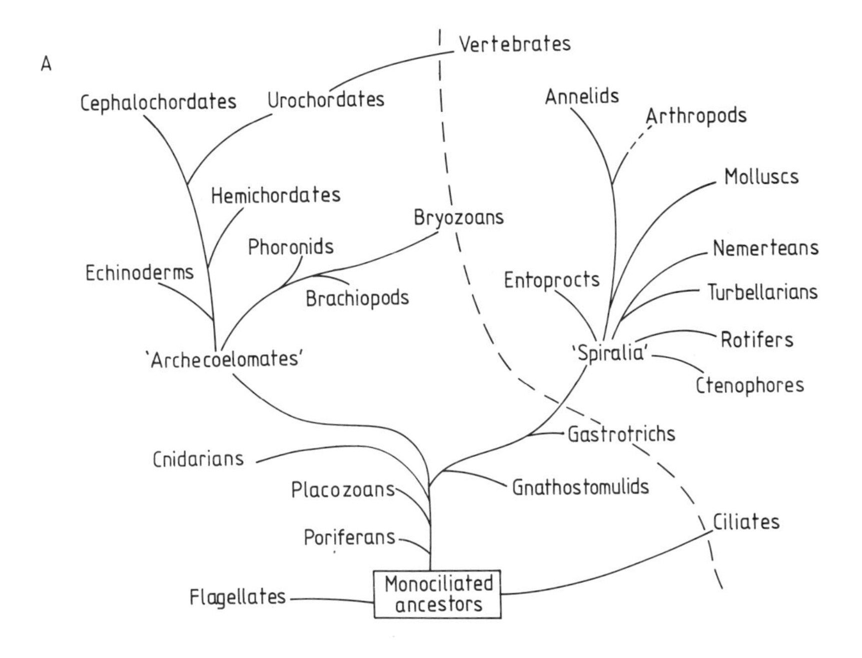

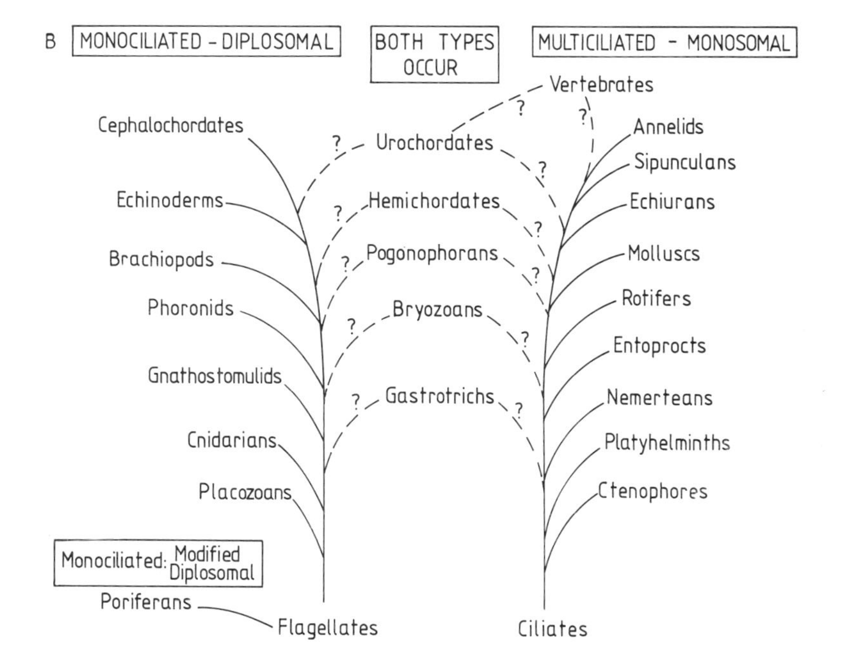

Fig 6.2. Possible significance of ciliation types in relation to animal phylogeny: *A*, the traditional dichotomy of the animal kingdom, with monociliate and multiciliate groups separated by the dashed line; *B*, a phylogeny based only on ciliation, without other preconceptions. (See also table 6.1).

Microvilli and collar cells

Many cell surfaces bear microvilli in the form of regular finger-like extensions, supported by a parallel central bundle of microfilaments. As they are the standard means of achieving greater surface area, particularly when arranged into a 'brush border', they are not unexpected in any cells involved in exchange processes whether absorptive, secretive, digestive or respiratory.

Table 6.1 *Occurrence of different ciliation patterns*

	Monociliated	Multiciliated
One centriole (monosomal)		Ciliates, Rotifers, Turbellarians, Entoprocts, Vertebrates, Ctenophores, Molluscs, Annelids, Nemerteans Some: Flagellates, Gastrotrichs*, Bryozoans (adults), Pogonophorans, Hemichordates, Urochordates*
Two centrioles (diplosomal)	Poriferans, Cnidarians, Placozoans, Echinoderms, Phoronids, Brachiopods, Gnathostomulids, Cephalochordates Some: Flagellates, Gastrotrichs*, Bryozoans (larvae), Pogonophorans, Hemichordates, Urochordates* One: Annelid Most excretory and sperm cells, many receptors	Some: Gastrotrichs*, Annelids (larvae), Urochordates*

(* Different ciliation patterns known even within one species)

There is however a particular cell type known as the collar cell (fig 6.3) where a regular ring of microvilli surrounds a single central cilium on the apical cell border. This cell type was originally known in metazoans only from sponges, where it was termed a choanocyte and taken as evidence for the unique status of the sponges. It was also believed to link them to the choanoflagellate protozoans, suggesting a separate ancestry of sponges as a sub-kingdom Parazoa, distinct from the Eumetazoa. Recent evidence has undermined this position (see for example Bergquist 1985), mainly by identifying very similar cells in a great many other phyla: Placozoa, Cnidaria, Echinodermata, Hemichordata, Gastrotricha and Gnathostomulida all possess them, and they also form the basis for many sensory cells and flame cell structures throughout the Metazoa (Rieger 1976; and see section 6.4).

As the classic collar cells are noticeably lacking from the main protostome groups they may have some phylogenetic significance. But within the

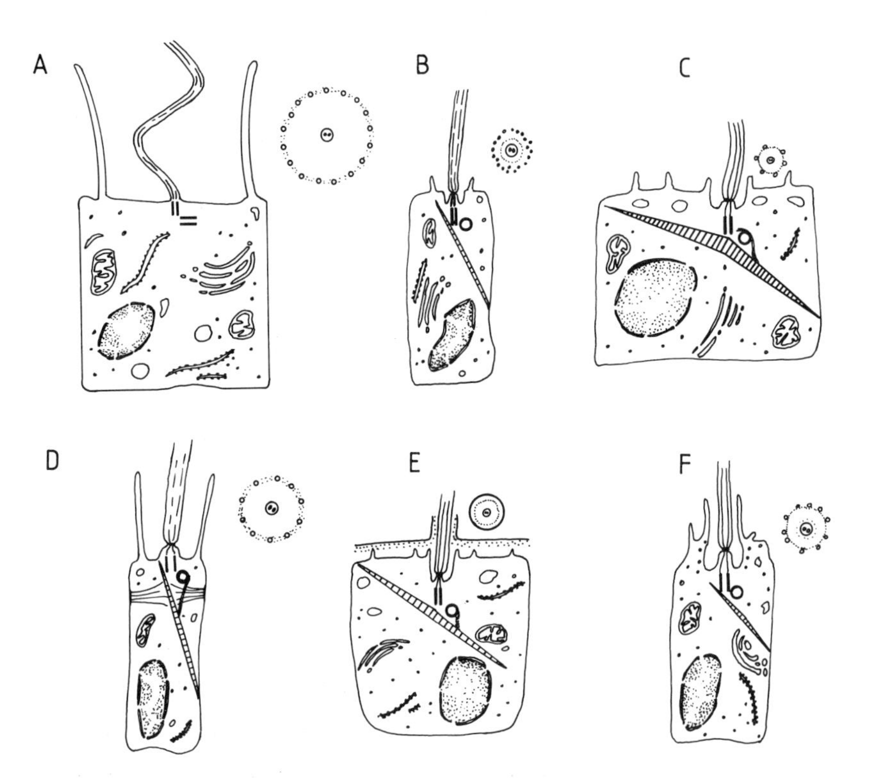

Fig 6.3. 'Collar cells' from various groups of invertebrates, based on Rieger (1976): *A*, sponge; *B*, placozoan; *C*, gnathostomulid; *D*, anthozoan cnidarian; *E*, gastrotrich; *F*, echinoderm.

remaining phyla Rieger believes they have probably had multiple origins from simpler ancestral monociliated cells, partly because they can be derived from any of the three germ layers, and can have multiple functions. Certainly their general structure suggests not unreasonable convergence on a form suitable for creating water currents and achieving orderly increases in surface at a suitable point in the fluid flow. While they may be useful indicators of links between groups their possibly convergent nature means that mere presence must be treated with due caution. However, they do represent a clear example of ultrastructural 'contributions' to phylogeny, having in many people's view first isolated the sponges, then reinstated them as Metazoa (though admittedly embryological consideration also played a part in these changing views; see chapter 7). Perhaps the sponges should still be set somewhat apart from other phyla in terms of ciliation though, because the basal apparatus of their choanocyte cilium (fig 6.3*A*) has proved to be unlike that of all other animals in centriolar orientation and lack of rootlets, and the microvilli are rather differently disposed.

Cell junctions

There is a considerable literature on the nature and role of the junctions that occur between membranes of adjacent cells, much of it reviewed by Staehelin & Hull (1978), Lane & Skaer (1980) and Green (1984). One attempt has been made recently to inject phylogenetic significance into these studies (Green & Bergquist 1982), by comparing the septate junctions of all invertebrates so far studied. These authors identify a total of thirteen variants on the basic septate junction, together with a special junction type in the sponges and tight junctions in tunicates and vertebrates, all these types having the common feature of forming a belt around the apices of epithelial cells. The views Green & Bergquist express are summarised in fig 6.4. While it must be admitted that the differences between junctions on which the authors base their analysis are often apparently minor, and are of uncertain significance functionally or phylogenetically, some interesting points emerge.

Porifera again appear to be set apart, with a very simple junction type. Amongst the cnidarians each class has its own unique junction, but it is that of the Hydrozoa that most nearly resembles the form taken in other lower invertebrates. These, including platyhelminths, nemerteans and annelids, as well as some of the lophophorate groups, share a pleated septate junction; and a slightly different version of this occurs in molluscs and arthropods. Smooth junctions also occur in arthropods, varying significantly between each of the major groups. In the hemichordates and echinoderms double-septum junctions occur, those of the latter group most clearly resembling the tight junctions so far described only in tunicates and vertebrates (and at a few special sites in arthropods). Thus the division of protostomes and deuterostomes seems to be supported, and the echinoderms are set nearer to vertebrates than are the hemichordates. The vertebrate\tunicate alliance is unequivocally supported. In addition sipunculans resemble protostomes (which seems to surprise the authors), and chaetognaths, possessing both septate junctions and pleated junctions, occupy a typically enigmatic position.

However, the phylogenetic tree produced by Green and Bergquist does give the impression that they set out to fit their findings into conventional phylogenies rather than using their data as a novel systematic tool, partly because having stressed the similarity of lower invertebrates they proceed (page 295) to split them into two groups for no obvious (inherent) reason. Their analysis clearly would be strengthened if the number of examples known from each phylum was greater and if functional constraints on cell junctions were better understood.

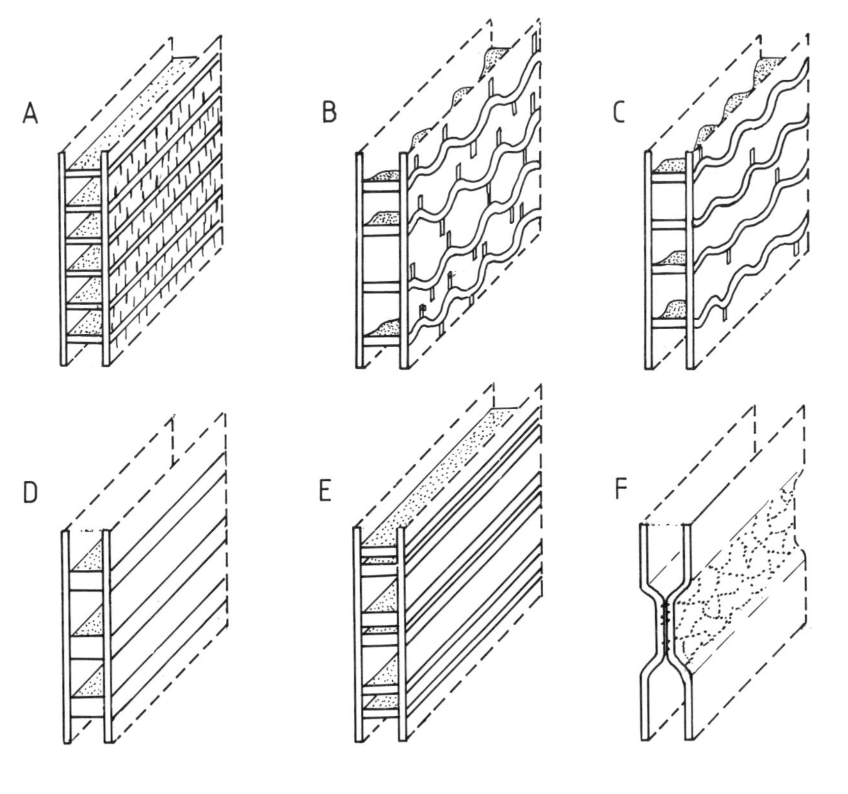

Fig 6.4. Structures of septate junctions between cells, drawn from descriptions and photographs given by Green & Bergquist (1982): *A*, the hydrozoan pattern; *B*, the pleated septate junction of lower invertebrates (including Platyhelminthes, Nemertea, Sipuncula, Annelida and some lophophorates); *C*, modified pleated arrangement in molluscs and arthropods; *D*, arthropod smooth junction; *E*, double septate junction in echinoderms and hemichordates; *F*, tight junction in tunicates (and similarly in vertebrates).

6.3 Epithelial secretions

Every animal's lifestyle and habitat are highly correlated with the properties of its outside surfaces, and there is always a conflict of interest at work here. 'Skins' need to protect the animal, and in physical, chemical and physiological senses they thus have to keep the world out. But at the same time they sometimes have to admit selected chemicals from that world (water or ions, gases or nutrients), and they always have to admit information about the outside world, by giving access to the animal's nervous system. In special cases skins also have to control the animal's movement (or stasis) relative to the world. It is virtually impossible to perform all these functions simultaneously with great efficiency, and epithelia and integuments inevitably represent compromises. One solution is a highly protective covering, with exchange functions concentrated in just a few areas; another is to remain effectively naked, with a cellular or even syncytial surface directly exposed to the world that may have to be equipped with defensive or adhesive specialisations.

The properties and secretions of epithelia are of course partly a phylogenetic issue, as representatives of particular phyla may have only a few options available; but the issue is greatly clouded by the effects of habit and habitat, because an optimal solution to the 'skin paradox' will depend so much on which requirements have most urgently to be met. Deducing phylogenies from properties of superficial cells and their products is made no easier by the apparently limited range of chemicals available (chapter 4) and the constraints upon their deposition. These problems will be met repeatedly in the reviews that follow.

Cuticular structure

A true cuticle, in the sense of a fibrous and/or granular material on the outside of the epidermis, may have been a very early acquisition in the Metazoa as it would confer potential protection, support and regulation on the newly multicellular animal. The nature of the invertebrate integument itself, and of its various cuticular secretions, has been recently and comprehensively reviewed (Bereiter-Hahn *et al.* 1984); and, as some details are now known for every phylum, cuticles have been a source of several phylogenetic speculations (Rieger & Rieger 1976, 1977; Rieger 1984; Storch 1979). The primitive cuticle may have been no more than a fibrous mass of mucoproteins or mucopolysaccharides, elaborated from the glycocalyx that protects most animal cells. Simple cuticles of this type can be found in most animal phyla (Rieger 1984), either all over the cell surfaces or between microvilli if these are present; there seems little phylogenetic pattern to their occurrence and indeed some examples may be secondarily simplified. However, some phyla (Ctenophora and Nemertea) have nothing beyond this simple (perhaps primitive) cuticle, whilst others have achieved marked specialisations of their surface coatings. The general trends of such changes are shown in fig 6.5.

In the platyhelminths and annelids the commonest cuticles are still variants on the microvillar prototype, but in some more specialised animals the fibrous matrix between the microvilli is strengthened with collagen, chitin or even calcium carbonate, and the microvilli themselves may be elaborated or reduced. It is not too difficult to see the much more elaborate cuticles of the various arthropod groups as modifications from the patterns found in these worms.

Rather similar trends in the nature of true cuticles are seen in other conventional invertebrate lineages; the various classes of Cnidaria, the deuterostomes, the pseudocoelomate phyla, and the lophophorates may all

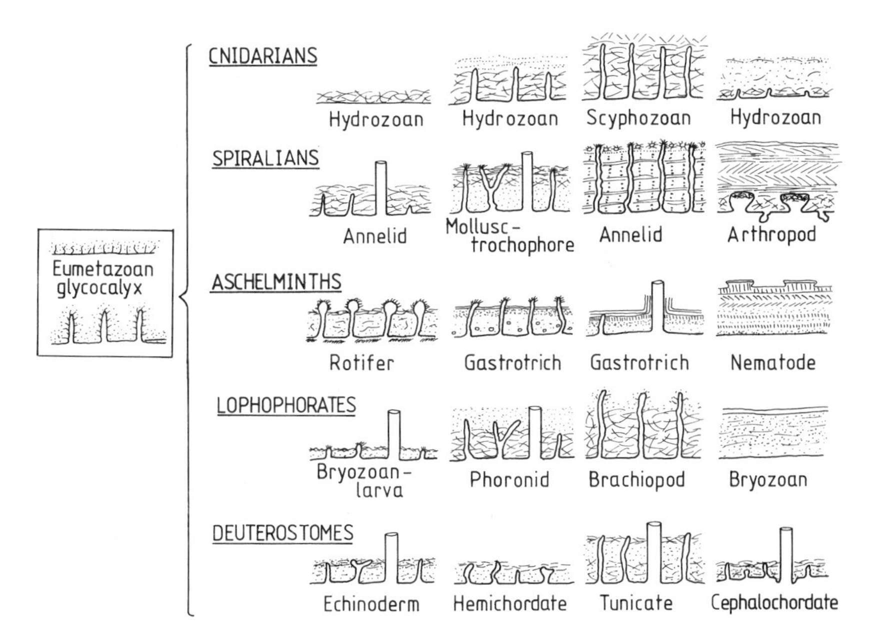

Fig 6.5. Cuticle structures in invertebrates, based on Rieger (1984) and data in Bereiter-Hahn *et al.* (1984). Most cuticles can be seen as modifications of the simple glycocalyx, with or without microvilli, found in lower invertebrate groups.

have proceeded along similar lines from a microvillar blueprint, though in the latter two groups the microvilli are suppressed more rigorously, and in lophophorates heavy calcification is particularly common. Rieger (1984) sees all this as evidence of multiple convergent lines in cuticle evolution. He suggests that primitive cuticles primarily functioned as molecular filters to aid the regulation of exchanges and the accumulations and uptake of dissolved nutrients, and that only later did supportive functions become important and result in the incorporation of collagens or minerals. Whatever the sequence of functional constraints on cuticles, their current structures often seem to do no more than indicate a general primitiveness of certain phyla, and give no real lead in understanding relationships between higher taxa.

There are a few special cases where cuticles have been used to establish a link between two phyla, however. The fundamental similarity of rotifers and acanthocephalans (see chapter 9) has been documented by Storch (1979); this seems particularly convincing given the very different habitats and problems of these animals. In this case though we are dealing with 'false' cuticles, as the stiffened laminas in both cases lie within the apex of an epidermal syncytium. More classically, the close similarity of ultrastructure in bryozoan and brachiopod cuticles has recently been reported (see Williams 1984) and cited as support for their close affinity in face of possible views to the contrary (chapter 13).

Finally we can only return to words of caution by mentioning the arthropod cuticle. This occurs in remarkably similar ultrastructural form in all the arthropod groups, yet as will be seen in chapter 11 there is a very strong school of thought that holds arthropods to be polyphyletic. Hence the affinity of any groups linked by their cuticle structures must remain open to doubt and to a convergent explanation.

Adhesive and defensive systems

Adhesive glands. Many lower invertebrates possess some form of adhesive system to give temporary attachment to surfaces, particularly where relatively tough cuticles are absent. These commonly consist of small organs with one or several cell types. For example in the turbellarian flatworms there are normally two kinds of gland cell, one producing the adhesive material and the other presumed to make a releasing factor, while a third cell type (a modified epidermal cell) provides support (Rieger & Tyler 1979). Rather similar structures occur in the gastrotrichs, and some archiannelids and nematodes also have a duo-gland system. By contrast a single gland type of adhesion is found in gnathostomulids, rotifers and nemerteans, and in a few turbellarians and nematodes, whilst in acoel flatworms modified secretory cilia provide adhesion. Homology seems likely within taxa, but the patterns of occurrence of these systems suggest convergent origins of the more complex duo-gland system; and once again the functional constraints are probably quite sufficient to account for this.

Cnidae. These are the everting organelles produced in the cnidoblasts of Cnidaria, and are perhaps the largest (up to 100 µm) and most complex intracellular secretions in metazoan epithelia. Mariscal (1974, 1984) identified three types, the nematocysts (fig 6.6*A*) being the best known and principally defensive in function. However some types of nematocyst, and all the spirocysts, are adhesive in nature. There are up to 30 kinds of nematocyst, generally simpler and less diverse in the Anthozoa than in the other two classes, but all having a capsule of collagen-like protein cross-linked with disulphide bonds. Cnidae appear to be unique to the cnidarians (despite many early claims to the contrary), occurring in other animals only where they have been 'stolen' intact from cnidarian prey. Their origins remain unknown, and they seem to provide no useful phylogenetic links, though a resemblance to extrusive organelles in protozoans has often been suggested (e.g. Hausmann 1978). But at least they give a clear coherence to the phylum Cnidaria, and may give some clues as to the relations between the classes of that group and their protozoan ancestors (chapter 7).

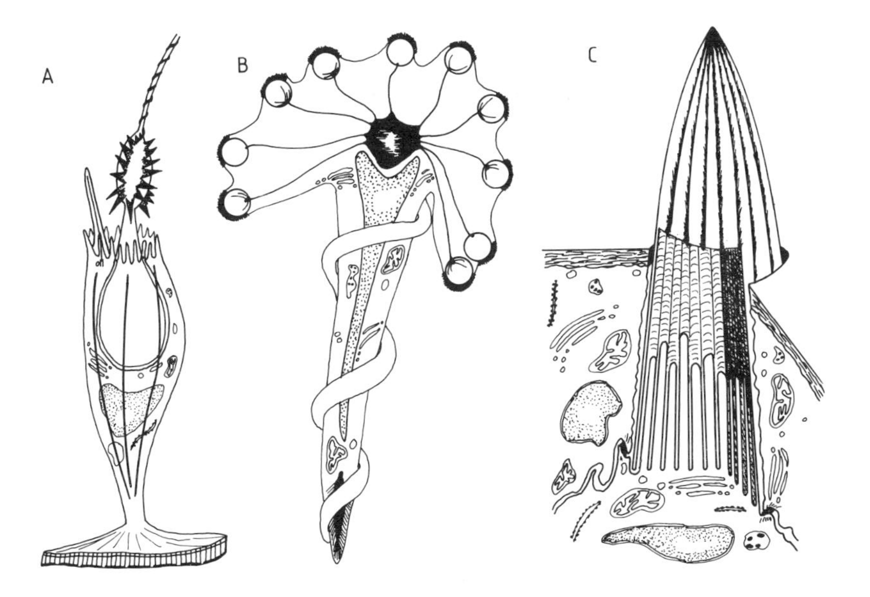

Fig 6.6. Secreted products of cells: *A*, the cnidarian nematocyst within the cnidoblast cell; *B*, the colloblast of ctenophores; *C*, the chitinous seta, formed around microvilli, found in annelids and other groups.

Colloblasts. Found only in the tentacles of Ctenophora, colloblasts seem to have no defensive function but are used solely to entangle prey adhesively (Hernandez-Nicaise 1984). Each is a complete cell (fig 6.6*B*), the upper spherical zone containing many dense granules that cause stickiness when released. The stalk of the cell links to synapses, and is anchored by a strong helical thread. Again this epithelial structure shows no links to any feature of other phyla, and can serve only as an intra-phyletic marker.

Rhabdites. Rhabdites are secretions found in the epidermis of turbellarians, but apparently originating from mesodermal cells. They produce a temporary gelatinous ooze when released, perhaps contributing to a protective coating and giving a noxious taste. They are unparalleled in all other phyla except nemerteans, though Graham (1977) points out a resemblance to glandular elements in the molluscan epidermis that are also produced by mesodermal cells. Smaller 'ultrarhabdites' also occur in turbellarians, within epidermal cells (Tyler 1984), and apparently these have similar functions.

Smith *et al*. (1982) have used these epidermal features as phylogenetic indicators within the free-living platyhelminth assemblage, but as they are absent from the other taxa of platyhelminths and from other phyla they cannot have any wider role as systematic characters. Early attempts to link them to cnidae (Reisinger & Kelbetz 1964) seem implausible in the light of more recent analyses.

Setae. One of the products of microvilli, occurring in many cells, is the seta (fig 6.6*C*). This stiff protrusion of chitin and tanned protein, sometimes mineralised, is secreted by one or more cells around a mould formed by a mass of microvilli protruding from its core, and is most familiar in soft-bodied worms such as annelids, where it was first described in detail by Bouligand (1967). Indeed it was initially regarded as diagnostic of this group, and its subsequent discovery in the pogonophorans was an important factor in a move to ally these two phyla (chapter 8) though the pogonophorans had formerly been ascribed to the deuterostomes. However the same type of seta is now known in echiurans, octopods, and brachiopods (references in Storch 1979). It may well be a convergent phenomenon, and cannot reasonably be used as evidence for close affinity of annelids and pogonophorans on its own, though it is taken by some as an essentially protostome character.

In general, defence and adhesion within relatively soft integuments is achieved by a variety of structures and glands, some of them unique to a phylum and others only of use in seeing very general trends, but occasionally suggestive of links between groups of phyla. Before leaving this topic one other mechanism should be mentioned, as the main means of acquiring both adhesion and a measure of impalatability in many invertebrates is secretion of mucus from specialised glandular cells. However comparisons on this front are particularly difficult, as mucus is chemically diverse and the terminology is still confused for invertebrates (see Richards 1984). Both cell type and mucus type are likely to be related primarily to function, however, as mucoid secretions are known to show extremely diverse properties and the glands producing them can be of many different forms even within one organism.

6.4 Specialised cell types

Spermatozoa

There are several good comparative reviews of sperm structure (Baccetti 1970; Baccetti & Afzelius 1976; Franzén 1977), and these cells have often been used to fuel phylogenetic debates, as besides the ready availability of data they offer the advantage of apparently clear homology in all groups of metazoans, and share an (almost consistent) common basic structure in the possession of a ciliary (flagellar) axoneme in the sperm 'tail'. Sometimes sperm cells have been of indisputable assistance in settling questions of affinity, yet since the work of Franzén it has been widely accepted that sperm type in part depends on the manner of fertilisation, and due caution must therefore once again be exercised (see Storch 1979).

The 'typical' tripartite spermatozoan is shown in fig 6.7*A*; it occurs with only minor modifications in a great many phyla where sperm are discharged freely into water. Most authors accept this as the primitive condition of the spermatozoan, from which all other designs are derived; though a useful discussion against this view can be found in Hanson (1977). Many groups that have divergent sperm types also have internal fertilisation (see fig 5.13), and the differences normally concern the shaft of the flagellum itself (see section 6.2). Many insect sperm, for example, have an extra set of outer microtubules, giving a 9+9+2 arrangement, and one or two large mitochondria that extend over much of the sperm length to give a thickened and flattened tail. In some orders of insect the sperm lack the central microtubules (9+9+0), and in others they have only one (9+9+1). Even more aberrant forms in insects are discussed by Phillips (1974) and reviewed by Baccetti (1979). Most of these patterns seem to modify the sperm's motility, and are presumably of some significance in permitting effective motion in the genital tract. Similar aberrations occur in other arthropod groups; for example spiders and uropygids have a 9+3 flagellum, but this is wound up inside the spermatid as it develops and may be non-functional in the mature sperm.

More profound deviations in sperm type, also shown in fig 6.7, occur in a few groups; either no flagellum is present (particularly in occasional arthropods, annelids and fish, and all the nematodes), or a biflagellate condition occurs (again conspicuous in some vertebrates and some aschelminths). The nematodes are especially noteworthy in having an amoeboid sperm that can squeeze in to the female genital aperture despite the high internal pressures characteristic of the phylum, thus underlining the functional aspect of sperm modification. A general phylogenetic review of sperm types (fig 6.8) certainly does not suggest that they give much indication of relationships at the phyletic level, though flagellar aberrations are reasonably consistent at the class or order levels as we have seen. Some interesting features are revealed by this phylogeny though. For example the primitive sperm type is absent in the platyhelminths (except in one primitive genus - see chapter 8), though this phylum is often regarded as close to the stock from which metazoans diversified. And the apparently more primitive status of the brachiopods amongst the lophophorate assemblage is an interesting indication.

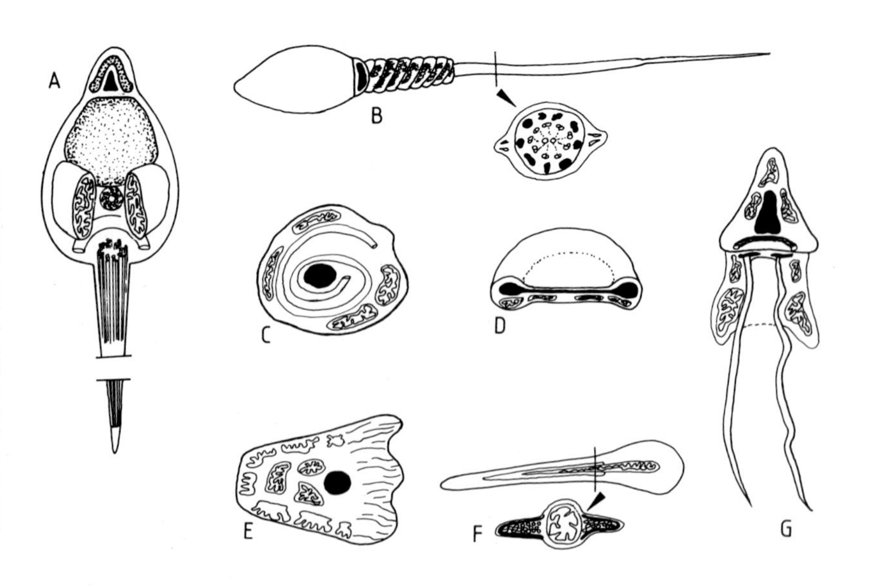

Fig 6.7. Spermatozoan diversity in animals: *A*, the classic sperm structure of most marine invertebrates; *B*, the modified tripartite sperm of some insects, with mitochondria extending down part of the shaft; *C*, arachnid sperm, with internal coiled axoneme; *D*, biconcave sperm from a proturan insect; *E*, amoeboid sperm from a nematode; *F*, non-flagellate insect sperm; *G*, biflagellate insect sperm; (mostly redrawn from Baccetti 1979).

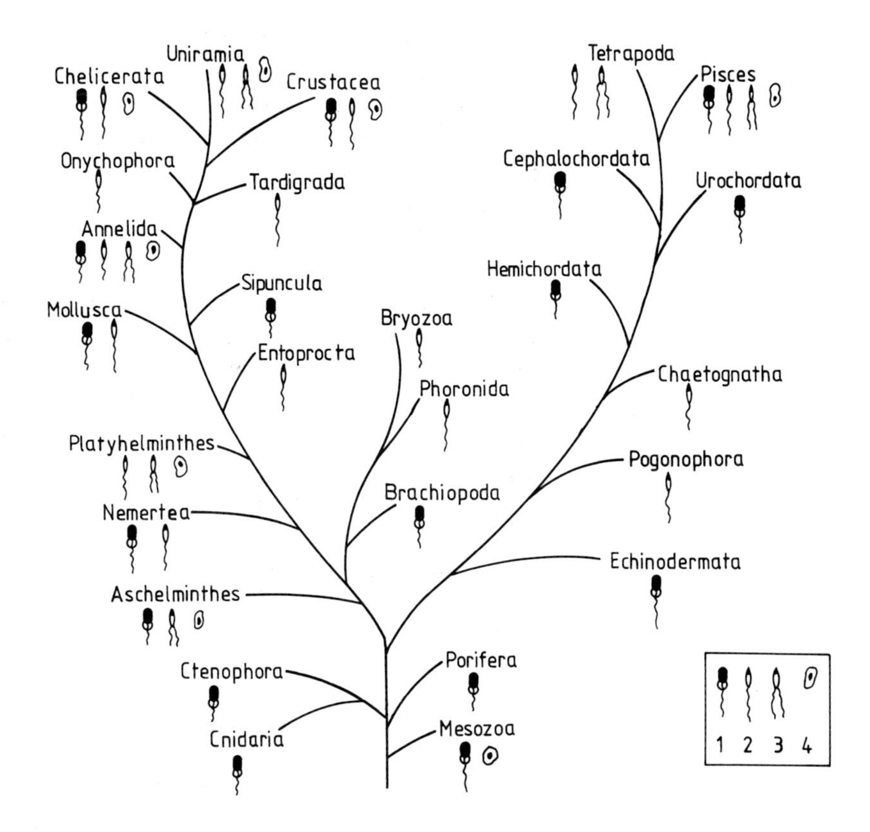

Fig 6.8. Distribution of spermatozoan types in invertebrates, based on Baccetti (1979): 1, classic marine sperm; 2, modified monoflagellate sperm; 3, biflagellate sperm; 4, non-flagellate sperm.

Nerve and muscle systems

The basic processes of nervous conduction in animals appear to have been established very early in their evolutionary history and to have undergone no major changes since their inception (see Shelton 1982). Indeed they share many features with the excitable properties of protozoan membranes. Thus it is unhelpful to look for phylogenetic indications in the mechanisms of invertebrate nervous systems, and their ultrastructure has proved similarly conservative. The basic neuron occurs essentially unchanged in form in all phyla except placozoans and sponges, which have at best a 'neuroid' tissue (Shelton 1982). The only differences to be found are in their disposition and concentration, the number of dendrites and axons, the complexity of synaptic patterns, and the nature of ensheathing cells. Cnidarians, ctenophores, hemichordates and echinoderms certainly have rather simpler arrangements of nerves than most other phyla, and retain nerve nets to a greater extent than any of the obviously bilaterally symmetrical groups, but this may just be a correlate of their effective radiality. The nerve nets in these phyla are in any case somewhat dissimilar, and echinoderm nerves seem particularly odd (for example in their apparently unique failure to stain with conventional silver treatments). Many 'simple' nervous systems are as yet too little studied to yield useful comparisons though.

In addition to true nerves, certain phyla have a system of epithelial electrical conduction; this is known in Cnidaria, Ctenophora, and (perhaps rather differently) Porifera. There cannot be much doubt that this is an essentially primitive feature, and it does not occur in any obviously 'triploblastic' phyla. However it serves as a link between the nervous and contractile systems of animals, as the latter is also firmly associated with epithelial properties in some invertebrates, and both nerve and muscle share certain properties in relation to excitability.

The myoepithelial cell is best known from the cnidarians, where the basal processes from both ectodermal and endodermal cells are aligned to make efficient contractile systems, and these cells were for a long time thought to be almost diagnostic of 'lower' metazoans. But it is now well established that similar myoepithelial cells are widely distributed in invertebrates (see Storch 1979; Hoyle 1983) and even in the chordates; and they may arise from any of the three germ layers. Their contractile mechanism is probably identical with that in true muscle, which is also identical with that in sponge 'porocytes' (independent contractile cells around the openings of poriferans), and in the 'fibre cells' of placozoans. Many authors have speculated that myoepithelial cells are the evolutionary precursors of more familiar muscle tissues, though Hoyle has pointed out that the deficiencies in our knowledge of comparative muscle function make any such assumption suspect. In any case true muscle cells, with no dual function, are also to be found in the cnidarians, and myoepithelial cells are unknown in ctenophores (which have true muscle); so there is little scope for imagining evolutionary sequences here.

Classifying muscle cells is difficult (fig 6.9), as traditional dichotomies derive purely from vertebrates and are not very helpful. Vertebrates have 'true' smooth muscles with no striation and no visible thick (myosin) filaments, anastomosing striated cardiac muscle, and classic skeletal cross-striated muscle. Invertebrates have a different type of 'smooth' muscle in which myosin filaments usually are visible and are organised into spirally displaced sarcomeres. Striations can be found in muscles in virtually all phyla, even in certain jellyfish. In many annelids and nematodes the oblique striations are particularly orderly, whilst in other groups they are disorganised and the muscle has often been confusingly termed smooth. Obliquely striated muscle may be advantageous in that it can shorten more drastically by actually shearing the heavy myosin (or sometimes paramyosin) filaments past each other as well as having the classic relative sliding of actin and myosin (Mill 1982). Distinct cardiac muscle also occurs in invertebrates, in almost all actively pumping hearts, and has anastomosing cells and properties convergent with the heart muscle of vertebrates. Classic cross-striated muscle is only found in the arthropod groups amongst the invertebrates.

In many cases it is possible on muscle morphology alone to assign an animal to its correct phylum (Hoyle 1983), but it is nevertheless very difficult to identify trends or possible evolutionary sequences in the design of musculature. The traditional view that myoepithelial cells gave rise to smooth muscle cells that in turn gave rise to increasingly highly organised degrees of striation is not really upheld by the distribution of these types, since cnidarians, flatworms and chordates all show both extremes. The idea that the fast-twitching, cross-striated vertebrate skeletal muscle is the peak of an evolutionary trend is deep-seated, but Hoyle points out that the type of muscle present is much better correlated with functional needs than with phylogeny, yet again. A typical invertebrate muscle has many advantages of its own, being much more versatile in resting length, contractility, and speed, and less readily fatigued. Muscle cell structure is in any case variable not only between phyla but between species, individuals, separate muscles, and even parts of the same muscle, and can vary within an animal's lifetime.

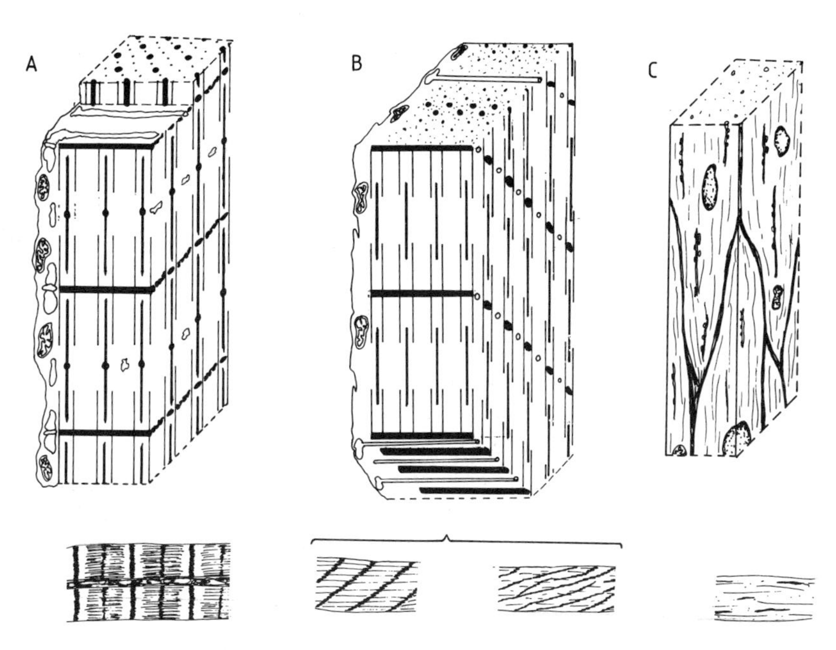

Fig 6.9. Muscle types in invertebrates, showing fibrillar patterns and overall appearance: *A*, cross-striated muscle of arthropods and vertebrates; *B*, obliquely-striated muscle common in many invertebrates; *C*, smooth muscle of vertebrates.

In fact the clearest correlations of muscle type are with locomotory and skeletal systems. Where there is a stiff skeleton, and hence a fixed contraction length, striation is most pronounced, with true cross-striation in vertebrates and arthropods and regular oblique patterns in nematodes and annelids. Where muscles contract in sheets of highly variable length and shape, less organised ('smooth') patterns occur. Within these generalisations, the most obvious feature of muscle design again seems to be that of convergence: the repeated occurrence of parallel, fusiform and pinnate arrangements of fibres, and the ubiquitous appearance of cardiac muscle types, are two clear examples of this.

Finally, having dealt with nerve and muscle separately their relationship to each other should be considered. Innervation patterns do vary considerably (fig 6.10) and there may be a genuine sequence here. In most relatively simple 'triploblastic' examples the muscle sends processes to the nerve (fig 6.10*A*), a pattern once thought to be rare and especially associated with nematodes, but now known in platyhelminths, nemerteans, gastrotrichs, onychophorans,

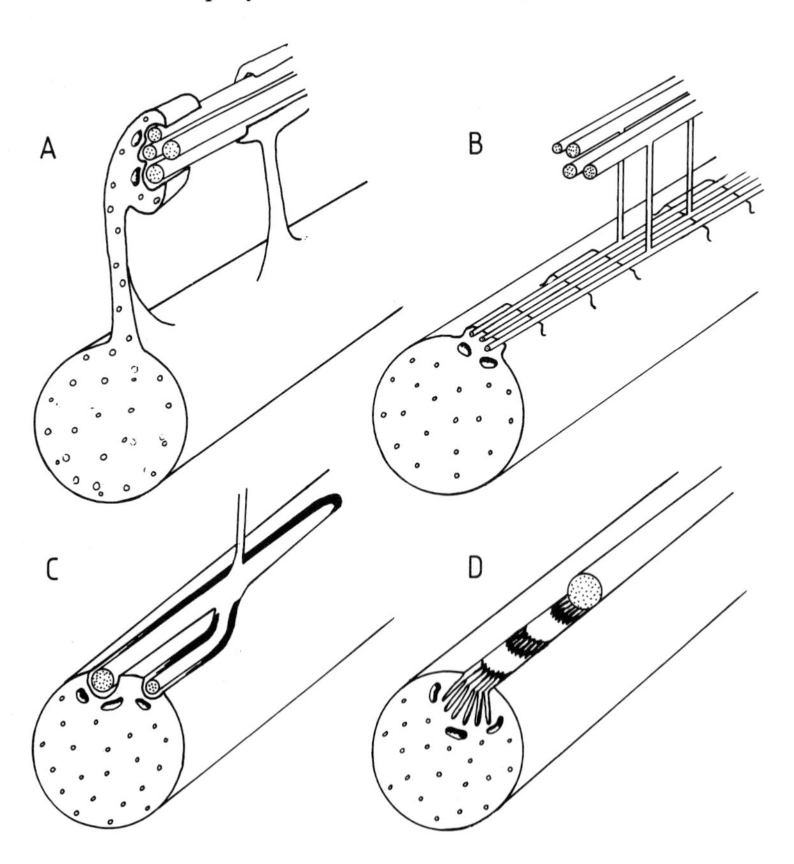

Fig 6.10. Patterns of neuromuscular innervation. In many invertebrates the muscle sends processes to the nerve *(A)*; in some annelids, molluscs and insects the nerves lie on protrusions from the muscle surface *(B)* or are sunk into shallow grooves *(C)*; only in vertebrates do complex invaginated grooves accommodate the nerve *(D)*. Based on Hoyle (1983).

echinoderms and other invertebrate deuterostomes as well. Some annelids, molluscs and insects have a system where short pillars protrude from the muscle, on to which nerves insert (fig 6.10*B*); but most members of these phyla have a further modification where the nerve lies in a shallow groove on the muscle surface (fig 6.10*C*), a situation found with increasing complexity in the vertebrates. Invertebrates also differ in the pattern of nerve-muscle relationships, usually having polyneuronal and multiterminal innervation, including peripheral inhibitory neurons, which gives much more scope for graded response; but there seems to be no obvious pattern of differences within the invertebrates themselves.

Receptors

The ultrastructure of invertebrate receptors has been established for a relatively long time as a source of phylogenetic conjectures, particularly where photoreceptors are concerned. Unfortunately other forms of receptor, though structurally well-described, are poorly understood in terms of physiology and function, and at best tend to be interpreted merely in terms of analogy with the only two groups where such matters have been properly studied, the vertebrates and arthropods. A summary of available knowldege on chemo-, thermo- and hygroreceptors is given by Altner & Prillinger (1980). Most of these, and also mechanoreceptors and special systems concerned with balance (statocysts), are simple epithelial bipolar cells serving as primary receptors, in some cases (cnidarians, gastropods, nematodes) linked by synapses to effector systems as well. The input region of the primary receptor always bears one or several cilia and microvilli, and the cilia are frequently unorthodox in their microtubule complement (having one or no central tubules, or unconventional bridges between the peripheral doublets) or in their insertion to the cell (see Barber 1974). Anomalies appear to crop up fairly randomly, and as yet there is no pattern to be observed that could help in interpreting phyletic relationships.

With photoreceptors the situation is different, as Eakin (1963, 1979) proposed two quite distinct evolutionary lines of ciliary and rhabdomeric (microvillar) types (fig 6.11), based on a survey of a good range of invertebrates. He effectively separated the 'classic' branches of the invertebrates in this manner, the protostomes and pseudocoelomate aschelminths having rhabdomeric photoreceptors and the deuterostomes plus lophophorates and diploblastic groups having light-sensitive cells based on a ciliary structure. These lines also seem physiologically distinct; most ciliary photoreceptors hyperpolarise in light, whereas most rhabdomeric types show a light-induced depolarisation.

Eakin's views have been attacked by Vanfleteren & Coomans (1976), who suggested that all photoreceptors were of monophyletic ciliary origin, by Clément (1985) who described a polyphyletic scheme with at least three different types interacting in their evolutionary histories, and by Salvini-Plawen & Mayr (1977) (see also Salvini-Plawen 1982*b*) who proposed at least 40 quite separate origins for photoreceptors, these being of at least four distinct basic types and giving a scheme best termed 'aphyletic' (Clément 1985). It is certainly true that there are a great many variations upon a theme in the realm of photoreceptors, as the figures given by Salvini-Plawen & Mayr show, and as is also clear from the Eakin classification shown in fig 6.11.

This variation would matter less if one could have confidence in underlying dichotomies. But it is now known that some 'rhabdomeric' polychaete and mollusc photoreceptors bear cilia, as do rare examples in flatworms, gastrotrichs and nematodes. Some echinoderm receptors and a few tunicate receptors, all classically part of the 'ciliary' lines, are clearly non-ciliary after all. The scallop *Pecten* has two kinds of eye, one of each type, while chaetognaths have recently been described as having lamellate eyes, forming a quite distinct third morphological type (Goto *et al.* 1984). Rotifers have at least three distinct kinds of eye (Clément 1985), one type only otherwise

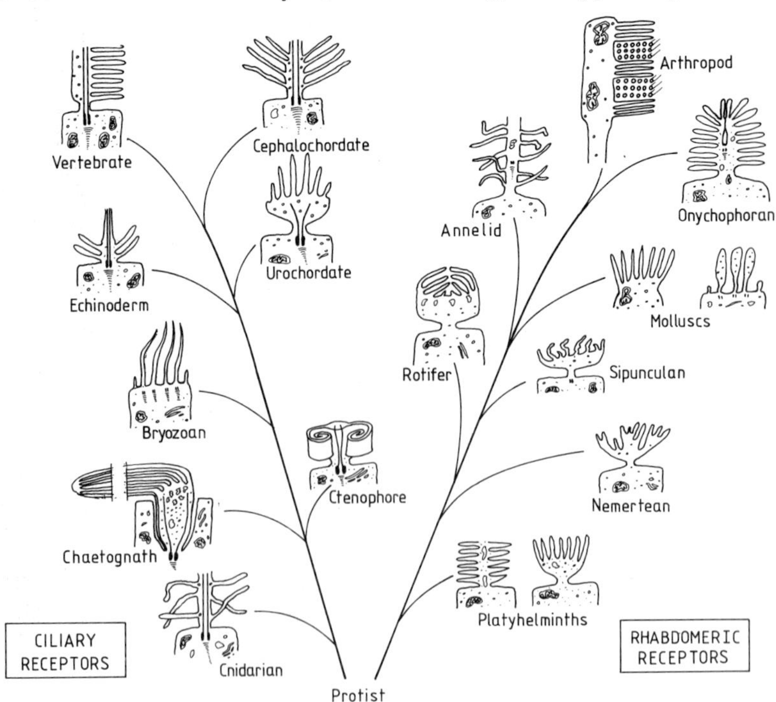

Fig 6.11. Photoreceptor structures in invertebrates, interpreted by Eakin (1963, 1979) as two major lines, in accordance with a dichotomous tree (but much criticised, see text).

found in phytoflagellates, and another rather similar to the eyes of some clitellate annelids, pogonophorans, and flatworms. Perhaps most damningly of all, the 'classic' rhabdomeric eyes of cephalopods and of arthropods have both recently been shown to have cilia present in their early development. A modern summary of photoreceptor occurrence is given in table 6.2.

Eakin defends his dichotomy despite recognition of these exceptions, asserting that while both cilia and microvilli may co-occur in receptors, and some switching of function may occur between them, the essential and primitive receptor (in which photosensitive pigment occurred in the ciliary membrane itself) gave rise very early on to two distinct and divergent lines. In the protostome line the rhabdomeric eye was thus a secondary development, but was perpetuated with just a few reversions; and at some point a few deuterostomes inconveniently invented rhabdomeres as well.

Table 6.2 *Photoreceptor types in invertebrates*

Ciliary receptors	Rhabdomeric receptors
Vertebrates Entoprocts Bryozoans Ctenophores Cnidaria Gastrotrichs	Echinoderms Pogonophorans All arthropod groups Urochordates Sipunculans
	Modified rhabdomeric
Embryos of arthropods and cephalopods Larvae of urochordates	Chaetognaths
Both types or mixed types	**Unknown receptor morphology**
Cephalochordates Hemichordates Molluscs Annelids Nemerteans Platyhelminths Nematodes Rotifers	Poriferans Brachiopods Phoronids Echiurans Priapulids Mesozoans Gnathostomulids Acanthocephalans Nematomorphs Kinorhynchs

Particularly useful summaries of the issues in photoreceptor evolution, by all the major contestants, are given in Westfall (1982); and possible evolutionary trends in photoreceptor evolution are discussed by Burr (1984), who considers functional reasons for their ultrastructure. While there may indeed be useful phylogenetic patterns to be found, it is clear that many authors now incline towards a multiple convergent origin of varying kinds of photoreceptors, usually from ciliary precursors. Once again, given that the possibly polyphyletic arthropods may have achieved all the components of their compound eyes independently (chapter 10), this degree of convergence cannot be dismissed as an unlikely hypothesis.

Excretory/osmoregulatory cells

Protonephridia, consisting of a terminal flame cell or 'cyrtocyte' and a proximal tubule, occur in most of the lower Metazoa, and have long been taken as the primitive 'excretory', or more correctly osmoregulatory, apparatus in animals. Goodrich (1945) developed a scheme that has held sway for several decades in which 'protonephridia' gradually became more sophisticated as true 'metanephridia', merging in the higher annelids with the primitive gonoduct to form a 'coelomoduct' osmoregulatory organ, and being lost altogether in many groups (notably molluscs and crustaceans); though many of the planktonic larvae retained protonephridial flame bulbs. Recently, attention has been largely focussed on the detailed structure of the protonephridium and the light this can shed on relations amongst lower groups; a review is given by Wilson & Webster (1974).

The protonephridia of gnathostomulids (close relatives of flatworms, and perhaps more primitive - see chapter 8) are amongst the simplest of all (Ehlers 1985*a*; Lammert 1985) and provide a useful model (fig 6.12*A*). There is a single and classic 'flame cell', with an apical nucleus, and simple perforations in the wall occur (where this terminal cell meets the tubule cell) through which fluid is forced by the beating of the central ciliary 'flame'. In this group there is only one cilium within the flame cell (Ax 1985; Lammert 1985), surrounded by eight microvilli; the resemblance to a collar cell as described earlier in this chapter is unmistakeable. Flame cells could therefore be seen as modified ectodermal collar cells. Beyond the collar cell/cyrtocyte only one tubule cell occurs, rolled up on itself and sealed by a desmosome to form the tube down which the filtered fluid passes to the exterior (Wilson & Webster 1974).

The evolution of this excretory system from an ectodermal cell is clarified by the present condition of platyhelminths (Ehlers 1985*a*). The two groups which may be most primitive (Acoela and Nemertodermatida - see chapter 8) do not possess flame cells, which may be a secondary loss. But an intermediate condition is found in another primitive family, the catenulid flatworms, where there are just two cilia and no microvilli (fig 6.12*B*); whilst 'higher' flatworms have multiple cilia, and acquire more complex filters as the terminal cyrtocyte and the tubule cell interdigitate and develop cytoplasmic projections to produce a 'weir'. Goodrich (1945) actually considered the multiciliate flame cells of certain flatworms to be the primitive form, and suggested that reduction to one cilium was a specialised condition; but ultrastructural analysis of the filtering apparatus seems to refute this idea. Indeed, Bartolomaeus (1985) has followed the development of flame cells in nemerteans and observed the derivation of a multiciliated flame cell from an initial monociliate condition; he therefore believes the latter to be ancestral for all the 'Bilateria'. In this context it may be of interest that gastrotrichs and priapulids (chapter 8) also have monociliate terminal cells in their osmoregulatory organelles.

Nemerteans are also set somewhat apart from flatworms by their not only multiciliate but also multicellular flame bulbs, and rather similar structures occur in entoprocts and in many archiannelids; Wilson & Webster suggest that though multicellular all these systems are structurally very like those of the flatworms. Rotifers have a superficially similar apparatus to that of the higher flatworms (Clément 1985), but the nucleus is no longer apical and the filtering weir is rather more elaborate (fig 6.12*C*); and something comparable may occur in acanthocephalans. According to Wilson & Webster's review this

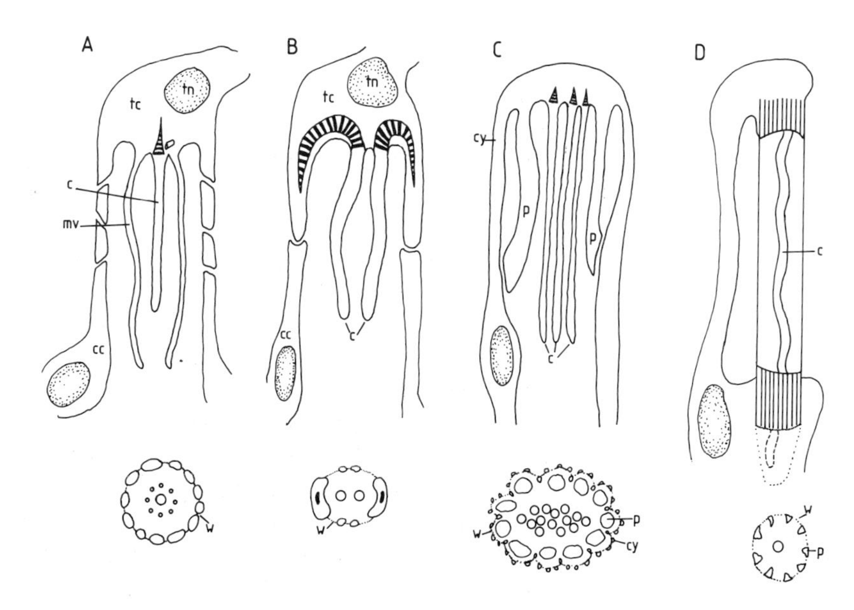

Fig 6.12. Flame cell structures. The monociliate condition of gnathostomulids *(A)* may be primitive; a 2-cilia cell occurs in catenulid flatworms *(B)*, and multiciliate cells ('flame bulbs') occur in other flatworms and in rotifers *(C)*, with more complex filters. In annelids and other groups a solenocyte may occur *(D)*, where numerous monociliate filtering tubules arise from each nephridium. (tc-terminal cell; tn-terminal cell nucleus; c-cilium; cc-canal cell; cy-cytoplasmic column; mv-microvilli; p-pillar; w-weir)

arrangement is distinctive and should be termed a 'flame bulb'. Other pseudocoelomate groups have neither flame cells nor bulbs (nematodes, for example), or have more typically flatworm-like multicellular organs (kinorhynchs, priapulids, and gastrotrichs).

Some annelids and related groups have a more specialised form of flame cell called a solenocyte (fig 6.12*D*), where bunches of fine tubules each containing a single cilium arise from the nephridial wall; this term could also be applied to the priapulids and gastrotrichs. In other adult forms of macro-annelids and most other 'higher' groups the nephridium opens as a funnel rather than having a closed flame bulb, and the term 'metanephridium' is usually preferred. The only major exception to this trend seems to come with the cephalochordate *Branchiostoma,* which is generally considered as a 'higher' animal close to the vertebrates (chapter 12), but which possesses solenocytes having closed ends unassociated with the coelom.

In general, then, the fine structure of these osmoregulatory organs gives substantial support for traditional views of the status of many of the bilateral worms. Inevitably the numbers and distribution of the organs vary with habitat and functional needs, and this alone would be unhelpful; but if examined on a fine enough scale they give clear indications of increasing structural complexity related primarily to ancestry, being simple in acoelomate worms, but increasingly elaborate in various coelomate and pseudocoelomate animals.

6.5 Conclusions

Repeatedly in the course of this chapter the problems of functional constraints on structure, and of possible convergences, have been met: these phenomena are no less applicable to ultrastructure than to gross morphology, and perhaps on the evidence of these surveys may be more obstructive for these smaller features, as some earlier authors suggested. Convergent origins have been suggested for collar cells, multiciliate cells, adhesive glands, setae, cardiac and striated muscles, eyes, and even some kinds of cuticle. In general, organelles and cells of particularly distinct type have been dealt with - cilia and cell junctions, sperm and muscle cells - but even there clear trends are hard to see. If other examples had been used, matters would often have been even worse. Intracellular organelles such as Golgi apparatus or mitochondria show either differences that are entirely associated with the function of their cell, or indeed almost complete constancy of ultrastructure; while cell types such as those specialised for active uptake and absorption, or for respiratory exchange, show such extreme convergence between tissues and phyla that they might lead to a rejection of ultrastructural criteria altogether.

But, on balance, while ultrastructure has not perhaps proved as decisive a tool as one might wish, it is only fair to recall its successes and some useful pointers raised here. Several ultrastructural features have turned out to be unique to one phylum; and for identifying relations within phyla many issues have been greatly aided by analysis of cells and their products. There is little doubt that at these levels, and especially for simpler metazoans, the electron microscope is a vital tool. For deciphering relationships between the invertebrate phyla the trends are often much less clear, but the evidence given by ciliation patterns and cuticular structures in particular can no longer be avoided, and other cell structures may also have value in showing general hierarchies of complexity. While retaining a healthy appreciation of the functional aspects of cytology, these features must be given due weight in establishing an overview of phylogeny.

PART III

Phylogeny of major groups

Based on the evidence and characters discussed in Parts 1 and 2 of this book, the chapters in Part 3 look at each of the traditional sections of the animal kingdom in turn. The sequence followed is a fairly classical one, so that particular groups can be found in the 'expected' places. The origins of the Metazoa, and relationships between the lower groups, are dealt with first (chapter 7); succeeding chapters are then devoted to the acoelomate and spiralian coelomate worms, and to the pseudocoelomates (chapters 8 and 9). The molluscs merit a chapter on their own (10); and then the arthropods are dealt with together (11), though they may not in fact be closely related to each other. Deuterostomes are dealt with collectively, with chordates included briefly (chapter 12). Finally the lophophorates are used in chapter 13 to highlight some problems with the underlying assumptions on which such a sequence of chapters is based.

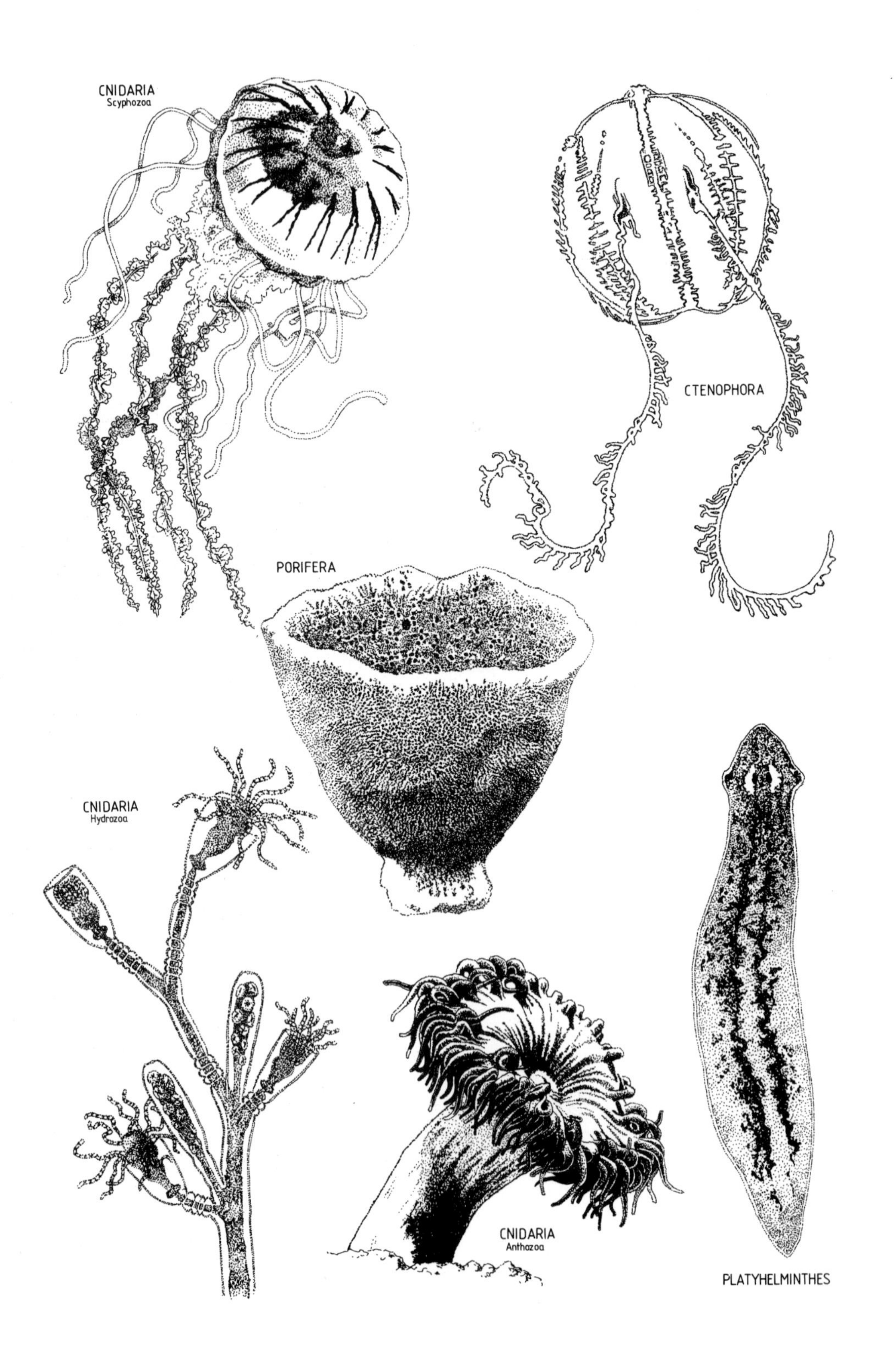
CNIDARIA
Scyphozoa
CTENOPHORA
PORIFERA
CNIDARIA
Hydrozoa
CNIDARIA
Anthozoa
PLATYHELMINTHES

7 The origin of the Metazoa

7.1 Introduction

The origin of multicellular animal life from single-celled ancestors is perhaps the most enigmatic of all phylogenetic problems, and the least likely to be 'solved' by additional evidence from any of the sources already discussed. Most obviously, it remains completely obscure in terms of the final arbiter of a fossil record. The issues with which this chapter is concerned are therefore inevitably highly speculative and any treatment of them is bound to be unsatisfactory. Nevertheless this area is gradually becoming amenable to a more synthetic approach, and it is now possible to bring together some of the problems already raised in part 2, since increasing knowledge of cell biology and biochemistry is beginning to give a fresh perspective to these murky areas of zoology. This is also an area where an increasing stress is being laid on the necessity of proposed ancestors being functionally plausible, and of stages in their evolution having obvious selective advantage, so that some early theories are gradually losing support.

The problem of metazoan origins has naturally exercised the imaginations of invertebrate biologists, and many theories have been advanced in the last 150 years. Most of these have been compounded with views on subsequent evolution from the earliest metazoan to the groups of existing lower Metazoa, and on relationships between these various phyla (particularly sponges, cnidarians and platyhelminths). In fact it is difficult to separate these issues, as any theory of the nature of the first multicellular animals inevitably has consequences for conceptions of 'what happens next'. Hence the origins and relationships of lower animals are dealt with together here, following the historical precedent; though making some attempt to deal with the issues sequentially rather than as a confusing whole.

Although there is a multiplicity of theories to be considered, most lucidly reviewed by Salvini-Plawen (1978), it is fortunate that there are in fact only a very few ways in which the initial multicellularity could conceivably have been achieved. Reduced to their very simplest, all the theories are of only two kinds, and *could* only be of two kinds (fig 7.1). The first possibility is that metazoan status was achieved by the coming together of two or more cells, forming a colony, whether by incomplete separation of daughter cells after mitosis (fig 7.1*A* i), or by independent aggregation of formerly separate cells in a symbiotic fashion (fig 7.1*A* ii), (which might be genetically rather difficult). The second major possibility is that a single cell became multinucleate without divisions of the cytoplasm, forming a 'plasmodial' organism (fig 7.1*B*), and that cell boundaries were established later to give a multicellular animal with many primitively syncytial tissues. (This should strictly be temed a 'coenocytic' condition, as syncytial implies a secondary loss of cell boundaries, but the use of the term syncytial in this context is already too well established to change.) All the proposals discussed here are variants on one (or both) of these themes.

The first true eucaryote Protista may have arisen as early as 1400 MYA (Schopf *et al.* 1973; Schopf 1975; Cloud 1976 *a,b*; and see chapter 3), and it is widely assumed that they represent a monophyletic entity given the uniformity of intracellular structures (see chapter 6), notably mitochondria and microtubular complexes. Most authors have concluded that autotrophic protists with aerobic and photosynthetic metabolism were the ancestral type, probably of an amoeboid form with a flagellar apparatus. Loss of the plastids would give primary heterotrophic protozoans (probably polyphyletically), having a naked cell membrane and a rather more complex flagellar/centriolar system. From these early single-celled 'animals' many further lines of heterotrophic protozoans would probably evolve before the first metazoans were to emerge. Which of these lines gave rise to those archemetazoans, perhaps 700-1000 MYA, is therefore inevitably a matter of pure speculation.

To qualify as a true metazoan, an organism must be multicellular, heterotrophic and potentially motile, with specialised gametic cells; some

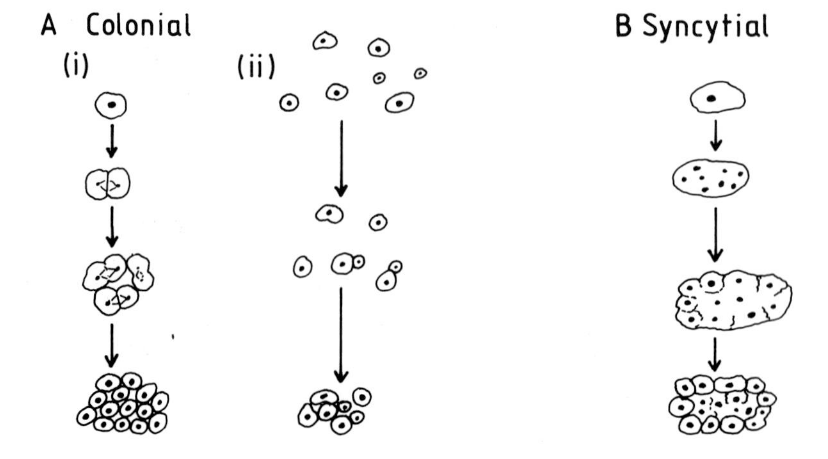

Fig 7.1. Possible mechanisms for achieving multicellularity, by aggregation of mitotically-related *(Ai)* or possibly unrelated *(Aii)* cells, or by incomplete subdivision of a single large cell *(B)*.

authors also require that it possess distinct tissue types other than the gametic tissue. An important stage in the development of metazoans from eucaryote protists must therefore have been the acquisition of meiosis and hence sexual reproduction; but the sporadic occurrence and degree of variation of this process in protists suggests that it may have had multiple convergent origins, as discussed by Margulis (1971, 1974) and Salvini-Plawen (1978). Hence the particular sexual pattern of metazoan cells is not necessarily inherited from any one group of protists (a view which has in the past contributed to numerous conflicts between possible theories) but may have evolved separately within the forerunners of the multicellular forms themselves. Following on from this argument, Salvini-Plawen suggests that the identical meiotic patterns, and diploidy, of all extant metazoans are convincing evidence for their phyletic unity - in other words, sex itself is polyphyletic, but the metazoans are not. To start with then, we should examine existing alternative theories of how this supposedly unitary evolutionary step could have occurred, and what the consequences are of each theory. Then it may be possible to evaluate the evidence for the nature of the first metazoans and the early stages of their evolution, and consider possible homologies within all extant metazoans that may unite them as a single phyletic entity, in the hope of being able to decide between these theories.

7.2 A survey of possible theories

Colonial 'blastaea/gastraea' theories

Commencing with the classic works of Haeckel (1874, 1875), a view of metazoan origins involving blastula-like and gastrula-like stages succeeding a phase of aggregation of protistan flagellates has long been influential. The version of this story proposed in the nineteenth century by Haeckel himself, summarised in fig 7.2, was heavily dependent on recapitulation theories, suggesting that the blastula of modern embryos recapitulates their ancestry as a

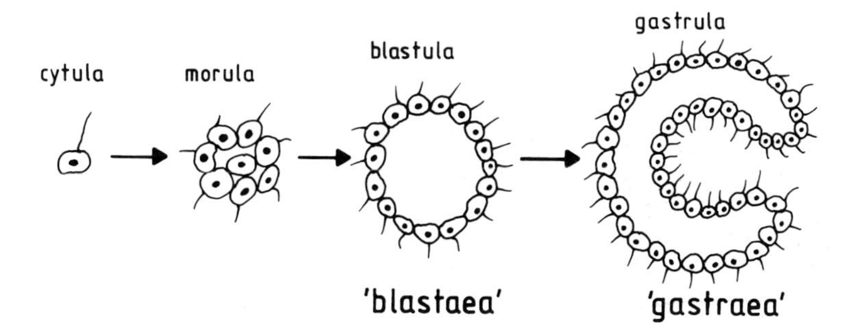

Fig 7.2. The idealised early stages of embryology, with the Haeckelian analogies of ancestral forms as 'blastaea' and 'gastraea'.

'blastaea' organism. The basic metazoan was thus a pelagic, radially symmetrical aggregation of flagellated cells, as currently represented by forms like *Volvox,* which have incipient division of labour between cells (even with a separation of somatic and gametic cells) and some coordination of flagellar activities. A 'gastraea' stage, which the gastrula of modern embryos recapitulates, would arise from the blastaea by a separation of locomotory and digestive regions, the latter being posterior and gradually tucking in to form an interior endoderm. This gastraea form, a simple two-layered, sac-like organism, readily gave rise to the similarly constructed aseptate cnidarians, from which other metazoans subsequently evolved.

Haeckel's theory has the great merits of simplicity and elegance, and presents a strong appeal in the orderliness of its explanations. But it is in some ways almost 'too good to be true', and in fact it is almost certainly invalid, being largely discounted now in the form in which he presented it. There are many objections, quite apart from the long-standing disenchantment with recapitulationist laws and theories (chapter 5). Most obviously, though there are many colonial flagellate protists (see Kerkut 1960), all the existing blastula-like types are unequivocably plants, and all have incomplete separation of constituent cells, with fine cytoplasmic bridges connecting the colony. The phytophagous volvocid organisms (possessing a curious mixture of colonial and syncytial features) are also primarily haploid. Clearly a zooflagellate colony with prolonged diploidy is needed as the root for this theory, yet the zooflagellates do not commonly show such a tendency to form colonies of the right blastuloid type. The nearest equivalents are the colonial choanoflagellates, some of which (e.g. *Sphaeroeca*) are pelagic and could provide suitable models for an archemetazoan; though the colonies in question are all fresh-water forms, and do not appear to show any division of labour as seen in some volvocids. Nevertheless, by invoking these as models of the ancestral forms, a version of Haeckel's theory can retain plausibility, and has been supported by authors such as Remane (1963*a,b*), Jägersten (1955, 1959) and Siewing (1969, 1976, 1980).

The development of this kind of theory has largely centred on the stages following the blastaea, and a variety of views has been spawned, in particular by German authors. The simple blastula form must invaginate to form a diploblastic gastraea organism, and in all phyla except sponges there must be a degree of internal elaboration of gastric surfaces. Remane took the view that this would occur in a pelagic animal, and proposed four original gastric pouches in his 'cyclomerism theory' (fig 7.3*A*) to account for archemetazoan origins. Jägersten, and subsequently Siewing, suggested that the ancestral form had become benthic before gastrulation occurred, giving a 'bilaterogastraea' (fig 7.3*B*) or a rather similar 'benthogastraea' respectively, each having three pairs of gastric pouches.

Even with these modified versions of gastraea theories significant controversies are raised. While some lower invertebrates do gastrulate by invagination (chapter 5), poriferans and cnidarians do not. Much more seriously, nearly all modern gastraea theories are inextricably linked with enterocoel theories of early coelom acquisition, for in the archemetazoans the fully formed gastric pouches of the cnidarian-like ancestor are supposed to transform rapidly into enclosed coelomic cavities (fig 7.3). All current acoelomates must therefore have regressed from a coelomate state; and the archemetazoan becomes synonymous with the archecoelomate. Such views have been persistently popular in the European literature, as discussed in chapters 1 and 2. But the various gastraea-type phases that these authors support, being so closely tied to enterocoelic theories, have difficulty in accounting for the diversity of modern coelomate embryology (chapter 5) and for the very existence of many acoelomate forms; all the arguments advanced in previous chapters on these points could be reiterated here. Once divorced from the enterocoelic overtones, the gastraea is a purely recapitulationist construct, and has little plausibility as a metazoan ancestor. Many authors have therefore retained the idea of a simple choanoflagellate/blastula colony as a plausible early metazoan, but looked elsewhere for the development of the next 'grade' of multicellular forms.

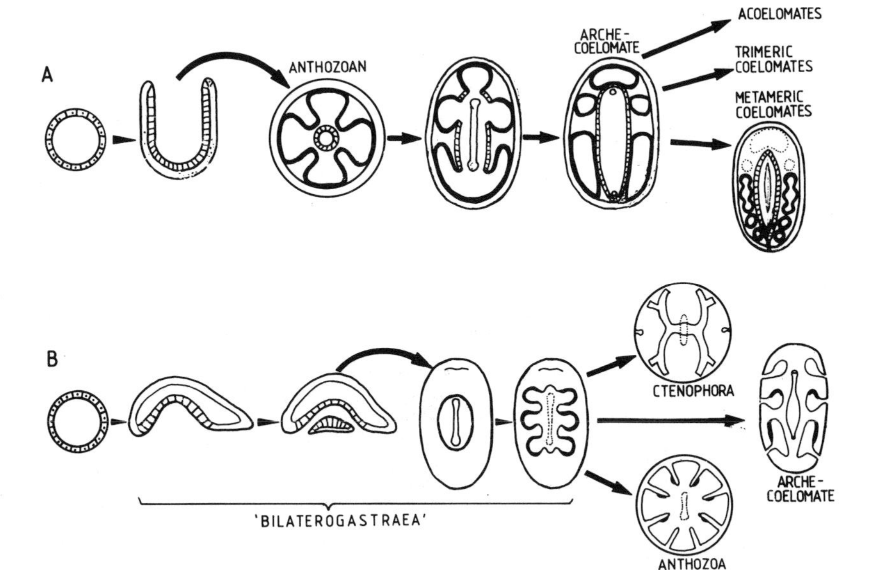

Fig 7.3. Archecoelomate theories of metazoan origins: *A*, Remane's cyclomerism theory, where the gastric pouches of a cnidarian are transformed into a trimeric coelomate condition; *B*, Jägersten's bilaterogastraea theory, where a bilateral and benthic organism gastrulates and acquires gastric (enterocoelic coelomic) pouches.

One variant of gastraea theories that does avoid some of the problem areas has recently been advanced by Nielsen & Nørrevang (1985). They suggest that a common monociliate gastraea stock gave rise directly to Cnidaria, and also via a 'trochaea' stage, with an equatorial ring of multiciliate cells, to the remaining bilateral phyla. This occurred through trochophore larvae on the spiralian side, and through a 'tornaea' larva on the deuterostome side, the latter form losing its multiciliation again (fig 7.4). While avoiding explicit archecoelomate postulates, this theory does require awkward and functionally inexplicable gains and losses of ciliation patterns. It follows Jägersten in placing perhaps excessive stress on the primacy of pelago-benthic life-cycles and the primitive stature of larvae, all of which is interesting dogma but untestable (see chapter 5); in fact, most lower Metazoa do not have larval stages (Steinböck 1963). The resultant theory seems over-elaborate on the one hand, and insufficiently supported by a functional rationale on the other.

In passing, one more theory related to but somewhat different from the blastaea/gastraea story should be mentioned. This is the 'plakula' theory, invoking a flattened disc of cells (rather than a hollow ball) as the plakula stage to precede a gastraea. The theory (again of nineteenth century pedigree) was revived by Grell (1971, 1981) with particular reference to the placozoans, peculiar animals that have a dorsal layer of flagellated cells and a ventral absorptive layer with little in between (see section 7.4). However, since they do have a certain amount of intervening parenchyma, it seems preferable to refer them to a planula grade as discussed below, and to do without the confusing plakula grade; especially as *Trichoplax* in fact passes through a perfectly good blastula stage to arrive at its adult morphology.

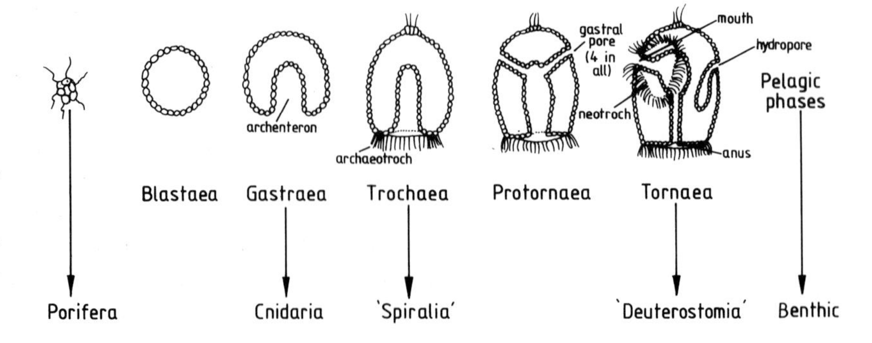

Fig 7.4. The trochaea theory, in which the pelagic trochophore larva (trochaea) succeeds the blastaea and gastraea, to form the ancestor of spiralians, and in turn gives rise to a pelagic tornaea as the ancestor of deuterostomes; (based on Nielsen & Nørrevang 1985).

Colonial 'blastaea/planula' theories

One way to avoid the problems discussed above is to retain the blastula as a first stage in metazoan origins but combine it with a non-invaginated, solid stage as the next step in evolution of the archemetazoans. This second stage has been termed the 'parenchymula', or 'planaea', or most simply the 'planula' (referring to the planula larval stage of cnidarians). The planula ancestor is taken to be a small, pelagic, ovoid but radially symmetrical solid mass of cells, the internal cells acquired by immigration from the blastula stage. From it radiated the cnidarian and ctenophoran lines, and when it became benthic and differentiated its dorsal-ventral and anterior-posterior axes it gave rise to the primitive Bilateria such as flatworms - the turbellarian acoels are therefore assumed to be derived from sexual planulae. The theory, partly foreshadowed in the late nineteenth century by Metschnikoff when he proposed a solid gastraea as a metazoan precursor, owes much of its recent popularity to Hyman (1940). She favoured it specifically as an attempt to update and render acceptable the colonial theories of Haeckel, avoiding most of the problems thereof. It has been added to and modified by Hand (1963), Ivanov (1968), Beklemishev (1969), and Reisinger (1970). A thorough review of these theories and their strengths is given by Salvini-Plawen (1978), and a summary of likely stages in metazoan evolution based on his views and to a lesser extent those of Ivanov is given in fig 7.5.

Several points can be made in support of this theory. It embraces both pelagic and benthic forms, and radial and bilateral forms, in a single theory, permitting all the Metazoa to be monophyletic from flagellate ancestors. It provides a good starting point for most of the 'lower' phyla, as planuloid larvae occur not only in many cnidarians but also in a few sponges and at least one ctenophore (Greenberg 1959). It neatly provides a plausible status for the placozoans, as rather solid forms of blastaea forming incipient planulae (Ivanov 1973). On the whole it makes ecological and functional sense, with varieties of planulae diversifying according to their different modes of life as planktonic, surface-dwelling or thiobenthic forms (though it is not clear how the initial planula stage was supposed to feed, having no mouth or gut - existing planula larvae are transient non-feeding stages). The theory also establishes the platyhelminths firmly near the rootstock of the bilateral phyla, with little modification required to derive them direct from a planula; indeed, the acoel turbellarians are superficially most convincingly like large planulae. All that is required to complete their transformation is the increasing potency of the endoderm to form mesodermal parenchyma and actual gut tissues leading from a new ventral mouth; and the acquisition of a (somewhat elaborate) reproductive system.

There are also, of course, some problems. As we saw before, having as a first stage a colonial flagellate ancestor does involve certain difficulties, even if the choanoflagellates are used as a model. At the next stage, the theory in its

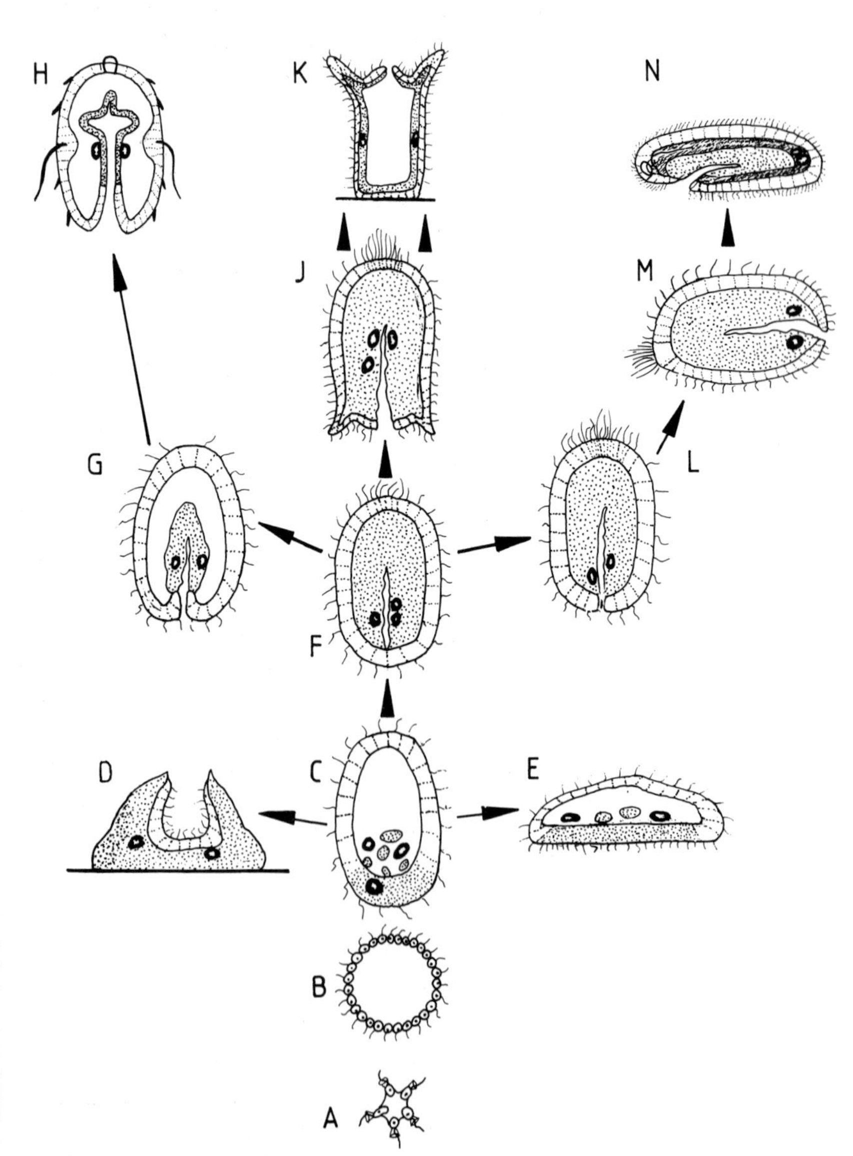

Fig 7.5. A scheme of invertebrate origins in which the planula is central to all further developments beyond the first acquisition of multicellularity. Colonial choanoflagellate cells *(A)* give rise to a blastaea *(B)*. Separate gametic cells develop (black circles), and ingression *(C)* gradually produces an internal layer and thus a true planula *(F)*. Sponges *(D)* and placozoans *(E)* are derived from the very early planula, with ctenophores *(H)*, cnidarians *(K)* and flatworms *(N)* as offshoots from the fully developed planula stage; (modified from Salvini-Plawen 1978).

original versions suggested that the medusa form of cnidarians would be the more primitive (Hyman 1940, 1959; Hand 1959, 1963), as being more easily derived from essentially radial forms, so that Hydrozoa and Scyphozoa (having medusoid stages) were more primitive than the solely polypoid and more pronouncedly bilateral Anthozoa. Although this has long been (and in many text-books remains) the conventional view of cnidarian relations, various authors have recently taken the opposite view, as we shall see below.

However, Salvini-Plawen's review (1978) combines a planuloid theory with a revised view of cnidarian evolution, invoking a settling planula with a single pair of tentacles as the first cnidarian and so starting with biradial polyps. This phase actually occurs in the embryology of Scyphozoa and some hydrozoan polyps; and as a theory it explains the differences between the medusae in these two groups, proposing that their pelagic medusoid phases must have arisen convergently. This seems to provide the most broadly acceptable (or least problematical) colonial flagellate/blastaea type of theory to date.

But it is worth noting two other issues here. Firstly the 'planula' is an extremely simple architectural construct, and is almost an obligate stage in any metazoan's embryology given the way in which cell aggregations behave; so it may well have developed independently several times (Greenberg 1959). Secondly, the problems with the occurrence of cilia and flagella, referred to in chapter 6, should be recalled. Monociliated cells, as derived from a flagellate, are represented amongst lower metazoans in cnidarians, placozoans, gnathostomulids and probably sponges, but multiciliated cells occur in ctenophores and platyhelminths. If the whole origin of Metazoa is monophyletic from flagellates, the presence of monosomal multiciliation in the very early planula leading to the turbellarians and other spiralian animals is somewhat difficult to account for.

For these and other reasons, a number of authors have developed theories for metazoan origins that do not rely on colonial (monociliated) protistan flagellates, forming blastulas, as their starting point; can such theories avoid some of these difficulties?

Colonial amoeboid/acoeloid theories

Reutterer (1969) proposed an archemetazoan origin amongst the amoeboid protozoans that exhibit some tendency to aggregate, and are essentially diploid. The ancestral forms would therefore be benthic, with nutrition by ingestion of surface deposits. Lacking primary surface ciliation, such acoeloid ancestors could perhaps readily give rise to either modern multiciliated groups like the Acoela, or to monociliated lineages like the cnidarians.

Hanson (1977) also advocated a colonial amoeboid form as the likely ancestor of cnidarians, as part of his effectively polyphyletic theory on metazoan origins (platyhelminths being derived from syncytial ciliates, as we will see below).

Syncytial ciliate/acoeloid theories

The possibility of metazoan origins occurring through growth, nuclear division, and cellularisation of a single protozoan, rather than by aggregation of several, was realised in the nineteenth century, but owes its modern formulation largely to Hadzi (1953, 1963). His version of the theory, also advocated by Steinböck (1958, 1963) and in part by Hanson (1958, 1963, 1977), envisaged a process of cell boundary formation in a multinucleate (plasmodial) protozoan of a ciliate type (perhaps one like *Paramecium*), or possibly of an aberrant form such as *Opalina* (an odd protozoan usually given either class status on its own or assigned to the periphery of the flagellate taxon). A bilateral ancestor of this kind could very directly give rise to the acoel turbellarians (fig 7.6), which lack any gut lumen. In particular, Hadzi believed the 'midgut' tissue in Acoela to be primitively syncytial in nature,

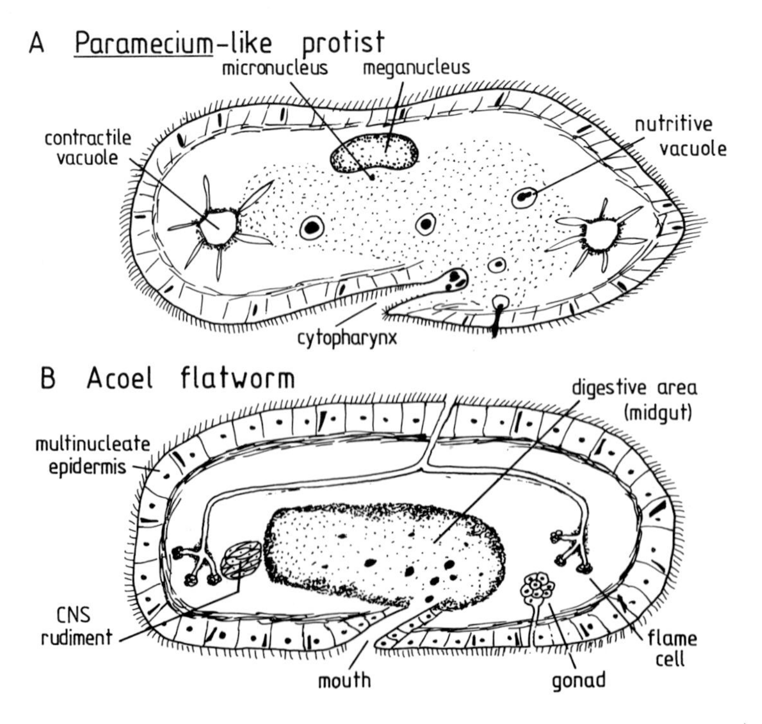

Fig 7.6. The resemblance between a large ciliate protozoan such as *Paramecium* and a small acoel flatworm, as interpreted by Hadzi (1953, 1963).

standing as testimony to a genuine link with plasmodial protists. Acoels would then be ancestral to all other metazoans, except possibly sponges which could have a separate origin from flagellates (as in so many other theories).

Hadzi drew heavily on the simple fact of resemblance between ciliates and acoels, and was imprecise in considering which features could actually be described as homologous; indeed he advocated somewhat rash parallels in pursuit of this theory (trichocysts with rhabdites, cytosome with pharynx, and the like). All this has allowed his views to be extensively criticised by the German school, accustomed to applying the rigid rules laid down by Remane (1952) to questions of homologies. As Hadzi's views of the later evolution of the metazoans were peculiarly linear, with all phyla derived from a single stem (fig 7.7; and see fig 1.3*C*), his work has often been ridiculed and has certainly not been widely accepted. Remane (1963*a*,*b*) and Ivanov (1968) further discredited it with attacks on the plausibility of cellularisation itself. Nevertheless Hadzi's ideas on archemetazoan origins have been accorded some serious consideration. The planuloid/acoeloid form leads readily to flatworms, which are now almost everyone's favourite starting point for metazoan radiation, but which are in fact multiciliate (like ciliates) and not monociliate (like flagellates). The order Acoela includes animals that are of a similar size to large multinucleate ciliates, and of an appropriate bilateral form; they have a similar feeding habit, ingesting bacteria and small protists; and no

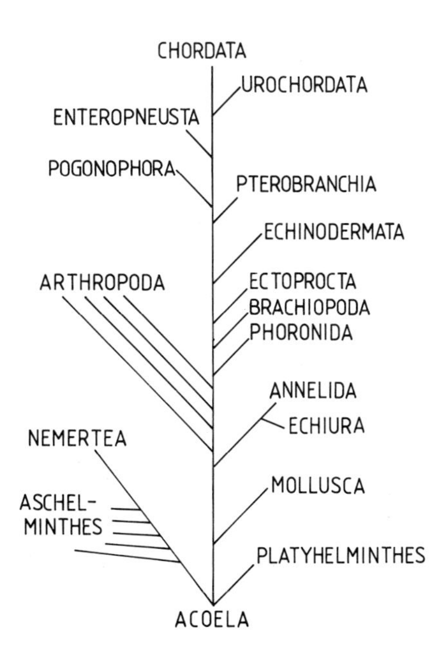

Fig 7.7. Hadzi's (1963) phylogeny of the animal kingdom, above the Acoela level.

awkward hypothetical intermediate stages need be invoked to make the transition from protist to archemetazoan.

However, there are problems. One major stumbling block must be the gnathostomulids, a group of flatworms now given phyletic status on their own (chapter 8) and thought to be closely allied to but primitive relative to platyhelminths. They are actually a monociliate group, so it is hard to see where they could fit into a ciliate/acoel scheme. Furthermore, the turbellarian starting point as envisaged by Hadzi requires the so-called diploblastic groups to be regressive, and bilateral anthozoans to be ancestral to the other cnidarian classes. Hadzi proposed (fig 7.8*A*) that the ctenophores evolved from neotenous polyclad larvae, and that cnidarians derived more directly from adult rhabdocoel-type platyhelminths. This is rather difficult to envisage, necessitating a link between the flatworm gut diverticula and the septae of cnidarians. Alternatively, Steinböck (1963) suggested that cnidarians branched off from a very early acoel form (fig 7.8*B*) similar to the planula invoked by other theories; the acoel preceded the planula, rather than vice versa.

Hanson's version of this theory (1977) is perhaps the most acceptable. He presents plausible selectionist arguments for the ciliate ancestry of turbellarians, and for cellularisation as a process, rejecting the more esoteric aspects of Hadzi's views. Overall he invokes a plasmodial/syncytial ciliate origin for platyhelminths and the spiralians, but a choanoflagellate origin for sponges and perhaps an amoeboid origin for cnidarians (see above); his views are summarised in fig 7.8*C*.

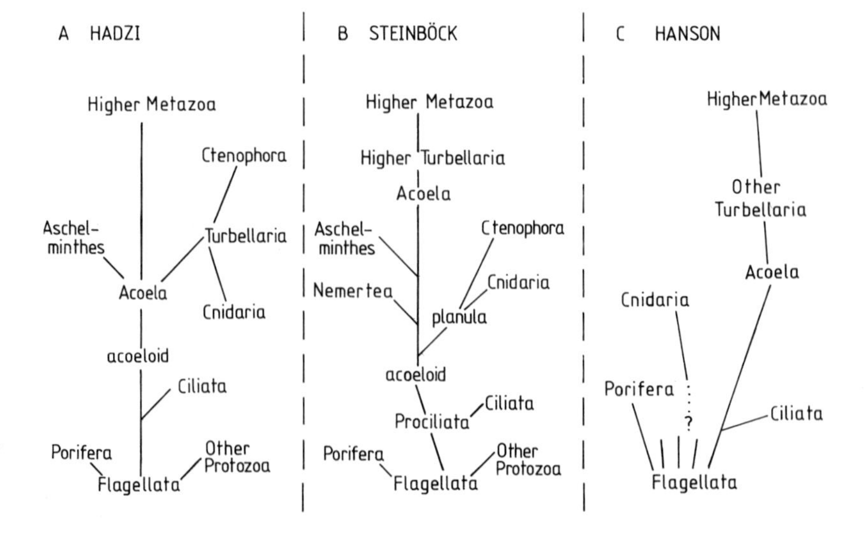

Fig 7.8. Different interpretations of the syncytial theory of metazoan origins, all of which regard the acoeloid form as primitive and derived from a ciliate-like protistan stock, but vary in their views on the status of cnidarians, ctenophores and other lower groups.

Other possible theories

Metazoans originated from procaryotes. A theory of this type was presented by Pflug (1974), when he proposed that animals actually arose from multicellular procaryotes that developed a nucleus. He based this view on certain fossils known as Petalo-organisms, from Pre-Cambrian (Ediacaran) rocks (see chapter 3). Some of these seem to be plants but others have an apparent internal cavity that could have had a digestive function, and may have been cnidarian-like. Apart from the difficulties of proposing diphyletic acquisition of complex eucaryotic features (mentioned in the introductory section), this theory also seems incompatible with any known embryological and growth patterns of animal life, which is never filamentoid or meristem-like as in the Petalo-fossils. Salvini-Plawen (1978) dismisses the petaloid organisms as an irrelevancy, mostly referred more plausibly to a plant or fungus lineage. Many other authors would concur with this view, and as was discussed in chapter 3 Glaessner (1984) suggests that some of the more animal-like petaloid forms are in fact perfectly good anthozoan cnidarians, whilst Seilacher (1984) believes they are evidence of a totally separate and extinct 'type' of metazoan life. They certainly do not seem to be good material on which to base such a radical theory of the origins of extant Metazoa.

Metazoans originated from plants. Consideration of the classic blastaea/gastraea theories of Haeckel raised the difficulty that all existing spherical hollow flagellate colonies are actually plants and not animals at all. Most authors have therefore discounted volvocid colonies as models. But Hardy (1953) proposed instead that such colonies could plausibly have given rise to animals by a series of logical steps. The colony gradually specialised its leading cells as flagellated locomotory structures, retaining their photosynthetic capacity; while trailing cells began to invaginate, forming a saucer-shaped creature. Due to the currents created by the flagella, debris including food particles would collect in the depression of the saucer, and cells bordering this region would lose their plastids and resort to heterotrophy instead. Thus a simple gut was formed and eventually photosynthesis was no longer required. Though less implausible than some, this theory has received little support, and does seem to propose somewhat superfluous steps given the presence of so many already heterotrophic single-celled protists as possible metazoan ancestors.

Metazoans originated anaerobically. A holobenthic, interstitial and anaerobic origin for metazoans is advocated by Boaden (1975, 1977), following analyses by Fenchel & Riedl (1970) and others of the thiobenthic organisms living today. These anaerobic meiofauna in the sulphide-rich deposits of most marine substrates are principally small bilateral organisms from many phyla, but exclude segmented forms, coelomates and cnidarians (Platt 1981). Their distribution in relation to depth and hence the anaerobicity of conditions possibly reflects an evolutionary sequence (fig 7.9), those forms found deeper being the earlier in evolutionary terms, and having evolved at low levels of atmospheric oxygen (fig 7.9*B*). It is true that all aerobic animals today also have anaerobic metabolic pathways but that the reverse does not hold, so there is a possibility that anaerobicity did come first; the geological information available would also support this view, as there is a preponderance of strongly reduced minerals in the earlier Pre-Cambrian rock. This story would also provide reasonable backing for an acoeloid type of ancestor (called a 'thiozoon' by Boaden), as it places acoel and catenulid turbellarians, gnathostomulids, and some aschelminths at the base of the metazoan radiation. However, the theory requires metazoan origins to be set very much earlier than is usually allowed - as long as 2000-2500 MYA - and there is no particular evidence for such a view. In fact the evidence amassed in favour of this thiobenthic sequence is all extrapolations from modern animals and their distribution, and could equally reflect a later radiation (perhaps one of many) from an acoeloid form derived by any of the described routes from protists. It is very likely that some of the modern thiobenthic representatives are probably the result of secondary invasion of the habitat.

Boaden's theory has little to say on the actual mechanisms of metazoan origins. He in fact postulated a separate origin for non-bilateral groups that are absent from the thiobenthos (poriferans, cnidarians and ctenophores), though this would probably not be an essential part of a thiobiotic ancestry for

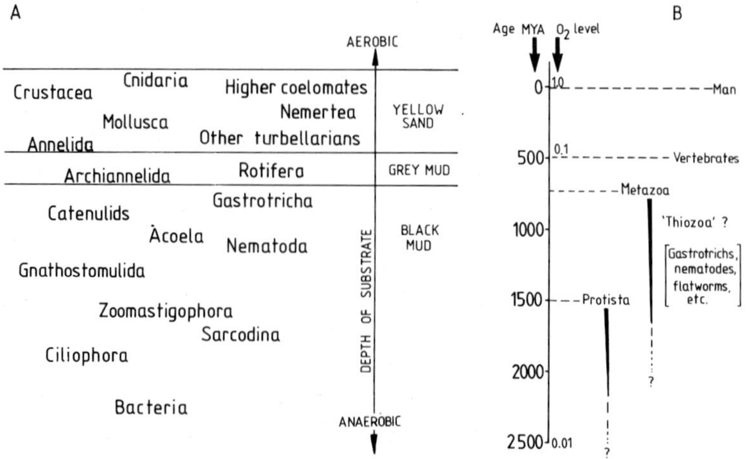

Fig 7.9. The 'thiozoon' theory of metazoan origins, assuming the modern inhabitants of anoxic muds to be relic representatives of an early anoxic metazoan radiation: *A*, the present-day distribution with depth of various protist and invertebrate groups, based on Boaden (1975); *B*, the geological sequence, setting metazoan origins very much further back in time than most other theories (from Platt 1981).

metazoans. Further discussions of the thiobenthic theory are given by Platt (1981) and by Glaessner (1984), whilst a severe critique is presented by Reise & Ax (1979).

There are, then, almost a surfeit of possible theories as to the origin of the animal kingdom from the Protista, summarised in fig 7.10. The process may have been mono-, di-, or polyphyletic, from ciliate or flagellate stocks. The first metazoan may have been benthic or pelagic, radial or bilateral; and it may have been a blastaea, gastraea, plakula, planula or acoeloid. The figure (7.10*B*) indicates that there are certain additional key controversies, centred on the relationships amongst the cnidarians, and between 'higher' Bilateria and the flatworms. Since the aim here is to make sense of the various theories and relate extant groups of lower metazoans to them, some of these controversial issues, and the relevant evidence such as it is, are examined in more detail below.

7.3 The first metazoans

Radial or bilateral?

In some older classifications of the animal kingdom, phyla have actually been segregated according to their primary symmetry, three of the groups of lower metazoans - sponges, cnidarians and ctenophores - being united as the Radiata, as was discussed in chapter 2. In this way they are distinguished from all the remaining bilateral groups (wormlike forms, and others which though apparently radial are clearly derived secondarily from bilateral ancestors). Often this view carried with it the suggestion that radiality in the three 'lower' phyla preceded the evolution of bilaterality, implying that the first metazoans were radially symmetrical.

Many texts (e.g. Barnes 1987) retain the idea of fundamental and primitive radiality in sponges (though many species have become irregular) and in diploblastic cnidarians and ctenophores (though a degree of biradiality is allowed as a secondary development). And in effect, all the theories that were outlined above, with the exception of the syncytial ciliate/acoeloid theory of Hadzi, begin with a radial ancestor. Hence for most authors, only the timing of the appearance of bilaterality is in dispute: it arises somewhat earlier in Jägersten's bilaterogastraea theory, and in the planula theories, than in Remane's and Marcus' versions of the gastraea theory.

Clearly there is no direct evidence as to the symmetry of the earliest metazoans. But as noted in chapter 2 virtually all extant metazoans do show some element of bilaterality, or at least biradiality, in their organisation. The

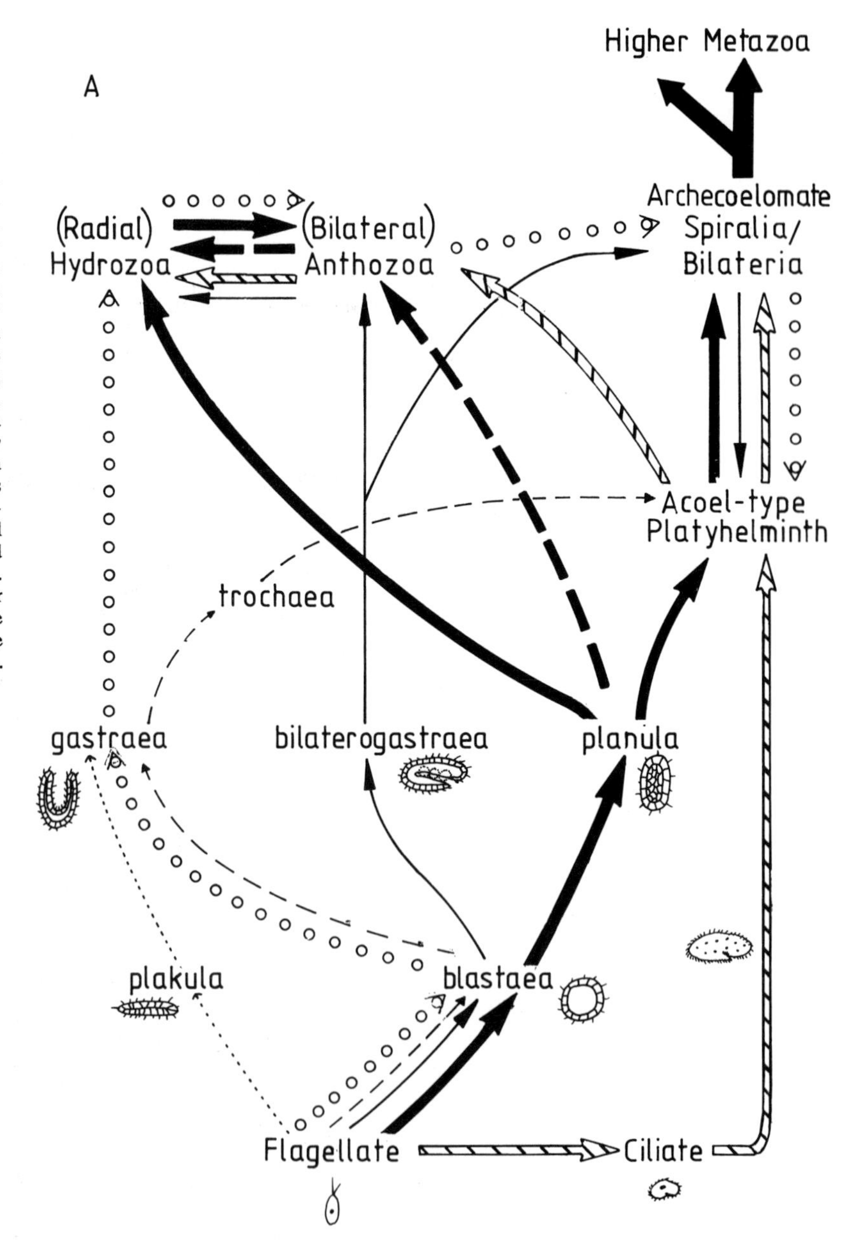

Fig 7.10. A summary of the different theories of metazoan origins. On the left, varying pathways between hypothetical ancestral forms and modern groups are shown: *Solid black arrows*, the planula theory of Hyman, Salvini-Plawen and others, with Hyman's variation for cnidarian evolution shown dashed; *circles*, gastraea theories derived directly from Haeckel, with archecoelomate ancestors (Remane, Siewing, etc); *continuous narrow lines*, Jägersten's bilaterogastraea version; *dashed narrow lines*, Nielsen's trochaea theory; *dotted line*, Grell's plakula theory; *diagonally-barred arrows*, syncytial theories such as Hadzi's and Hanson's. On the right, the major questions raised by all these opposing theories are clarified.

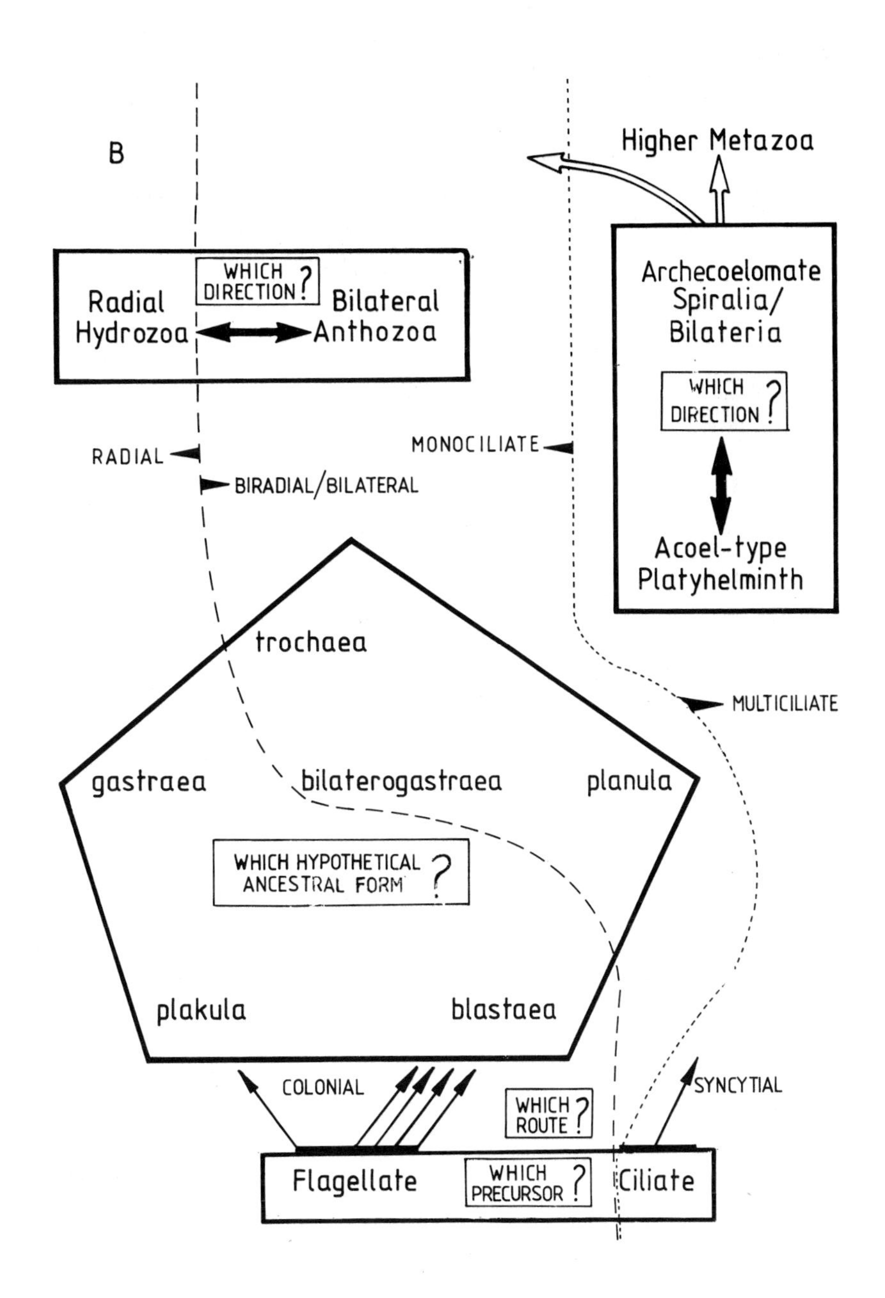
B
Higher Metazoa
WHICH DIRECTION?
Radial Hydrozoa
Bilateral Anthozoa
Archecoelomate Spiralia/ Bilateria
WHICH DIRECTION?
Acoel-type Platyhelminth
RADIAL
BIRADIAL/BILATERAL
MONOCILIATE
MULTICILIATE
trochaea
gastraea
bilaterogastraea
planula
WHICH HYPOTHETICAL ANCESTRAL FORM?
plakula
blastaea
COLONIAL
SYNCYTIAL
WHICH ROUTE?
Flagellate
WHICH PRECURSOR?
Ciliate

supposed Radiata are certainly a somewhat suspect group. Sponges are rarely symmetrical in any sense at all, but have elements of biradiality in their cleavage and in the amphiblastula larva (Tuzet 1973; Bergquist 1985); and ctenophores are very often clearly biradial (fig 7.11*A*). Even the cnidarians (fig 7.11*B-F*) show clear longitudinal axes of symmetry in respect of the disposition of their mesenteries and muscles in all cases except some hydrozoan polyps, and the latter are among the most sophisticated of their kind (see below), unlikely candidates as ancestral types. For these reasons, the planula/acoel type of metazoan precursor, of essentially biradial form, seems the most logical starting point; this is very much the conclusion reached by Salvini-Plawen, in his review supporting a blastaea/planula origin, though for the present there is no need to offer a judgement on whether the planuloid form arose by Hadzi's or by Hyman's route.

Pelagic or benthic?

This question relates very closely to the preceding one, in that it is widely taken for granted that pelagic forms are commonly radially symmetrical whilst epibenthic forms - living on the surface - tend to be bilateral (unless they are actually sessile or become interstitial in which case radiality may be resumed).

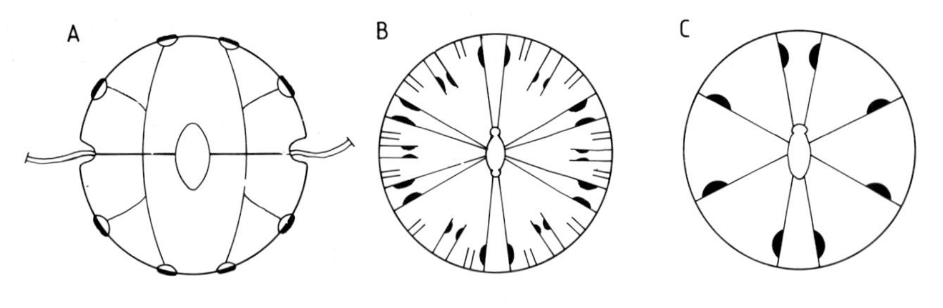

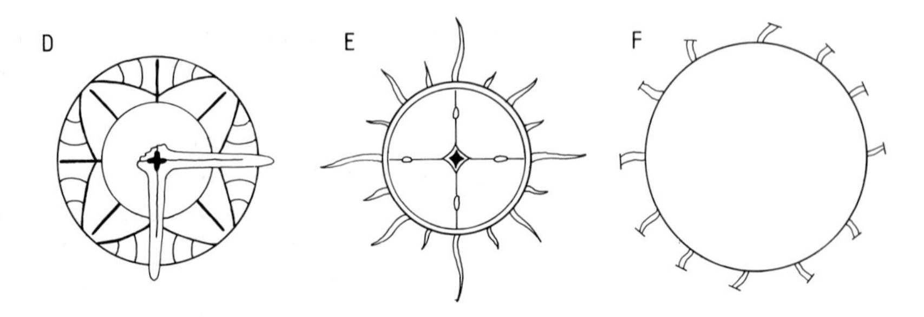

Fig 7.11. Patterns of symmetry in supposedly radial groups of lower metazoans. The ctenophores *(A)* show clear biradiality, whilst many cnidarians *(B-F)* also have biradial axes, particulary in the disposition of mesenteries in actinarian anemones *(B)* and octocorals *(C)*, only becoming truly radial in the advanced hydrozoans *(F)* where mesenteries are absent.

Pelagic ancestry is therefore commonly assumed, because in most theories radiality has already been taken for granted. Such an assumption was inherent in the views of Jägersten and Siewing reported earlier, where a pelagic radial blastaea form gave rise to a bilateral phase on becoming benthic. Indeed Jägersten's whole theory of life-history evolution (1972) revolved around this principle, with the newly benthic adult retaining a pelagic larval stage. So it might be assumed that if the ancestor was a radial blastaea/gastraea it was probably pelagic, and only if it was a bilateral acoel or planula by the direct routes envisaged by Hadzi or Reutterer was it likely to be benthic. Boaden of course also adopted a benthic ancestor, but for quite different reasons.

The basic assumption underlying all this is clearly an over–simplification (see also section 2.2), and will not bear close scrutiny. The ciliate forms from which a syncytial metazoan might have grown are not noticeably benthic, whilst the many small pelagic larvae now extant are rarely fully radially organised themselves, even in groups which may perhaps be primitively radial. Either type of ancestry seems equally possible: the phytoplankton may have formed a good food source favouring pelagic forms, as so many early theories required, but the surface deposits and associated decomposing bacteria and protists available to benthic animals are frequently more concentrated and easier to collect. Benthic animals would perhaps achieve higher growth rates for this reason, and since size increase may be an important factor in favouring multicellularity a benthic origin could be strongly advocated. But it is never going to be possible to decide on this issue; it must be left in abeyance, merely noting that whether the ancestral forms were benthic or pelagic is not so simply allied with their symmetry as some would like to believe.

Flagellate or ciliate?

The concern here is with the issue of monociliate (commonly called flagellate) cells, with a second centriolar structure in the basal body of the cilium giving a diplosomal condition; and multiciliate cells, with monosomal basal bodies. This question was considered in some detail in chapter 6; it is central to the issue of metazoan origins, for within the classic groups of lower metazoans both types of cell occur. Monociliated cells are characteristic of sponges, placozoans, cnidarians and gnathostomulids, and of course of flagellate protists; multiciliated cells occur in ciliates, and in all the spiralian groups including platyhelminths. Both types of cell occur in various kinds of gastrotrichs. Hence it must either be assumed that the ancestor was of one type and that in some derived lineage a switch to the other type occurred (e.g. fig 6.2*A*), or that there were at least two separate origins for the Metazoa, some phyla deriving from a flagellate/monociliate ancestor and some from a ciliate form (fig 6.2*B*).

Since most protozoologists agree that flagellate/amoeboid protists are ancestral to all others, the arguments in fact centre around whether multiciliation was established before (Hadzi) or after (most others) the evolution of multicellularity. In the latter case, it must have arisen independently in the ciliates and in the early Metazoa. Chapter 6 gave some of the pertinent arguments; for example, there can be little doubt that transformations between the two conditions *have* occurred at various points in evolutionary history (e.g. amongst the gastrotrichs and bryozoans, and en route from early chordate forms to the vertebrates). So it is not unlikely that a rather fundamental divergence into the different cell types should occur after a unitary origin of the metazoans. The most likely hypothesis would of course be that this unitary origin was from a flagellate, as the radiate groups and sponges are normally taken to be most primitive (though they are *not* all flagellate/monociliate). The planuloid ancestor would probably have to be the point at which multiciliation then arose; some planulas could lead to the acoel flatworms, perhaps via gnathostomulid and gastrotrich-like groups (cf. Rieger 1976, and chapters 8 and 9), and others (perhaps independently) to the ctenophores. This view resembles that advocated in Salvini-Plawen's (1978) review, where he maintains that uniflagellar (= monociliate) diplosomal cells are a fundamental character of all true metazoans (making no mention of the rather obvious exceptions).

The alternative view would require a multiciliate monosomal ancestor derived from a ciliate protist, as Hadzi proposed. In fact modern ciliates have unusual and specialised basal bodies and seem implausible as direct ancestors for the multiciliate/monosomal condition. Instead an opalinid form might be invoked, or some other group of now extinct multiciliate protists derived from the flagellates - which would also avoid the criticism that modern ciliates are implausible ancestors because of their two types of nuclei and curiously complex conjugation-based sexual systems. This 'ciliate' ancestor would directly give rise to acoels and higher spiralians. But it would have to convert subsequently to the monociliate state and gain the second centriolar structure at least three times: in cnidarians (and hence perhaps in the deuterostome groups if these were descended from the same stem); in gnathostomulids, close to and probably primitive to the platyhelminths (which as discussed above poses some problems for Hadzi's views); and in sponges (though Hadzi believed these to have been of independent origin - which is reasonable given that they differ from all other diplosomal metazoans in the disposition of their second basal apparatus centriole (chapter 6)).

Hadzi did not particularly concern himself with the ciliate or flagellate nature of cells in the lower metazoans, and other authors have not taken up seriously the implication that the first metazoans were multiciliate forms. If they were, the problem of explaining how all metazoans appear to have primitively uniflagellate spermatozoa is at first sight insuperable. But Hanson (1977) has reconsidered this issue, and points out that the widespread assumption of a single 'primitive' sperm type is itself largely dependent on underlying assumptions and preconditions about early metazoan phylogeny, and may not be tenable.

There remains the possibility that the two types of cell occur in the Metazoa because this is a polyphyletic assemblage, and that different ancestors were either mono- or multiciliated independently. This would at first sight be no more unlikely than either of the views so far considered; in fact Hadzi's phylogeny is already effectively diphyletic, with just sponges arising from flagellates, and many other theorists have in the past proposed the sponges to be a separate lineage from protists. It may only be necessary to extend these views somewhat, to ally cnidarians, sponges, gnathostomulids and placozoans with flagellate ancestors (perhaps the first pair from a gastraea type, and the second pair from planulas?), and platyhelminths and ctenophores with ciliates, to be able to side-step many of the awkwardnesses imposed by a monophyletic phylogeny. The constant occurrence of a 'ciliary necklace' at the base of the axoneme in metazoans (chapter 6), a feature shared with flagellates but absent in ciliates, may seem damaging to such a polyphyletic view; but again an ancestral multiciliate protist derived from flagellates but not part of the highly modified Ciliophora could provide a way around the problem.

On this issue then, a reasonable conclusion would be that ancestral metazoans may have been just flagellate, are rather unlikely to have been just ciliate, but could quite possibly have been of both types (though not actual ciliophorans) independently.

Diploblastic or triploblastic?

The theories considered in section 2 have very different implications with respect to the nature of the first metazoans. Either they were simple 'diploblastic' gastraeas with ecto- and endoderm, readily transformable into classically 'diploblastic' cnidarians, or they were solid planula/acoeloid forms, perhaps with little internal differentiation (Hyman) but clearly with the immediate possibility of producing a three-layered state (Hadzi). One of the main criticisms of Hadzi's views has always been the implausibility of commencing with a triploblastic group (the platyhelminths) and trying to derive diploblasts from it by an apparently regressive step (see fig 7.10).

Certainly the progression from unicell to diploblast and on to triploblast has a more immediate logic and the appeal of orderliness, as with so much of Haeckel's theories.

There are several important points here. Firstly, the actual 'diploblast' nature of the cnidarians and ctenophores has long been a matter for dispute anyway, as was discussed in chapter 2. Both groups have a layer of non-cellular mesogloea (often extensive) between ectoderm and endoderm, but in both this is invaded by cellular elements from the two epithelia, particularly in the anthozoan cnidarians. Nerve cell processes traverse the mesogloea, and the contractile tails of the cnidarian myoepithelial cells (chapter 6) extend into it. Wandering amoeboid cells transport nutrients through it. In some respects it forms a perfectly good third layer of tissue; and the distinction between diploblasts and triploblasts becomes blurred.

Secondly, the old idea that cnidarians and ctenophores formed a close-knit group, within the Radiata or even as a phylum together (Coelenterata) has been largely abandoned recently (see Harbison 1985). They have nothing more in common than a very general plan, and no true synapomorphies exist (Salvini-Plawen 1978). Cnidoblasts and colloblasts may perform similar functions but are clearly no more than analogous (see chapter 6). Both groups have tentacles, but in cnidarians these are often hollow and are formed from both ectoderm and endoderm, whilst in ctenophores they are solid and purely ectodermal growths. Ctenophores have true muscle, not myoepithelial cells (chapter 6). And cnidarians are clearly monociliate/flagellate, whilst ctenophores have complex multiciliation.

Thirdly, it is perhaps not so implausible that triploblasts could appear to 'regress' into an almost diploblastic state. The mesogloea of cnidarians and ctenophores could readily be derived from a true mesoderm rather than from an acellular blastocoelic matrix. Neither the nerve nor muscle systems of these two groups are now seen as being so different from (and regressive relative to) other metazoans as was once the case (see chapter 6). Apparent radiality is itself a suspect character and of doubtful primitiveness, as discussed above. However, one issue cannot be readily explained away: flatworms have orderly spiral cleavage, whereas cnidarians show little or no trace of this, and it is difficult to see why this process should have been abandoned if cnidarians are indeed derived from platyhelminths.

Nevertheless, the cnidarian/ctenophoran designs suit particular ways of life, and the unique cnidoblasts and colloblasts respectively give benefits that may render more sophisticated architectures, locomotory systems and nervous systems (and perhaps even embryological patterns) superfluous. Thus regression from an apparently 'superior' triploblastic flatworm-like ancestor does not appear as incongruous in the light of some detailed modern knowledge as most earlier theorists maintained; and the whole question of whether ancestors were diplo- or triploblastic may be a rather unimportant matter of semantics.

Syncytial or colonial?

Either system of producing the first multicellular animals would be feasible in theory; perhaps the best approach here is to examine the processes as they actually occur in the protists, and in lower metazoans.

Colonial aggregations occur quite commonly in flagellate protists (Mastigophora), particularly in the plant-like phytoflagellates. Choanoflagellates also aggregate, often as stalked colonies, though *Sphaeroeca* is pelagic and the genus *Proterospongia* includes forms where the cells are embedded in a central gelatinous mass. Amoeboid protozoans (Sarcodina), though closely related to the flagellates, do not have living colonial representatives; and examples are rare and usually rather specialised in the ciliates. Hence if a colonial origin is to be proposed, a flagellate ancestor certainly seems the most likely starting point; though it is a pity that extant protist colonies are all either plants or fresh-water animals.

A multinucleate condition that might precede multicellularity by subsequent cellularisation is quite common in protists. Several groups of zooflagellates contain two or more nuclei (notably diplomonads and oxymonads, though both are primarily symbionts or parasites). Many amoebas and heliozoans are multinucleate; and in foraminiferans the multinucleate protoplasm often gets divided up into a form of cell by membranes, though there may be no correspondence between the nuclei and these 'cells'. Ciliates have both macro- and micronuclei, only the latter being concerned with reproduction. The micronucleus is often single, but up to 80 have been reported in some species. A few, such as *Stephanopogon*, possess *only* micronuclei, and in considerable numbers. And opalinids (perhaps a phylum on their own, or a super-class within the Mastigophora) are particularly clear examples of multinucleation, having many identical nuclei within a large flattened multiciliated body. They would appear good candidates as plasmodial precursors for a syncytial metazoan stock, via Hadzi's route, but for their curious habits as commensals in the guts of frogs. However, since flagellate protists may have been the source of amoeboid and ciliate groups, as well as opalinids, we could return to the suggestion that another multiciliate and multinucleate group may have been derived from flagellates that in turn gave rise to metazoans. Thus either a colonial flagellate or a multinucleate 'derived flagellate/protociliate' seems plausible as the rootstock of metazoans.

However the latter proposal relies heavily on Hadzi's assumption that the most primitive living metazoans also exhibit ancestral syncytial organisation. This was long held to be true of the gut area of acoels, though their sexual, glandular and nervous cells appeared distinct (e.g. Beklemishev 1963). However, recent electron microscope studies have blurred the whole issue of syncytial/cellular status, and some authors have denied that the Acoela are any more syncytial than other invertebrates. In any case, it is now clear that their syncytial tissues are secondary phenomena, as the gut region is clearly cellular during its early ontogeny. Secondary syncytialisation is a condition also found in other small interstitial metazoans, where it seems to be a derived condition relating to size and lifestyle. If acoel guts are not primitively syncytial, then Hadzi's views certainly lose a good deal of their force, though we cannot expect modern acoel turbellarians necessarily to reflect the condition of the first metazoans.

On balance, there are clearly problems with either theory given the present occurrence of colonial and multinucleate conditions in protists and simple metazoans. But at least one of these processes must have occurred to produce multicellular animal life; the only conclusion can be that both are possible and neither is inherently more unlikely than the other.

Coelomate or acoelomate?

A number of the theories outlined in section 7.2 require the archemetazoan to be a small coelomate form with enterocoelic pouches derived at the gastraea stage from a cnidarian precursor. Acoelomates, and particularly flatworms, must then be secondary. This issue was considered briefly from functional and embryological viewpoints in chapters 2 and 5; here some of the relevant points can be briefly reiterated.

Archecoelomate theories do not bear up well in terms of functional considerations, as Clark (1964, 1979) clearly demonstrated. The earliest metazoan must surely have been very small, and would have moved efficiently with ciliary action; the acquisition of one or several fluid-filled cavities would not give mechanical benefits at this size range. Thus if the main benefit of a coelom is mechanical, as a hydrostatic skeleton (Clark 1964), these ancestral forms should not have required or benefitted from their enclosed pouches. Alternatively, if the coelom serves mainly as a means of permitting size increase, as was discussed in chapter 2, then its acquisition at such an early stage is less implausible. But in that case its subsequent loss, with a return to ciliary-based locomotion and a loss of circulatory system and anus, by ctenophores, platyhelminths and some other groups not of notably minute size or interstitial habit, seems illogical and inconsistent with any conceivable selective pressures. And whatever the primary function of the coelom, it would have had to be a powerful necessity that overcame the disadvantages of actually sealing off gastric pouches whose initial function must have been to increase the absorptive surfaces available to the animal - particularly in an enlarging animal still lacking an organised through-gut! Enterocoely seems functionally most implausible in such early animal forms. The number of pouches proposed in the various theories also seems to raise rather than solve difficulties; whether there were four, five, six or eight initially, it is hard to explain why extant forms display such varieties of single paired, oligomeric or metameric coeloms nowadays (Clark 1979). By contrast, if the coelom is indeed polyphyletic (chapter 2), as seems only logical, then this diversity of modern coelomic organisation is only to be expected. Given polyphyletic coelom origins, a theory for metazoan origins that proposes an early and unitary development of coelomic pouches cannot be upheld. The platyhelminths probably preceded other worm-like groups and were the ancestral stock in which coelom formation was initiated; enterocoelic gastraea ancestors are not acceptable postulates.

The ancestral metazoan, then, remains enigmatic. Biradiality, and a relatively solid acoelomate form, seem most likely, and either a pelagic or benthic lifestyle would be perfectly plausible though the latter might offer greater food resources for growth. A colonial, flagellate/monociliate condition is most likely if all groups share descent from one ancestor, but multiple origins by different routes, including multinucleate ciliates, may be at least equally plausible. Now we need to pursue these possibilities in terms of their implications for evolution amongst the lower metazoan groups, and test them against available evidence from fossils and living animals.

7.4 Evolution of the lower Metazoa

Porifera

It is now almost universally assumed that sponges are derived from choanoflagellate protozoans; the resemblance between their choanocyte cells and these protists is thought to be too great to allow other possibilities. Colonial choanoflagellates such as *Proterospongia* represent an obvious intermediate, lacking the cellular differentiation and limited coordination of a complex sponge but differing very little from the simplest ascon-type poriferans. The question therefore comes down to whether the sponges are a separate evolutionary line, making the metazoans at least diphyletic, or whether all multicellular animals actually derive from choanoflagellates.

The former view has long been popular, with supporters of all possible theories, and it gave rise to the concept of the sub-kingdom Parazoa reserved solely for the sponges. But chapter 6 made it clear that some of the features formerly supposed to be unique to sponges are in fact common amongst the lower metazoans - notably the collar cells (see also Simpson 1984). And sponge embryology, widely quoted in texts as being aberrant and irreconcilable with other metazoans, can be reinterpreted to allow poriferans a legitimate place amongst other metazoan groups (Salvini-Plawen 1978; Bergquist 1985). Bergquist also cites membrane junctions, cell recognition mechanisms and immune systems as thoroughly metazoan attributes of sponges and refutes the need for a sub-kingdom; she in fact splits sponges into two phyla within the Metazoa, with the glass sponges (Hexactinellida) given separate phyletic status. For similar reasons, Salvini-Plawen has strongly supported the view that sponges and all other metazoans have a common origin amongst choanoflagellate protozoans, though he retains their sub-kingdom status distinct from Eumetazoa (Histozoa). This makes a very plausible story, and of course requires a colonial flagellate/planuloid theory to be adopted.

Two points might be borne in mind though. Firstly, the monociliate condition of sponges differs somewhat from that of all other metazoans in terms of basal body architecture (chapter 6). Secondly the existence of collar cells (relatively simple in design) in many metazoan groups may be a case of convergence for functional reasons (chapter 6), rather than of common ancestry necessitating a belief in choanoflagellate origins. Hence Salvini-Plawen's arguments lose a little of their conviction; sponges may be a very early offshoot of a common metazoan stock, or they may still be taken as something of an anomaly, one of several phyletic lines from protozoan stocks.

Cnidaria and Ctenophora

The supposed relatedness, diploblastic nature and radiality of these two groups are problems that have already been discussed and doubted. It remains to consider their specific ancestry and status amongst lower metazoans.

Cnidarians form a clearly defined phylum because of their unique cnidoblasts, and the underlying design of this phylum could be readily derived from either a conventional gastraea stage (which is architecturally similar anyway) or from a solid planula (as happens in their ontogeny). By either route, it is possible to arrive at either the polyp or the medusa stage first, though it has been conventional to regard the medusa as primitive since pelagic and radial forms are preferred as ancestors by most proponents of a colonial gastraea or planula theory. However, the two essentially opposed theories set out earlier in this chapter have differing predictions as to the direction of evolution in the Cnidaria. If the traditional variety of colonial gastraea or planula theory is correct, a radial ancestor is needed and evolution must have been from the more radial groups to more bilateral forms; whereas if the syncytial acoel view is to prevail, then a bilateral form must have been antecedent to more radial types. Here, then, is some hope of testing between the two theories.

Cnidarians are represented by three relatively distinct classes (though some modern schemes split them into four). Of these, the Anthozoa are distinctly bilateral, in the disposition of mesenteries, muscles and siphonoglyphs, having either a six-rayed (Zoantharia - fig 7.11*B*) or eight-rayed (Octocorallia - fig 7.11*C*) pattern. The remaining groups show an essentially tetraradial pattern (fig 7.11*D-E*), with little evidence of any simple bilateral axis; and in some hydrozoans complete radiality is achieved (fig 7.11*F*) with no septa present (*Hydra* being an obvious example). Thus Hyman's views would predict (as she indeed believed and strongly advocated) that hydrozoans were primitive and anthozoans advanced, with medusa preceding polyp; whilst Hadzi's views (and also Jägersten's theory) suggest the reverse - that anthozoans (entirely polypoid) are the most primitive group of cnidarians, the conventional tendency of a group to become more radial as they become sessile (and commonly colonial) being demonstrated in them and even more decidedly in the derived hydrozoans.

Despite the claims of earlier cnidarian experts (Hyman 1959; Uchida 1963) that anthozoans and bilaterality were clearly *not* primitive in the Cnidaria, most of the recent evidence from extant and fossil cnidarians must be said to give some tentative support to this latter point of view. Anthozoans have simpler life-cycles and an apparently lesser ability to cope with physiologically difficult environments. They have markedly less elaborate and less diverse cnidoblasts (Mariscal 1974), hydrozoans having 23 types of which 17 are unique. Hydrozoans also have more complex musculature, ectodermal glands, and nervous systems, all indicative of a high degree of secondary differentiation. As regards fossils, it is strictly impossible to decide whether polyps or medusae came first, as many of the early radial cnidarian-like traces could equally be interpreted either way, according to the degree of flattening invoked, and both forms are perhaps present in the Pre-Cambrian fossil beds (chapter 3; Glaessner 1984). 'Hydrozoans' and 'pennatulid-like anthozoans' are supposed to be particularly well-developed in these deposits, with only very primitive 'scyphozoan' forms, suggesting a later origin for this class. The anthozoans have been even more pronouncedly bilateral in the past than they are today though, as shown in forms such as the Cambrian-Permian Rugosa (rugose corals) where there was a strong axis of symmetry (fig 7.12). This

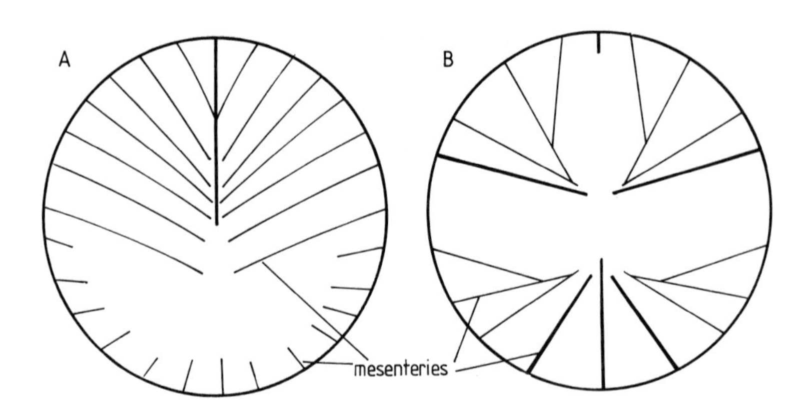

Fig 7.12. The extreme bilaterality in early fossil cnidarians such as the rugose corals, showing the mesentery pattern in two different species.

would suggest a trend from primitively pronounced bilaterality towards a secondary radial condition in the Anthozoa, and hence a bilateral origin for the whole phylum. (However, given the long Pre-Cambrian history of cnidarians there would have been time for many secondary changes of symmetry planes as conditions and lifestyles altered before these rugose corals evolved.)

The evidence then seems to be in favour of a bilateral ancestor, usually associated with a plasmodial route such as Hadzi's. However, Hadzi actually required a complex adult platyhelminth to be transformed into a cnidarian, which seems peculiarly unlikely; a route via an early acoeloid form (e.g. Steinböck, Hanson) requires fewer regressive steps. And in fact Salvini-Plawen's version of the planula theory embraced all the points just raised and incorporated an anthozoan-hydrozoan (biradial-radial) progression, via a planuloid ancestor becoming benthic and sessile with two initial tentacles. Hence this updated account of cnidarian relationships cannot be said to be inconsistent with either main theory, though it renders the older Haeckel and Hyman versions of the colonial flagellate story somewhat implausible.

One extra point may be made; if evolution has indeed been from anthozoan ancestors to later hydrozoans, the greater endowment of cells in the mesogloea of anthozoans, and their near absence in hydrozoans, may actually be indicative of a trend away from a triploblastic state to approximate diploblasty, partially supporting Hadzi's view that diploblastic animals are secondarily modified. However, as hydrozoan polyps are nearly always small, their lack of mesogloeal cells may be unrelated to phylogeny.

Cnidarians certainly represent an early and highly successful radiation, probably from a bilateral, flagellated, planuloid form, and as a group they probably had no further major descendants (Glaessner 1984). They must be accorded a rather isolated position in the animal kingdom. Since the diet of modern forms is almost entirely carnivorous, their evolution may have significantly post-dated that of other Metazoa; the supposedly 'cnidarian' Ediacaran fauna may have been a very early, separate, and unrelated radiation, as discussed earlier (chapter 3).

As regards ctenophores, their supposed close affinity with cnidarians can no longer be accepted (see above), and their evident biradiality is most readily derived from a planula form; they do not appear to hold any comfortable place in a gastraea-based theory. With a planuloid scheme, their origin could be either direct, from a sexual proto-planula as Hyman and Salvini-Plawen would propose, or rather more circuitously from neotenous larval forms of later spiralian groups (Hadzi and others). They may even be the intermediate stage between planula and more advanced Bilateria (Bonik *et al.* 1976). In whatever direction the changes may go, the ctenophores are probably closer to platyhelminths, with whom they share features such as ciliation, gonoducts, mesodermal muscles and determinate cleavage, than they are to cnidarians.

Platyhelminthes

Most currently popular models for the origin of metazoans suggest that flatworms can readily be derived from a planuloid/acoeloid form; that is, a small, ovoid, solid ciliated creature. The differences are that by Hyman's or Salvini-Plawen's route this planula would have (inappropriately) monociliate cells and no gut, whereas by Hadzi's more heretical route the acoel ancestor would be multiciliate as found in modern flatworms and would have an incipient digestive system. In either case, the planula/acoel is assumed to acquire greater bilateral symmetry and cephalisation as it settles to a benthic lifestyle, and the beginnings of a proper gut with three organised cell layers also appear.

Various views of relationships within the phylum Platyhelminthes will be considered in more detail in the next chapter. Most authors support a primary small acoelomate condition for the group, though differing as to whether the actual Acoela are themselves primitive or derived; a useful discussion is given by Hanson (1977), though for dissenting views, see Rieger (1985) or Ax (1987). Even if the Acoela themselves are secondarily syncytial, it is of course not implausible to invoke an 'acoeloid' form as ancestral.

However, a major difference in all the principal groups of platyhelminths compared to any proposed ancestor is the marked elaboration of their reproductive systems; indeed, the presence of unusual biflagellate spermatozoa throughout the group (quite different from the classic and presumably ancestral uniflagellate condition) seemed a particularly awkward feature in a phylum thought to be near the roots of the Metazoa, until 'normal' sperm were located in *Nemertoderma* (Tyler & Rieger 1975). As seen above, it is possible that the

Gnathostomulida also have a role to play here. They are monociliate forms, so are readily derived from a traditional planula, and some have monoflagellate sperm (Sterrer *et al.* 1985). They could represent offshoots of an early metazoan stock that would reconcile multiciliate spiralians with the colonial flagellate/planula route into multicellularity (see also Ax 1985).

Acoelomates therefore seem attractive candidates as being close to the ancestral metazoan; a proto-platyhelminth derived from a planula can serve as a good starting point for most other phyla, as later chapters will show. The alternative view necessitated by some derived gastraea theories (section 7.2), that require metazoan ancestors to be coelomate and the acoelomate platyhelminths to be a regressive state, is improbable on various grounds, discussed elsewhere in this book.

Placozoa and Mesozoa

Finally brief consideration should be given to two phyla that have at different times been considered important elements in the story of metazoan origins. The views of Grell (1971), concerning *Trichoplax* (Placozoa) were mentioned in section 7.2, in relation to a 'plakula' grade of organisation. But this enigmatic animal (fig 7.13*A*) can equally be accommodated in a planula/acoel theory, as a very simple planuloid form, for it consists of outer monociliate cells, resembling the collar cells of other phyla, with a small amount of internal contractile mesenchyme. By the definition given in the introduction to this chapter it is doubtful if it should be accorded true metazoan status at all, since the rarely formed gametes seem to be derived from normal somatic cells; so at best it can be given a status as a very early offshoot from metazoan stock. As such it would appear to tie in better with views such as those of Hyman and Salvini-Plawen than with any of the other theories yet advanced. But it may equally plausibly be regarded as a separate experiment in multicellularity from some colonial flagellate precursor.

Mesozoans (fig 7.13*B*,*C*) have also been claimed to be very close to the ancestral metazoan, but here the evidence is more equivocal; because they are all parasites of other invertebrates, and have complex life-cycles, their simplicity of structure may well be secondary. Since they are multiciliated, some authors propose that they are merely degenerate flatworms (e.g. Stunkard 1972); while others (Hyman 1940; Lapan & Morowitz 1972) place them as an early offshoot of Metazoa (presumably having developed the multiciliate condition independently). Again an alternative view would be to accord them a unique place as independently multicellular animals derived from ciliates. It is even possible that the two classes of mesozoans are unrelated and have separate ancestries. The Orthonectida (7.13*B*), less

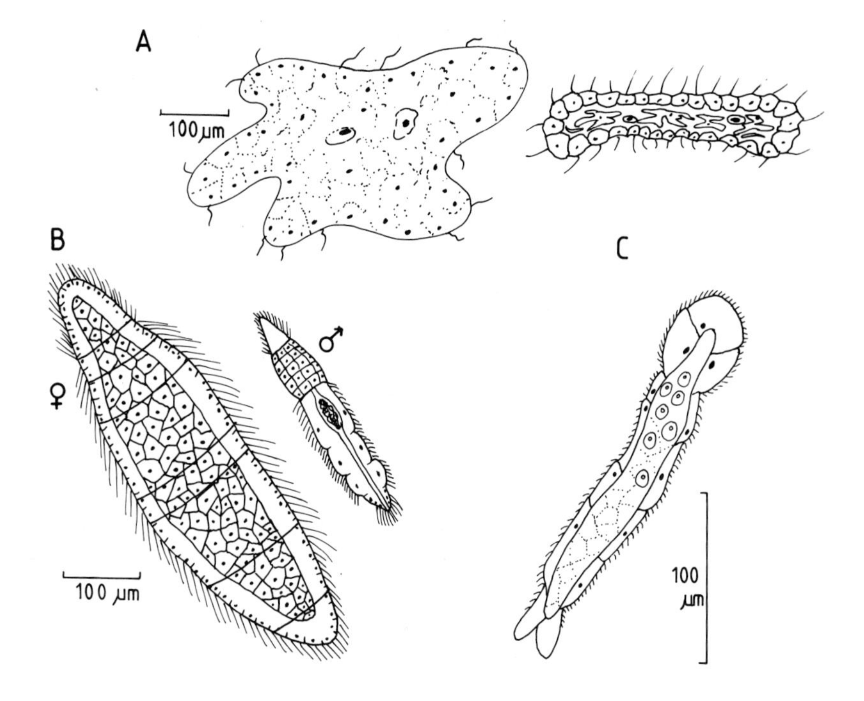

Fig 7.13. Structure of the least familiar lower metazoans: *A*, the placozoan, *Trichoplax adherens*, surface view and side view; *B*, the mesozoan group Orthonectida; *C*, the mesozoan group Dicyemida or Rhombozoa.

specialised forms living unattached in a range of invertebrates, do seem more flatworm-like than the Dicyemida (7.13*C*) that live only in the excretory systems of cephalopods; the latter have meganuclei in their axial cells and curious developmental patterns in the reproductive cells, thus being surprisingly reminiscent of ciliate protists.

7.5 Conclusions

With a certain degree of over-simplification, the various theories for metazoan origins can be summarised as colonial theories (sub-divided into gastraea and planula varieties) from a flagellate ancestral type, as opposed to syncytial theories with a multiciliate (ciliate or opalinid protist) ancestral type. To give a brief summary of the evidence is nearly impossible, but the following points have emerged as useful indicators.

Colonial/flagellate theories. Many flagellate protists do show coloniality; though extant examples are all either plants or are fresh-water animals with no apparent trend towards division of labour. Many lower invertebrates do show a flagellate/monociliate condition. Monociliate spermatozoa (which are in themselves very like certain zooflagellates) appear to be the ancestral condition for all Metazoa (but see Hanson 1977). However, multiciliation must have evolved at least once and probably several times independently within the metazoans if they had a flagellate origin. Most versions of the theory use a radial ancestor, and therefore predict that hydrozoan cnidarians preceded the anthozoans; but much of the evidence refutes this.

In relation to gastraea versions, most current proposals are tied in with enterocoelic coelom formation and require coelomic pouches in a very early ancestor. The problems mentioned above are not avoided, and new ones are created in providing functional explanations for such pouches in small cnidarian-like forms and for the loss of a coelom in all modern acoelomates. These theories (Remane's cyclomerism, Jägersten's bilaterogastraea, and Siewing's benthogastraea) seem to have little to recommend them. Nielsen & Nørrevang's 'trochaea' theory avoids the enterocoely complication, but substitutes an unnecessarily fussy sequence of changes in larval ciliation patterns and appears functionally unconvincing.

In relation to planula versions, all lower metazoans can be conveniently derived from such a planuloid form, but the monociliate/multiciliate issue is not averted, and it is by no means clear that one single planula ancestor can be invoked. Salvini-Plawen's theory adopts a biradial ancestor and thus at least avoids the problems with cnidarians, allowing anthozoans to be more primitive; it appears to be the most plausible available theory starting from a single colonial flagellate form.

Syncytial/multiciliate theories. Many protists are multinucleate and therefore candidates for syncytial metazoan status and subsequent cellularisation; this is particularly common in ciliates, and most clearly represented in the opalinids. Many lower metazoans are multiciliate, including flatworms and all the spiralian/protostome groups commonly assumed to be derived from them. But cnidarians, sponges, placozoans and gnathostomulids are monociliate and would be awkward to derive from the multiciliate acoel ancestor; and the occurrence of basically monociliate sperm in all metazoan groups is hard to explain. Cnidarian anatomy would have to be derived from acoeloid anatomy, though this is not such an implausible 'regression' as has sometimes been maintained in the past. Bilaterality would be primitive, in accord with the probable relations within the Cnidaria.

Both the major theories quite clearly present difficulties, although the more acceptable varieties of each are in essence converging on a planula/acoel form as the first real metazoan and make similar predictions thereafter. This is therefore a good moment to revert to the issue of whether this ancestor was invented only once, making the metazoans monophyletic, or whether we can in fact avoid some of the problems outlined here by proposing a polyphyletic ancestry. Perhaps a simple planula form arose twice or several times. One or more such forms could give rise to the ctenophores, flatworms and spiralians, possibly from a multiciliate stock but more plausibly initially monociliate to take account of gnathostomulids. Another (or several others), from flagellates, could evolve into the essentially flagellate/monociliate sponges, placozoans, and cnidarians (and conceivably ultimately the main deuterostome lines as well).

Salvini-Plawen's review (1978) firmly rejected the possibility of polyphyly of the Metazoa, citing a number of common features that unite them: diploidy, with identical meiosis; diplosomal basal bodies and identical flagellar structure; homologous spermatozoa; presence of collar cells; biradial symmetry; potent regeneration; and the presence of primitive nervous and muscular tissue. To this list might be added the unifying presence of collagen (chapter 4) and of the ciliary necklace. However, the discussions in chapter 6 cast grave doubt on some of these features. Most spiralian groups do not have diplosomal cilia, and do not have collar cells unless their flame cells (cyrtocytes) are assumed to be derived from these sources. Both nervous and muscular mechanisms may have their roots back in the protists and cannot therefore be used as evidence for monophyly; the same is true of the ciliary necklace. The biradial symmetry of some groups is a matter of dispute (section 7.3); and the potency of regeneration is a doubtful and somewhat nebulous character anyway. Collagen presence may be necessary for maintaining multicellularity, and could have evolved convergently from common precursors when oxygen levels rose high enough for its biosynthesis to be possible (see chapters 3 and 4). We are left with the reproductive features; apparently homologous meiosis, and similar, primitively uniflagellate, spermatozoa. As we have seen, this last feature is a problem with any possible theory, as so many lower metazoans are basically multiciliate yet have monociliate sperm.

To set against these reservations, the points in favour of a polyphyletic status for metazoans need some thought. Firstly, there is the central issue of selective pressures for multicellularity; the advantages of increased size in itself, and of divison of labour/increased specialisation, and of possible protection for the vulnerable gametic tissues within a casing of somatic cells. Since these advantages are profound, it is surely very likely that the condition of multicellularity would have arisen many times, and perhaps by different routes; if it happened once successfully, there is every reason to suppose it would happen over and over again. There do seem to be rather fundamental differences between even the more familiar phyla of the animal kingdom, largely tying in with the recurrent theme of cell ciliation. Sponges do seem to be rather separate from other monociliate groups and might have had their own distinct ancestry; cnidarians show no convincing links to other phyla; and the placozoans and mesozoans might be yet more manifestations of attempts at a planula grade from protistan beginnings. And a number of different planula origins for the 'proto-platyhelminths' seems not implausible given the mixed suite of features (especially embryological) found in modern Platyhelminthes (Smith *et al.* 1985; Inglis 1985; and see chapter 8). Given the strong pressures for one set of gametes to be mobile, and the very limited biological ways of achieving motility, it would be by no means inconceivable that all the planuloid-acoeloid groups separately derived from protozoans should have flagellate sperm, and it is not entirely clear how far beyond this simple fact the supposed homology of all spermatozoans actually goes. Reference back to chapter 6 will underline this point, and also remind us of the considerable diversity now found in sperm type, such that some groups do not have, and may never have had, monoflagellate sperms at all.

We shall never know the actual origin of the metazoans, and whether this was a unitary or multiple occurrence, though it is becoming increasingly difficult to believe a true metazoan state was only reached once in view of the lack of conclusive similarities amongst the phyla of lower invertebrates. It is certainly worth remembering that a polyphyletic approach is every bit as plausible as other theories and has had a perfectly respectable history, dating back again to the nineteenth century, and being retained by default by many twentieth century theorists who derived sponges but no other groups from the choanoflagellates. It has recently been advocated more explicitly (see fig 7.14) by Greenberg (1959), Nursall (1962), Hanson (1977), Cloud (1978), Sleigh (1979) and Anderson (1982). Nursall presented a clear case for the plausibility of separate phyletic ancestry from protists in support of his exceptionally polyphyletic views (see fig 1.8). Having reviewed the theories and the evidence, probably the best we can do is to remain agnostic, but with suspicions of polyphyly, on these issues. The first metazoans must have

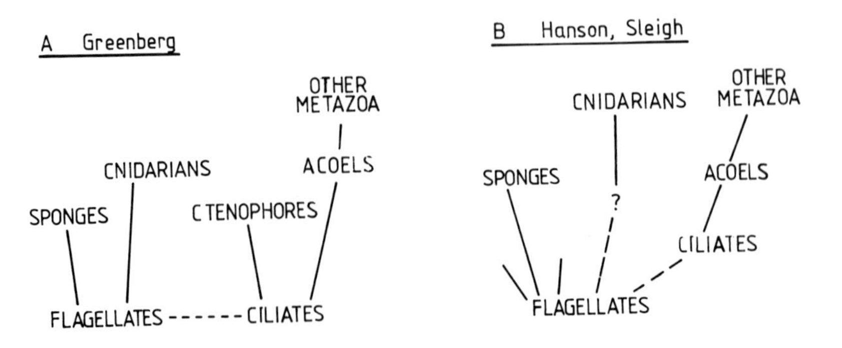

Fig 7.14. Two alternative polyphyletic views of metazoan origins.

evolved considerably before their first preservation in the Ediacaran rocks at least 650 MYA; it is really not surprising that there are no firm conclusions as to the course their early evolution followed.

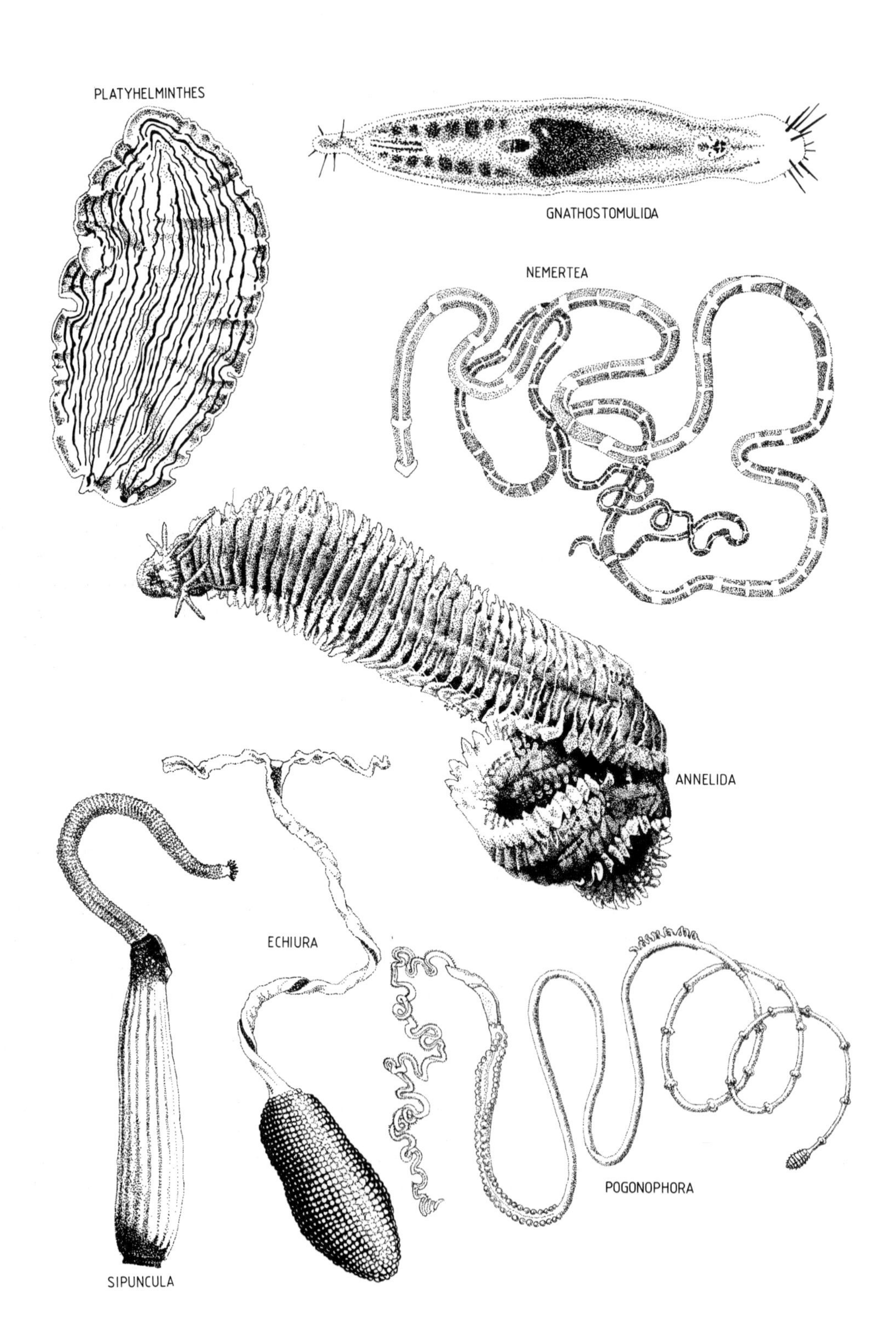
PLATYHELMINTHES
GNATHOSTOMULIDA
NEMERTEA
ANNELIDA
ECHIURA
POGONOPHORA
SIPUNCULA

8 Acoelomates and other lowly worms

8.1 Introduction

In the preceding chapter, some attempt was made to decide where the first metazoan animals came from and what they might have been like. The most plausible answer takes as a starting point a planula/acoeloid form, small, ovoid and ciliated. As discussed in section 7.4, this can give rise both to the cnidarians, which have a planula larva, and to the bilateral acoelomate worms. It also provides a sensible starting point for radiations of pseudocoelomate and coelomate phyla, as the rest of this book should make clear. In effect, the planula can be seen as the early worm, and most of the designs it gave rise to are also elaborations on the theme of being a worm. This chapter deals with a subset of these variations on a theme: the worms that are conventionally regarded as part of the protostome or spiralian assemblage. All are supposed to share the classical patterns of spiralian development set out in chapter 5, so that in comparing them it should also be possible to get a perspective on the Protostomia/Spiralia super-phylum as a whole.

8.2 Acoelomate worms

Platyhelminths

Apart from a predominantly German school of thought requiring flatworms and most other simple worms to be secondarily reduced from a coelomate state, the majority of currently popular theories suggest that flatworms are direct derivatives from the planuloid/acoeloid form. This early metazoan is assumed to settle to a benthic lifestyle, thereby acquiring greater bilaterality in association with moving directionally in two dimensions, and the beginnings of a proper gut as the tissues differentiate fully into three cell layers. By this stage it must also possess or acquire spiral cleavage, if it is to become a forerunner of the protostomes.

Most views of relationships within the Platyhelminthes clearly support these views, requiring a primary acoelomate condition derived from an acoel-type worm, with guts becoming more elaborate and the division into distinct germ layers more pronounced as the phylum evolved. Conventionally the turbellarian group Acoela (virtually solid, with a ventral mouth and central 'syncytial' area acting as the gut) do show many features indicative of primitiveness, and have in the past been taken as nearest to the ancestral group within the phylum (e.g. Hyman 1951*a*; Hanson 1977). Catenulida and Nemertodermatida, with fully developed guts and more distinct mesodermal matrices, were assumed to represent the next level of sophistication.

But recently Ehlers (1985*a*,*b*) Smith & Tyler (1985), and Ax (1987) have argued that the last two groups may in fact be the more primitive, in particular the nemertodermatids. Acoela are odd in having modified cilia (chapter 6), a lack of any intercellular matrix (Rieger 1985), and a peculiar form of 'duet' cleavage (fig 8.1) somewhat like that of gastrotrichs and nematodes (Joffe 1979). They might therefore be thought to be a derived group, a case of secondary reduction (see also Ax 1963, 1987), (though some authors see the duet cleavage pattern as ancestral to fully organised spirality: see Salvini-Plawen 1978; Inglis 1985). In common with most turbellarians, acoels also have unusual biflagellate spermatozoa (quite different from the classic uniflagellate condition that is usually taken to be ancestral - see chapter 6). The location of 'normal' monoflagellate sperm in the genus *Nemertoderma* (Tyler & Rieger 1975) therefore tends to support the view that the order Nemertodermatida is the most primitive extant group of flatworms. However other authors (e.g. Ax 1985,1987) have pointed to the low number of cilia in catenulid epidermal cells (and flame cells - see chapter 6) as a primitive feature, and therefore take the Catenulida as a starting point within the flatworms. Meanwhile, Hanson (1963, 1977) reiterates the traditional view by pointing out that Acoela are simple in virtually *all* aspects of their biology relative to other flatworms, and in *all* their various habitats; this would not be expected in a case of secondary reduction, and it may therefore be reasonable to retain the acoels as a genuinely simple ancestral group. It is actually rather difficult to see why the Acoela should have lost their proper gut secondarily, as they still need

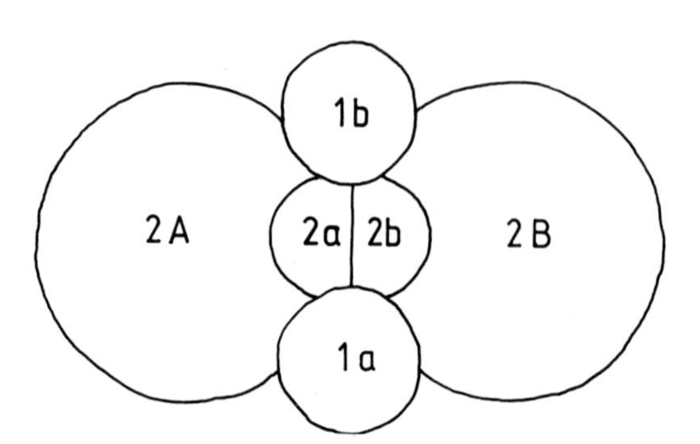

Fig 8.1. Duet cleavage in Acoela (Platyhelminthes).

and use the alternative syncytial arrangement. Perhaps the safest view would be to regard all three of the orders discussed here as derived in parallel from even simpler acoelous ancestors. Smith *et al.* (1985) consider that the platyhelminths as a whole may actually be polyphyletic, so that each of these groups could be primitive.

Whichever way this point is taken, the consensus amongst recent workers is clearly that the 'simpler' orders of flatworms are the more primitive, supporting the trend from small and unspecialised to larger and more highly organised turbellarians such as the polyclads, rather than taking the larger forms to be primitive and closer to a coelomate ancestor. The view that metazoan ancestors were coelomate, and the acoelomate platyhelminths a regressive state, thus seems not only counter-intuitive but still contrary to any evidence from within the platyhelminths (despite the efforts of Rieger and others to find such evidence, discussed in chapter 2). This chapter therefore persists in regarding the platyhelminth design as a primary acoelomate condition, the product of a direct planuloid ancestry; and as the starting point for most of the subsequent evolution of metazoans.

From the early turbellarians, the other groups of flatworms can fairly readily be derived (fig 8.2). Three main lineages are commonly identified; the nemertodermatid/acoel group, the catenulid group and the macrostomid-polyclad group. This last group includes the most familiar planarian worms, and its members often show the traditional protostome features of spiral quartet cleavage and 4d mesoderm particularly clearly. Some have the free-swimming 'Müller's larva', possibly related to a trochophore (see fig 5.10) but with a simple mouth opening that never forms the slit-like blastopore of a

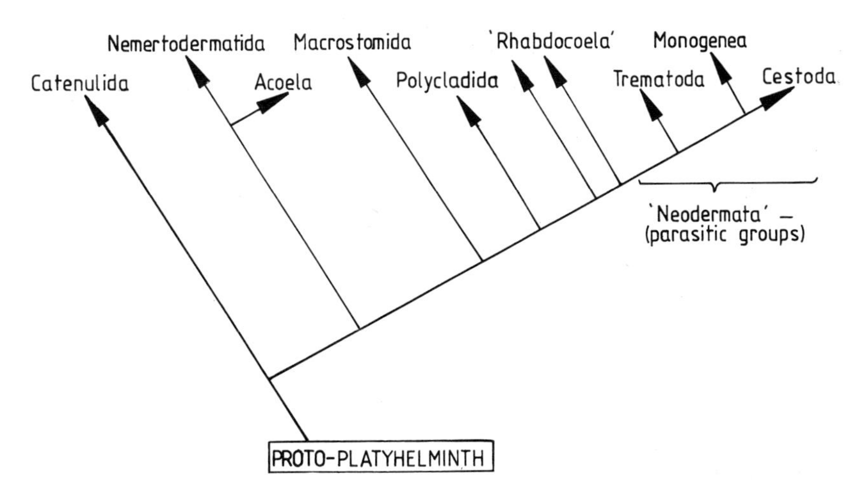

Fig 8.2. Phylogenetic relationships within the Platyhelminthes (based on Ehlers 1985*a*).

trochophore, and lacking the necessary features to be a true trochophore according to Salvini-Plawen's (1980*b*) analysis.

From within this assemblage, and probably from ancestors close to the group traditionally called rhabdocoels, the trematodes, monogeneans and cestodes evolved (Ehlers 1985*a*,*b*). These parasitic groups all have a non-ciliated and syncytial epidermis (present in only a few turbellarians), derived from stem cells or 'neoblasts' that progressively replace the original ciliated epidermal cells. Taking this to be a homology between the three parasitic groups, Ehlers unites them in the new taxon 'Neodermata', within which the monogeneans and cestodes are more closely related to each other than to the digenean trematodes. The three traditional classes of parasitic flatworms are thus nearly allied to each other as offshoots of just one branch of the whole flatworm assemblage, and it is probably unreasonable to give each of them class status alongside a Class Turbellaria embracing all the free-living forms. If Ehler's views are correct (and they do closely follow other recent phylogenies of the phylum) then the Turbellaria in any case becomes a paraphyletic group, technically unacceptable; and the term is gradually being dropped as a strict taxon, though it remains extremely useful in a descriptive sense.

Once established, a free-living platyhelminth stock could reasonably be taken as the starting point not only for parasitic platyhelminths, but also for other bilateral acoelomates such as nemerteans and probably for molluscs (see chapter 10), and for subsequent coelomate and pseudocoelomate radiations (Clark 1979). However, there are a few problems here, in that there are so many varying degrees of expression of the protostomian embryological traits within the platyhelminths that it is difficult to decide just where higher groups might have sprung from. Perhaps the other acoelomate groups may help in sorting out these anomalies.

Gnathostomulids

The Gnathostomulida were once regarded merely as specialised carnivorous flatworms, but are now accorded phyletic status in their own right as a sister group to the platyhelminths (Ax 1985). They are monociliate forms, also possessing collar cells (chapter 6), and could therefore be readily derived from a conventional planula of the sort found in modern cnidarians. Some of them have monoflagellate sperm (Sterrer *et al.* 1985), another probably primitive feature, and all have spiral quartet cleavage of the standard type. So in many respects they could be considered closer to the ancestral bilaterian form than the more familiar flatworms, as table 8.1 (modified from Ax 1985) reveals. There are only a few respects in which they might be seen as more advanced.

Table 8.1. *Primitive and advanced features in flatworms and gnathostomulids*

Primitive state	Gnathostomulida	Platyhelminthes
Monociliate with two centrioles	YES	NO Multiciliate, with one centriole
Protonephridium with single cilium and with microvilli	NO	YES Protonephridium with several cilia and no microvilli
Monoflagellate sperm	YES (Common)	NO (Only one known case)
Simple blind-ending gut	YES (but traces of an anal pore)	YES
Simple mouthpore	NO Complex pharynx with jaws	YES

One example is their possession of traces of an anal pore (Knauss 1979; Sterrer *et al.* 1985), though at least one rhabdocoel flatworm also shows this phenomenon (Karling 1966). Gnathostomulids also have complex jaws and pharyngeal apparatus, and perhaps more pronounced cross-striation of most of their muscles, though this latter point is of somewhat dubious value in the light of discussion in section 6.4.

If the gnathostomulids are taken as the most primitive of all extant bilaterian phyla, they serve to link several of the other lower phyla in a most convenient fashion. Their monociliate condition links them to the cnidarians, but also very clearly to the gastrotrichs which have both mono- and multiciliate cells and may be one of the most primitive of the 'pseudocoelomate' groups (see chapter 9). And since the catenulid flatworms are characterised by unusually low numbers of cilia per cell they may perhaps be primitive within their own phylum as discussed above, representing a transition from the gnathostomulid condition to the fully expressed multiciliation of all other platyhelminths. A set of relationships as shown in fig 8.3 seems possible on the basis of present knowledge; though the suggestion that the platyhelminths are themselves polyphyletic complicates this issue.

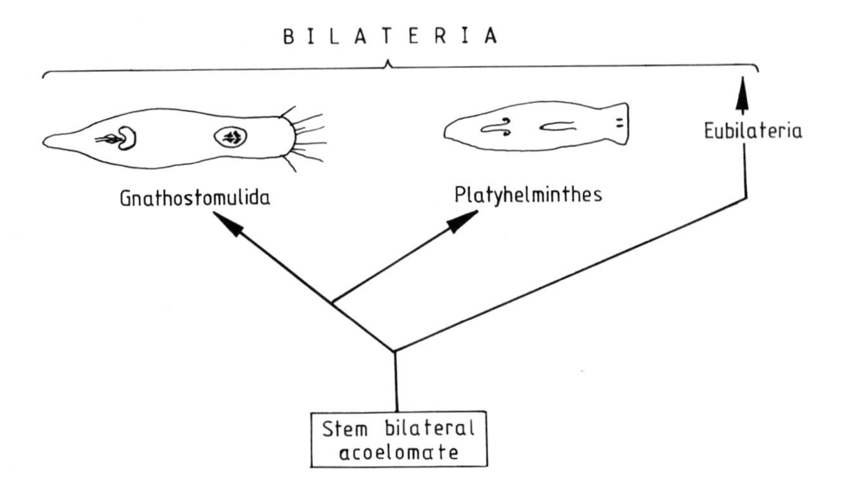

Fig 8.3. Relationships of platyhelminths and gnathostomulids to the rest of the bilateral metazoans, according to Ax (1985).

Nemerteans

The carnivorous 'ribbon worms' that constitute the Nemertea (or Rhynchocoela) have in the past generally been designated as acoelomates, taken as representing a slight advance over the flatworm condition but falling short of the status of worms with 'real' body cavities. The discussion of body cavities in chapter 2 should already have cast some doubt on this picture; and many recent authors have come round to seeing the nemerteans as a unique design of worm, having invented a special cavity to meet their own needs which nevertheless by any reasonable criteria should probably be included in the term 'coelom'. Since this cavity - the rhynchocoel - serves principally as a hydrostatic skeleton for the proboscis, rather than for the whole body which remains relatively solid, these worms hold the peculiar position of being acoelomates that nevertheless possess a coelom. This of course raises the possibility, particularly for many authors subscribing to archecoelomate views of metaozoan ancestry, that nemerteans are in fact regressed from a coelomate state rather than primitively acoelomate.

The anatomical situation is clarified in fig 8.4. The proboscis of nemerteans is quite separate from and dorsal to the gut (though there are secondary connections in a few cases); and it is everted by muscles operating on the walls of the rhynchocoel in which it lies. Separate retractor muscles withdraw it back to this cavity after use in feeding or defense (in some species it is armed with spines). The cavity in which it lies is intra-mesodermal, its walls being muscular, and it has a distinct lining; it is by all reasonable definitions a coelom, and arises as a schizocoel. In some cases the proboscis is short and the rhynchocoel is small in proportion, but in other worms it runs most of the length of the body and can be of considerable volume. In these latter cases it probably has some role in the locomotion of the whole worm when the proboscis is not in use, since it does then give a spacious enclosed fluid-filled system against which the body wall muscles can act (see Gibson 1972).

Even where the rhynchocoel is relatively small, the nemerteans commonly show an improved capacity for muscular locomotion compared to the acoelomate flatworms. Their body is usually somewhat larger and less flattened, and the central space is filled with less dense mesodermal tissue; cells are sparser, and there is considerably more fluid present. The extent of cellularisation of the middle layers varies with the size and locomotory mode of the species (Clark 1964), nemerteans that creep slowly with ciliary aid being rather solid but those which are capable of reasonable peristaltic movement having nearly cell-free areas filled with a thinly gelatinous fluid material. In this way, the nemerteans can be seen as part of a continuum (in a functional rather than phylogenetic sense), away from the restrictions of a solid body and towards the acquisition of a fluid-filled hydrostatic body cavity.

Apart from their proboscis and body cavity, nemerteans have acquired two other major features that set them apart from the flatworms: the gut has an anus, giving a through-flow system, and there is a circulatory system of blood vessels. These two features are probably linked, the blood vessels serving chiefly to distribute nutrients received from the increasingly specialised and localised parts of the straight gut. Hence the major longitudinal blood vessels are sited very close to the intestine. However, flow in the vessels is erratic in direction and force, and the system bears no clear resemblance to the blood vascular system of 'higher' worms; it can best be seen as an independent development. Recent analyses (Turbeville 1986; Turbeville & Ruppert 1985) suggest that it is in fact a specialised 'coelomic' system, the mesoderm splitting to give an array of circulatory tubules, lined with peritoneal cells,

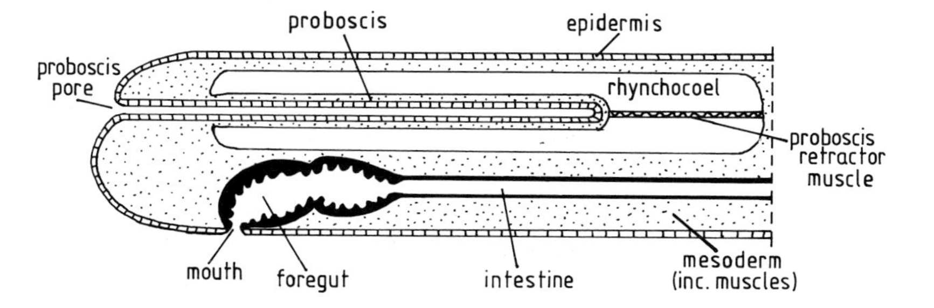

Fig 8.4. Transverse section of the front end of a nemertean, showing the position of the rhynchocoel (proboscis cavity) within the mesoderm.

meeting immediate functional needs. However, the authors concerned come to this conclusion partly because they wish to see nemerteans as regressive coelomates; it remains unclear whether the nemertean blood vessels are ontogenetically homologous with any kind of coelom, and it is probably unhelpful to use the term for such restricted cavities unless one accepts the premise that the acoelomate condition is secondary. In the absence of any evidence indicating such a regressive state in nemerteans, this premise may be rejected; and the blood vessels cannot then sensibly be termed coelomic, since that term has been reserved as a catch-all phrase for large secondary cavities (chapter 2). However, the nature of the nemertean blood vessels does raise the interesting possibility that vascular systems in the animal kingdom are more polyphyletic than traditional analyses might allow.

Nemerteans show a number of other features that give indications of their relationships, in most respects seeming to be rather similar to the flatworms, so that despite the differences discussed above close affinities are always accepted. The multiciliated microvillar epidermis is similar in the two phyla (Norenburg 1985; Turbeville & Ruppert 1985), with comparable gland cells present in both and with some nemerteans even having a form of rhabdite (chapter 6). The nervous systems are similar, with lateral and ventral nerve cords and a number of lesser longitudinal nerves; and the protonephridial systems also bear strong resemblances, though often associated with the blood vessels in nemerteans. Their muscle layers are similar, though more highly ordered in nemerteans, varying in arrangement between the classes and orders but always with distinct circular and longitudinal layers, and some dorso-ventral fibres.

In most aspects of adult morphology, then, nemerteans seem likely to be derived from platyhelminths and to be somewhat advanced in comparison. In their embryology the nemerteans again display clear links with the flatworms, having typical spiral cleavage and a classic gastrula. In some the blastopore forms the mouth, though in one class (Hoplonemertea) it closes and both mouth and anus form independently at some distance from it (Jensen 1963). Most nemerteans show direct development, but many heteronemerteans pass through a larval stage called the pilidium that is often interpreted as a modified trochophore. In fact it bears only limited resemblance to the normal annelid trochophore, or to the free-living larva of flatworms (see fig 5.10); (though there may be links between the frontal organs in the polyclad Müller's larva and the nemertean pilidium). The young nemertean worm actually metamorphoses in a complex fashion inside its larval epidermis, leaving the helmet-shaped pilidium form behind when it hatches as an adult; there is little likeness here to the classical protostome development of trochophores. Since nemerteans also have primitive spermatozoa relative to most of the flatworms

(Franzén 1967, and chapter 6), and have a very simple genital apparatus (Beklemishev 1963), many authors find it hard to see them as derivatives of anything like the extant platyhelminths.

So, whereas conventionally the nemerteans are often shown as a sidebranch of the line to annelids (fig 8.5*A*), there is no particular morphological or embryological evidence for this scheme, notwithstanding the occurrence of one partially 'segmented' nemertean (Berg 1985). The more acceptable view is that the nemerteans are not closely allied to a direct line from flatworms to the coelomate spiralians, but rather an early and specialised independent branch derived from some other group of flatworms. In fact the primitive spermatozoans of nemerteans indicate a separation from flatworms almost before the present platyhelminths were established at all, perhaps from a proto-flatworm rather like a nemertodermatid as discussed above. A position as shown in fig 8.5*B* therefore seems most likely; the nemerteans have no close relatives amongst extant flatworms, and no direct descendants.

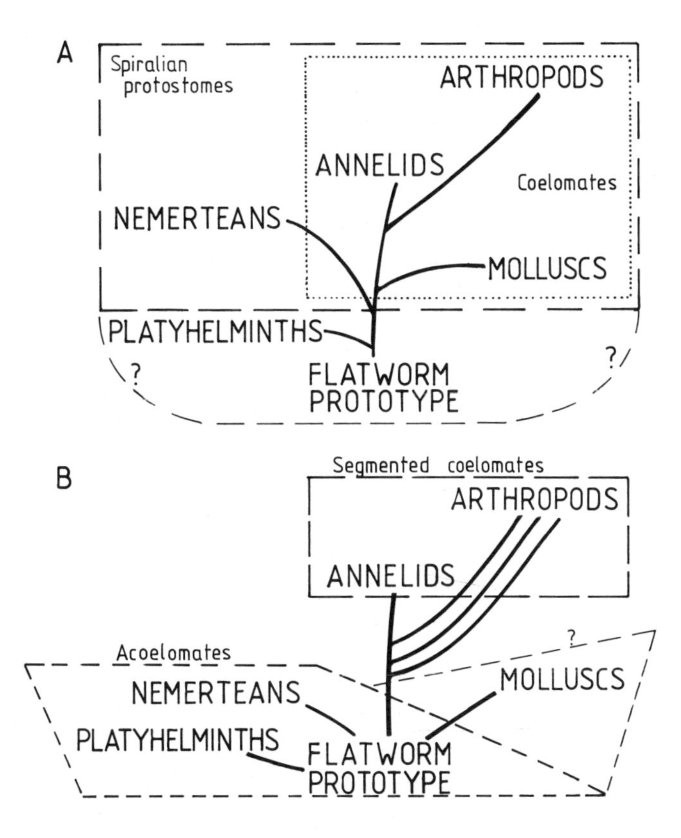

Fig 8.5. Relationships of spiralian groups, particularly with reference to the nemerteans: *A*, traditional position of nemerteans as an offshoot on a line from the acoelomate flatworms to the annelids; *B*, alternative status as a quite independent lineage from the early acoelomates.

The status of acoelomates; protostome and spiralian assemblages

Taking the acoelomates as a whole, their relationship to other worms usually designated as spiralian or protostome seems somewhat enigmatic. If we refer back to the summaries of protostome features given in chapter 5 (table 5.1), it is possible to assemble the evidence more easily.

(1) The three phyla considered above do all show spiral cleavage, mostly of a rather typical form though of a modified duet pattern in acoels and some other groups.

(2) They generally have a primary mouth, formed from the blastopore. In the flatworms and gnathostomulids this is hardly surprising since there is no anus, and the blastopore, if it persists at all, can have no other possible fate. And exceptions do occur: in hoplonemerteans the blastopore closes and forms neither mouth nor anus.

(3) Some of them, particularly the more advanced flatworms and groups above them in the usual hierarchy, show clear derivation of the mesoderm from the 4d cell; others have less defined modes of mesoderm formation, especially where duet cleavage occurs.

(4) They have no conventional coelom, so the schizocoelic/enterocoelic dichotomy cannot be applied. The nemertean rhynchocoel forms as a split in the mesoderm, but not at the early stage of a split in the 4d mesodermal bands; it cannot reasonably be taken as homologous with normal annelid-type schizocoely. The nemertean blood vessels are not likely to be homologous with any traditional coeloms either.

(5) Larvae of flatworms and nemerteans occur rarely, and in neither case bear particular resemblance to the protostome trochophore beyond that which is dictated by earlier embryological events.

In the light of these comparisons, it seems unhelpful to lump the acoelomates in a broad category 'Protostomia'; that term is better reserved for groups with distinctive and possibly homologous forms of coelom. The three phyla of acoelomate worms could however be allied with an assemblage to be termed the Spiralia, taken in a broader context, as they do show spiral cleavage. But since the earlier discussion of this phenomenon (chapter 5) suggested that it was an unhappy phylogenetic character, perhaps being 'negative' (arising inevitably as the mechanically obvious way to organise cleavage in embryos lacking surface coats to restrain the blastomere movements) and certainly being sporadic (occurring in many classically non-spiralian animals), we cannot put too much weight on it. However the acoelomate worms do usually also share a degree of determination of cell fate in the embryo, with matching cell destinies at least for the mesodermal blastomere, and this cell-by-cell parallel did survive as a useful character in the analyses given in chapter 5. Hence the scheme in fig 8.5*B* remains plausible, even if in effect it says little more than that the acoelomates are a primitive metazoan stock where each of the three extant phyla can best be derived from a preceding proto-platyhelminth form (or from several parallel such forms). The modern platyhelminths have left no obvious descendants, and we will have to go back to this proto-flatworm to find the ancestors of other spiralian worms.

8.3 Spiralian coelomate worms

Annelids

The annelids are the classic coelomates in the animal kingdom, with a highly organised and spacious secondary cavity, and with standard protostome/spiralian embryology. Their traditional position between the unsegmented acoelomates and the higher invertebrates (notably arthropods) makes them almost the reference point with which all other groups are compared in deciphering phylogenetic affinities, and in that sense one cannot doubt that they *are* spiralians, *are* protostomes, and *are* coelomate. The only problem here is therefore to decide quite where they came from, and which other groups if any share the same lineage.

Within the phylum, attempts to identify the 'proto-annelid' have never been at all satisfactory (Mettam 1985). There are in effect four possibilities, and each has been upheld at different times. Any one of the three familiar classes - Polychaeta, Oligochaeta or Hirudinea - could be closest to the ancestral form. Or the simpler meiobenthic forms that make up the group termed 'Archiannelida' could be important; it has traditionally been held that these represent an ancestral stock from which the polychaetes arose (Clark 1969), with oligochaetes and leeches as later developments. The taxon Archiannelida is quite clearly unsatisfactory, as some of its usual members are certainly related to the main classes of macro-annelids (Westheide 1985) and have gradually been removed from the assemblage and put in other classes. However, other 'hard core' members of the group do appear genuinely simple, showing only rudimentary annelidan features, and some even appear to be completely acoelomate (Rieger 1980). Many textbooks treat the taxon Archiannelida as a mixture of archaic and degenerate forms, and pass on to the established larger worms. Others (e.g. Clark 1969) reject them entirely as having no relevance to the problem of annelid origins. Yet other authors, such as Mettam (1985), still propose that some meiofaunal annelids could represent the ancestral form.

Taking this possibility first, it remains debatable whether the core members of the archiannelids (fig 8.6) are primitive or are secondarily reduced; opinions vary largely according to their owner's views on the archecoelomate issue. They do show a number of features indicative of a simple state (a lack of parapodia, reduced or absent chaetae, few and similar segments), and some that can be interpreted as genuinely primitive, for example the protonephridia, and the simple and highly microvillar cuticle (Rieger & Rieger 1976). Some Archiannelida (notably *Polygordius* and *Dinophilus*) possess particularly classic embryological patterns and trochophore larvae, whereas in larger polychaetes these may be highly modified. However, Westheide (1985), in reviewing current ideas, believes that there are no grounds for regarding even the core groups of archiannelids as closely related, let alone monophyletic. He regards the 'simple' characters of archiannelids as evidence for a neotenous state rather than for primitiveness; thus he links the various families of archiannelids to the larval stages of other polychaete families. (He does, however, point to a possile homologue in at least three of the archiannelid groups, in the presence of similar pharyngeal organs.)

Clark (1969) gives a clear analysis of the historical problems in holding any one particular archiannelid to be ancestral. He also doubts (1979) that the Archiannelida could be primitive on functional grounds, since coeloms are seen by him as an adaptation for hydrostatic burrowing and should not evolve in small animals where they would be unnecessary. But this recalls an argument in chapter 2; the body cavity can instead be seen as a means of

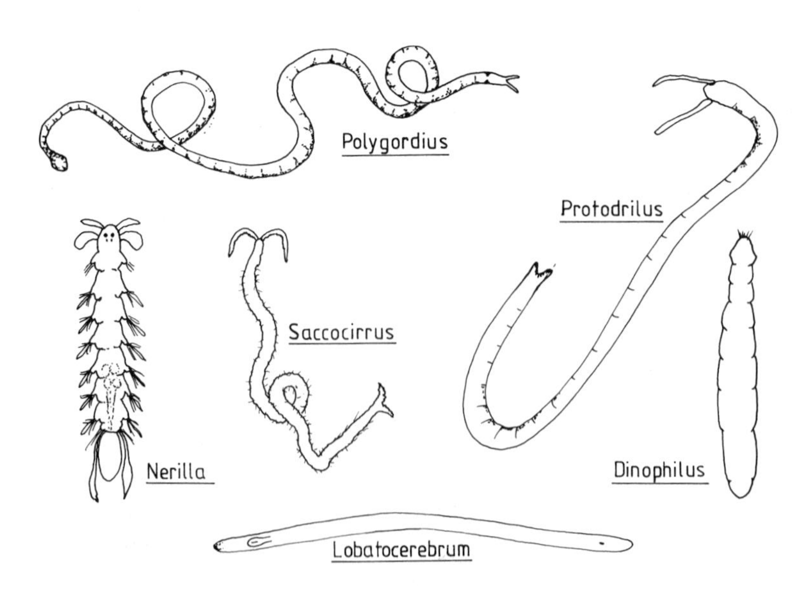

Fig 8.6. The groups commonly ascribed to the 'Archiannelida', discussed further in the text.

attaining larger size, and only thereafter does it become suited to Clark's mechanical function: it cannot have 'made its first appearance in large animals', as Clark demands, because the animal could not *be* large without already having a body cavity. In support of this, Fransen (1980) demonstrates a positive correlation between coelomic development and muscular activity even in these small annelids, so we cannot discount their primitive nature on the grounds that they are too small to 'need' a coelom and must have remnants of it only as a hangover from a larger ancestor. They could instead represent the evidence for a direct flatworm origin of the annelids, with progressive acquisition of a spacious cavity. Similarly the low numbers of segments and absence of septa cannot be dismissed as a secondary degenerative condition in these and other annelids. A non-septate condition offers distinct advantages in certain habitats, as discussed in chapter 2, and septation could surely have been acquired by degrees as the animal grew and became a faster and more continuous macro-burrower, as it is not one of those awkward features that must be present in its full unexpurgated form to be of any use. Hence some of these small meiofaunal creatures could indeed be close to the ancestral annelid form, each of the macrofaunal classes evolving from them in different ways as adaptations to different locomotory modes, some of which required increased septation and larger body cavities.

For those who regard the macro-annelid plan as primitive, there are only limited options. Polychaetes are in many ways the least derived form of annelid, from which the oligochaetes and leeches (collectively Clitellata) with their relatively specialised freshwater/terrestrial habitats could be descended; this is by far the commonest view in the older literature. Since arthropods are generally taken as reasonably close relatives of annelids, an ancestral form with incipient appendages seems attractive, despite the long-standing controversy over the homology of parapodia and arthropod limbs (Lauterbach 1978; Siewing 1978; and see chapter 11). Following such lines of thought, Bonik *et al.* (1976) start the annelid lineage with parapodial polychaetes, for in their view the coelom and segments were evolved primarily to give mechanical assistance with lateral undulations.

An ancestor already endowed with complex parapodia and the associated musculature seems to some zoologists less likely as a stem group than an oligochaete with simple peristalsis using circular and longitudinal muscle sheets, much more similar in pattern to the flatworm mechanical system. Clark (1964, 1969, 1979, 1981) regards the oligochaete form as primitive, since he invokes both the coelom and segmentation as mechanical adaptations to aid burrowing and requires simple muscle sheets in his ancestor. He has been followed by many other authors, and text-books therefore often tend to suggest an oligochaete-type prototype, with leeches as an early offshoot and polychaetes as a slightly later development. An example of this type of phylogeny is shown in fig 8.7 (note that the archiannelids are relegated to a status as part of the polychaetes). But it should be stressed that transitions between the two major annelidan forms in *either* direction run into functional and mechanical problems; and in particular the difficulties of deriving either leech or polychaete musculature (both of which are very different from all other worms) from the very simplified condition found in earthworms should not be underestimated.

The third possibility is raised by Pilato (1981), who quite reasonably regards the proto-annelid as a turbellarian-like form with flatworm musculature, but therefore takes the leeches (Hirudinea) as his ancestral model (having a solid body and marked dorso-ventral musculature). This leaves obvious problems in explaining the character of the coelom and the blood vascular system in the remaining annelids, and most authors persist in regarding leeches as the most specialised of annelids, derived from oligochaetes with whom they share a lack of parapodia and head appendages, and similar hermaphroditic reproductive systems.

On balance, some variation on the first of our four possibilities looks the most likely. Indeed many authors who adopt a macrofaunal ancestor would agree that annelids originated from a small acoelomate form, but would obviously be wary of regarding particular groups of extant 'Archiannelida' as actual ancestral forms. Annelids probably began from amongst small

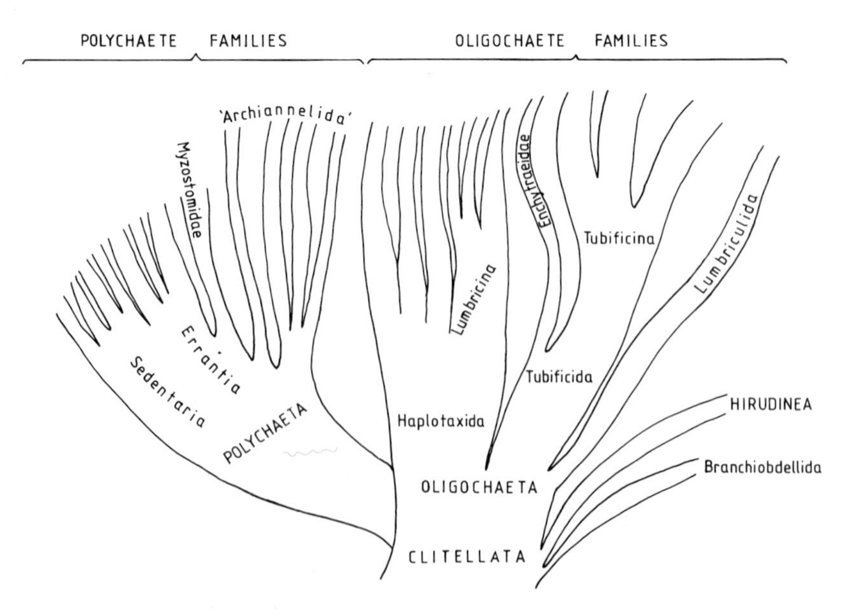

Fig 8.7. Possible relationships within the phylum Annelida, taking the clitellate design (complete septation and musculature, few chaetae, no parapodia) as primitive, and polychaetes as an early offshoot.

acoelomate forms that were surface creepers or meiobenthic in habit (the proto-platyhelminths that were discussed above); and evolved into larger more actively burrowing coelomate forms by progressive acquisition of a body cavity (fig 8.8). The group Lobatocerebridae (fig 8.6), recently described by Rieger (1980) and tentatively assigned by him to the Oligochaeta, may represent a stage in this transition, being effectively acoelomate and unsegmented but with a complex epidermis and cuticle, paired ganglionic nerve cords, and an anus and hindgut system. (Rieger of course argues that this animal is evidence for acoelomates being derived from annelid-like coelomates, supporting his archecoelomate theory; but it could equally represent the 'archiannelid' transition between acoelomates and more advanced coelomate annelids.) From amongst these early flatworm-derived coelomates, some group acquired segmentation for mechanical reasons, and chaetae probably for protective/mechanical reasons, and thus became truly annelidan. Some of these remained as meiofaunal animals, including some (but certainly not all) of the assemblage 'Archiannelida'. The three groups of extant macro-annelids diversified independently (fig 8.8) by virtue of different locomotory modes exploiting this early segmentation. These changes were clearly established very early on, since segmented worms are abundant in the Cambrian deposits and may be evident even in Pre-Cambrian strata (see chapter 3).

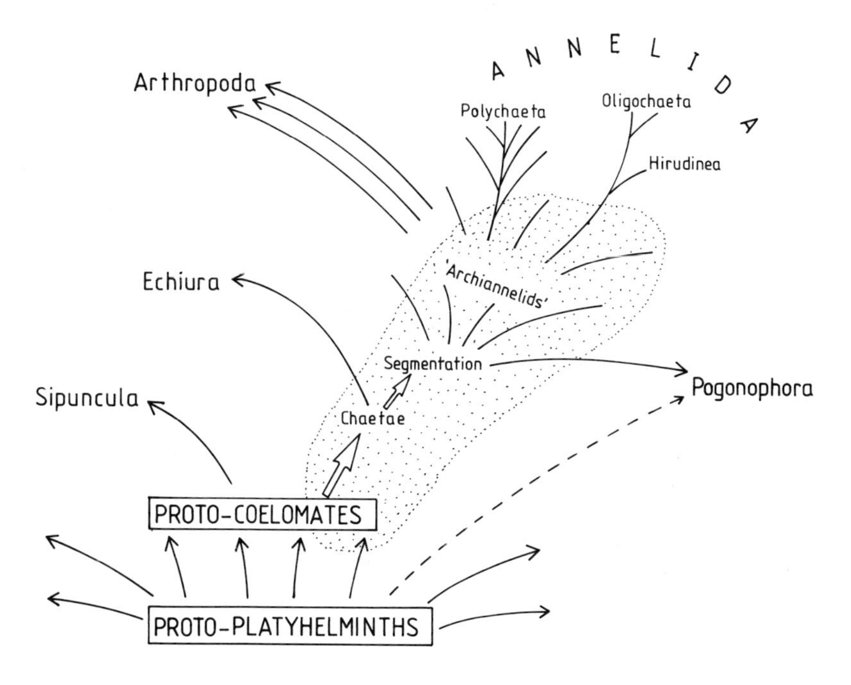

Fig 8.8. An overview of the phylogeny of annelids, deriving them from acoelomates by progressive acquisition of a body cavity, chaetae and segmentation, and with separate radiation of the three major classes. All animals within the shaded area could be included within the amorphous group 'Archiannelida' (including the effectively acoelomate family Lobatocerebridae).

Annelids, then, can be taken as the descendants of an early proto-coelomate worm that became segmented. How far can such an ancestor account for the other phyla of worms that have coeloms and a spiralian/protostome pattern of development?

Echiurans

One phylum that can be readily assimilated into the scheme outlined so far in this chapter is that of the echiuran worms. For once there is little controversy about their status; the only point of contention recently has been whether they deserve separate phyletic status or should be contained within the Annelida as Beklemishev (1969) and others have advocated. These worms differ visibly from annelids in a lack of adult segmentation and the possession of a large feeding proboscis rolled up to form a gutter. But in all other respects they are clearly closely allied with annelids. They have a large coelom that is continuous throughout the body, surrounded by a typically annelidan body wall. The cuticle is simple, with cilia present on the undersurface of the proboscis and with chitinous setae at the front of the trunk and sometimes in a posterior circlet as well. A closed blood vascular system is normally present, and there are one or many pairs of metanephridia.

It is in their development, though, that echiurans most clearly display annelid affinities. They have typical spiral quartet cleavage, and a free-swimming larva that is clearly allied to a trochophore (see fig 5.10). Even though the term trochophore has been much abused, as was discussed in chapter 5, there is little doubt that an echiuran larva should merit the term; indeed it is the *only* larva apart from that of annelids which Salvini-Plawen (1973, 1980*b*) includes within the designation 'trochophore'.

For at least some species of echiuran, the link with annelids seems to become incontrovertibly clear as development proceeds, for the larval 4d mesodermal bands become enlarged and show regular transverse thickenings. These bands produce two coelomic cavities which show partial segmentation. The echiurans could therefore be seen as segmented coelomate protostomes, differing from annelids only in the loss of the segmentation in adult forms. An origin amongst the earliest 'proto-annelids' would then seem likely, with segmentation presumably being lost as the worms became increasingly sedentary deposit feeders. However, Clark (1969) points out that the suggestions of segmentation are very limited and do not persist for long, and that they are in any case quite absent in some genera of echiurans. The acceptance of echiurans as embryologically segmented animals rests on limited observations from the nineteenth century, and more recent analyses have described the mesodermal thickenings and coelomic 'compartments' as indistinct and irregular, and having no correspondence with the developing nerve ganglia. The resemblances between annelids and echiurans may thus be rather fewer than some earlier authors have accepted, and the echiurans should not be placed within the same phylum as segmented annelids. But the two key features of development to identical trochophores and the presence of identical chaetae (together with a number of other similarities listed by Clark 1969) should still be enough to keep the two phyla very closely allied in any phylogenetic scheme.

Pogonophorans

The deep-water pogonophoran worms have been known for a relatively short period relative to most other phyla, but within that time have changed their status more radically than almost any other group of animals. In most texts dated before 1964 they were treated as oligomerous deuterostomes, having a classically tripartite enterocoelic coelom and a certain resemblance to hemichordates. Since then, when the back end of a worm was dredged up from deep sea muds (Webb 1964), they have been transformed into classic protostomes, and some authors even include them as a sub-phylum of the annelids (e.g. van der Land & Nørrevang 1975). That zoologists can change their view of a phylum so radically is perhaps in itself something of an indictment of the super-phyletic views that have held sway for so long.

Pogonophorans have a remarkably long and thin body, often encased in a chitinous tube, with an anterior tentaculate cephalic zone, a collar or forepart, and a long trunk, giving the tripartite effect. However at the far end of the trunk we now know they also have a short 'opisthosoma', a bulbous swelling that normally lies buried and is perfectly segmented and septate in the annelid fashion. When examined without this part they do indeed seem to be deuterostomes, if the criterion of tripartite organisation is given weight; but when seen in their entirety they become metameric. Clearly we need to know more of their internal organisation and their embryology to make real decisions about their status.

Unfortunately internal anatomy is rather unhelpful, partly because it is very difficult to know which way up a pogonophoran goes! They have no gut, and the blood vessels and nervous system could be interpreted either way; when the group were deuterostomes, the nerve cord was dorsal and the 'dorsal' blood vessel ran backwards (see chapter 12), but once they became protostomes the nerve cord defined the ventral side (Nørrevang 1970) and blood-flow directions therefore fitted with the classic annelid pattern, running forwards dorsally. The 'coelom' is spacious, but appears to lack a peritoneal lining (Gupta & Little 1975). The opisthosoma is septate and very like annelid organisation, though the segments lack repeated excretory or reproductive organs. A few aspects of the anatomy do give some clues though; there are setae on the trunk and in particular on the opisthosoma that closely resemble those of annelids (George & Southward 1973); and the cuticle is also annelidan in character (Gupta & Little 1975).

Embryologically, pogonophorans remain somewhat enigmatic. They often brood their eggs, which undergo bilateral cleavage, formerly said to be not obviously either radial or spiral, but recently interpreted (Bakke 1980) as roughly spiral. There is plenty of yolk, so no blastocoel appears and gastrulation is by epiboly. Hence (see chapter 5) there is briefly a posterior 'blastopore', but this forms neither mouth nor anus as there is no gut. Some authors maintain that the coelom forms by enterocoely (Ivanov 1975), though this is hard to envisage when there is no enteron; others claim to have observed schizocoely (Southward 1971; Nørrevang 1970). There is a brief and unique larval stage. So there seems to be little here that can really help in deciding relationship, though many embryologists do now concede a spiralian affinity (e.g. Anderson 1982).

For most recent authors, the segmented setiferous rear part of the body is enough to settle the matter, and the pogonophorans have taken up a position next door to, or even within, the phylum Annelida. However, opinion is perhaps swaying a little against this view once again; Siewing (1975) and Cutler (1975) both argue that pogonophorans should more properly be sited as an intermediate between the two super-phyla (see chapter 12), and the latter author even believes the group to be something like the antecedents of both main branches (the back end dominates to give protostomes, the front end to give deuterostomes). Jones (1985) argues that pogonophorans are not particularly close to annelids in respect of coelomic organisation; he doubts the coelomic status of certain of the cavities, and points out that others lack a medial mesentery which is inconsistent with a schizocoelic origin (see fig 2.6). Jones believes the pogonophorans should be split into two major sub-phyla anyway, the Perviata and the Obturata (commonly called vestimentiferans), the former being markedly more distant from annelids.

However, links between annelids and pogonophorans are moderately convincing, when viewed in functional terms. Both segmentation (chapter 2) and setae (chapter 6) undoubtedly *could* be independently acquired features, but in this case there are no good functional grounds for proposing such an effect. The opisthosoma of a pogonophoran is very small, and seems unlikely to have an especially important role in burrowing or anchorage that could explain the need for a segmented condition; it is therefore more likely to represent a vestige retained from a more fully segmented ancestor, and pogonophorans can be seen as an offshoot of some proto-annelid stock. Their deep-sea habitat (where planktotrophy is inappropriate and yolky eggs result), and their feeding mechanism with associated loss of a gut, would be enough to cause most of the anomalies in development that now obscure their presumed protostome affinities. A position not too distant from the main annelid/echiuran lineage therefore seems reasonable. But whatever status they are accorded, the pogonophorans must leave us with increased suspicions of the characters on which the protostome/deuterostome dichotomy is based.

Sipunculans

The 'peanut worms' are possibly the most clearcut case in the animal kingdom of unsegmented coelomates, and as such they seem to deal a body blow to archecoelomate theories suggesting that the coelom and segmentation arose in concert (see chapter 2). Sipunculans are not segmented as adults, and show no trace of segmentation in their embryology. In the past they have been allied to holothurians, phoronids, and echiurans with about equal frequency (see Hyman 1959), but have now apparently acquired a settled place close to annelids and echiurans (Clark 1969). However, as indisputably unsegmented forms they are perhaps likely to be less closely related to annelids than are the echiurans, and the extent of their similarity to these 'core' protostome groups remains controversial.

The sipunculan body is in two parts, a posterior trunk bearing an anterior introvert with tentacles around the mouth. The body wall may have complete sheets of longitudinal and circular muscle, and bears bristles; the coelom is spacious, and there are two classic metanephridia. However in some ways the sipunculans are unlike annelids; the gut is almost U-shaped, the anus lying in an anterior dorsal position on the trunk, and the nervous system though ventral is normally unpaired. Bristles are indeed present, but they are not homologous in structure or origin with those of annelids (Clark 1969; Rice 1975; and see chapter 6), and they do not contain chitin, a substance seemingly absent in the phylum as a whole (chapter 4).

Sipunculans appear to reveal clearer protostome affinities in their development (Rice 1985). They have spiral cleavage, but with a radial arrangement of the apical cross cells more like that of molluscs (see chapter 10, and fig 10.4) than like that of annelids. The micromeres are often larger than the macromeres, resulting in large and yolk-laden prototroch cells, often similar in size and pattern to those of some polychaetes. The mouth is formed from the blastopore (the anus opening separately at a rather late stage and in an unusually dorsal position), and mesodermal bands formed from the 4d cell later give rise to the coelom by splitting. The larva varies somewhat between genera, and usually bears some resemblance to a trochophore (fig 5.9).

However Salvini-Plawen (1973, 1980*b*) does not regard it as a true trochophoral type: the prototroch is often very extensive and can surround the mouth, and there are never protonephridia in the larval stage. In addition the whole structure is contained within an egg envelope, and never feeds. Many sipunculans have a more complex 'pelagosphaera' planktonic feeding larval stage instead (Rice 1985).

Thus the sipunculans have many features indicative of a protostome affinity but also show anomalies. They raise again the problem of evaluating embryological evidence; do we take their developmental patterns as confirmation of a fundamental unity in the 'Protostomia' due to common descent, or merely as further evidence for convergence due to limited practical possibilities and functional constraints? On balance the similarities of cleavage and cell fate are probably enough to link sipunculans with annelids and echiurans, and some similarities with molluscs suggest that all these phyla have common ancestors; indeed some of the schemes outlined in chapter 1 put molluscs and sipunculans rather close together as monomeric protostomes, sharing the 'molluscan cross' formation (see Inglis 1985). But since the acoelomates also share most of the developmental features at issue here, the common ancestor of all these groups is more probably right back at a turbellarian-like stage. Possibly, from amongst the 'proto-platyhelminths' that we have already invoked, one group that lost a typically protostome ability to synthesise chitin gave rise to the sipunculans. Evidence for any closer relationship of peanut worms with the other spiralian coelomate worms, or with molluscs, is insubstantial.

Origin of spiralian coelomates

In modern schemes, coelomates can either be derived from the acoelomates already discussed, in which case they must acquire a coelom and also in some cases segmentation; or they can be seen as direct descendants of the 'archecoelomate', which in many schemes is inherently segmented by virtue of the mode of coelom formation (see chapter 2). We have already criticised the latter hypothesis on functional and embryological grounds (chapters 2 and 5 respectively), and on the basis of evidence from acoelomates discussed in section 8.2. Since some classic protostomes are not segmented - notably the sipunculans, but also perhaps the molluscs discussed in chapter 10 - we are now also able to criticise the archecoelomate concept as inconsistent with extant spiralian coelomates. And analysis of the various coelomates considered here, whose cavities have different spatial organisations, different embryological patterns, and different linings, amply underlines the problems in considering all the 'coelomic' cavities to be homologous.

There are, then, altogether too many problems with deriving these coelomate groups from an archecoelomate ancestor. Annelids and the other worms must instead be seen as derived from an acoelomate ancestor. Annelids and echiurans probably share a common ancestry above the acoelomate level, in some proto-coelomate of moderate size that exhibited chitinous chaetae and perhaps segmentation; and pogonophorans *may* be a side-branch from the same stock, subsequently much modified in response to their peculiar habitat. However, sipunculans show only remote links to these groups, having no trace of segmentation, and are more readily placed as a separate development from the acoelomate groups. Similar reservations apply to the molluscs, as will be seen in due course.

8.4 Conclusions

A number of groups of invertebrates share features with the platyhelminths, particularly in embryological terms, that suggest phylogenetic links; and the

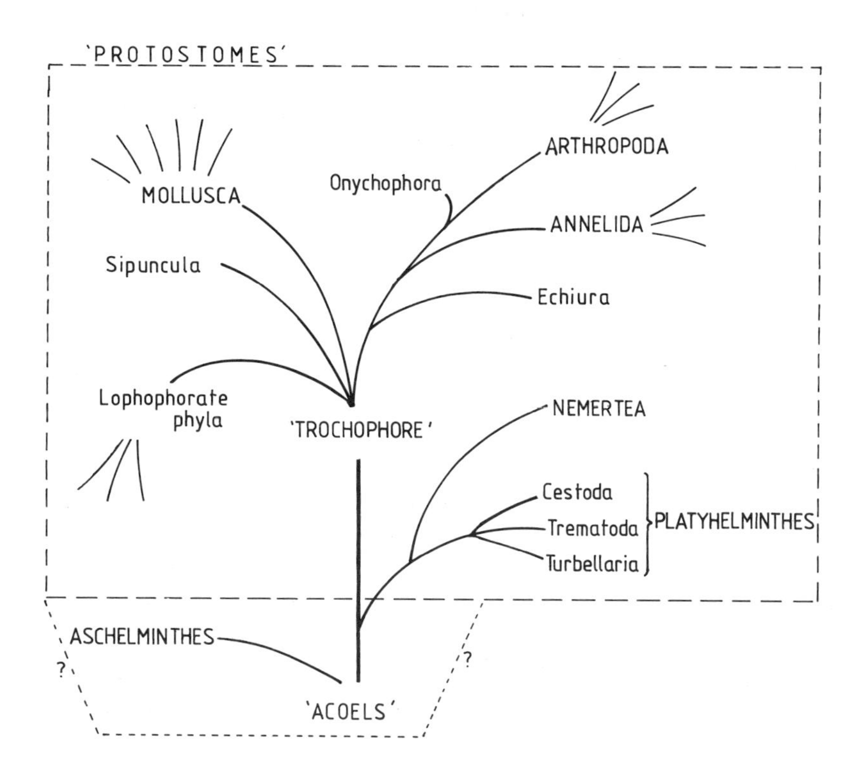

Fig 8.9. A traditional view of spiralian relationships, based on Hyman's works. All phyla with spiral cleavage are part of a single branch from the acoels, termed Protostomia or Spiralia.

whole assemblage is often united as the Protostomia or Spiralia, as shown in the traditional phylogeny of fig 8.9. However closer analysis reveals many anomalies and dissimilarities in embryological and adult patterns, and in many cases it is easier to explain similarities as convergence than to account for seemingly functionless differences. That is to say, given the developmental and functional constraints on animal morphology, when there are only a few possible mechanisms for achieving a particular effect (cleaving an embryo, or making a body cavity) quite unrelated groups can end up showing that mechanism.

This view suggests that many of the protostome groups should actually be derived independently from one or several early acoelomate 'proto-platyhelminths'. It renders the whole concept of the super-phylum suspect, because the common ancestral form is then no more than a slight advance on the planula/acoel that is probably the common ancestor of *all* the Metazoa. The idea of a broad protostome assemblage embracing all the worms discussed here therefore seems unhelpful; a set of relationships as shown in fig 8.10 (for the same set of phyla as in fig 8.9) may be more realistic. Do the traditional terms protostome and spiralian then retain any use?

Protostomia. Some authors have used this term only for coelomate phyla (traditionally annelids, arthropods and molluscs, plus all the lesser coelomate worms dealt with here). Others take it more broadly and attempt to fit nearly all animal phyla into either this or the Deuterostomia categories (see chapter 1, fig

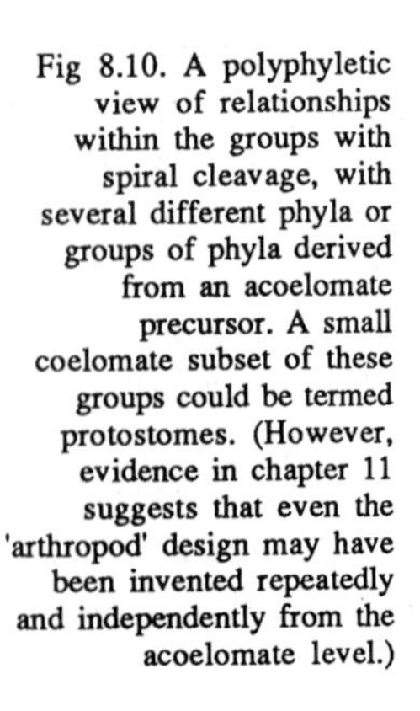
Fig 8.10. A polyphyletic view of relationships within the groups with spiral cleavage, with several different phyla or groups of phyla derived from an acoelomate precursor. A small coelomate subset of these groups could be termed protostomes. (However, evidence in chapter 11 suggests that even the 'arthropod' design may have been invented repeatedly and independently from the acoelomate level.)

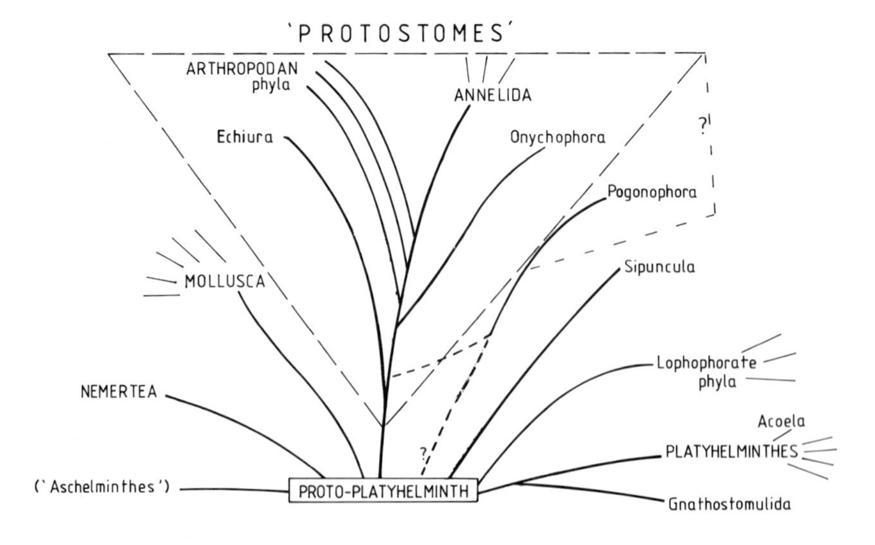

1.1). From the analysis here, it seems that even in the narrower of those senses the term and the concept it embodies are fairly meaningless. However where the similarities between groups extend to fine details of embryology and structure (for example the segments and chaetae found in annelids, in pogonophoran opisthosomas, and perhaps in echiurans), and cannot be accounted for on functional grounds, they should surely still be taken to indicate a common lineage. The two phyla Annelida and Echiura may therefore quite reasonably be taken as related, and the Pogonopohora may belong here too; and since some of the groups of arthropods (notably uniramians) are almost certainly descended from the same line (chapter 11) this whole array of phyla *could* be given the designation Protostomia. But it is not at all clear that this term then remains useful; partly because in the past it has been used in a much broader sense, and partly because within it the numerically dominant uniramian arthropods actually show very few of the classic protostomian features in their development.

Spiralia. This term has been used in a broader sense than the Protostomia, usually including all the acoelomates and perhaps pseudocoelomates that show spiral cleavage (Anderson 1982). As such it does concentrate the mind on a particularly clearcut similarity between an array of groups that are highly diverse as adults. The problems entailed in the embryological characters on which it rests have already been discussed, and there is clearly no certainty that spiral cleavage in itself was invented only once. However, the precise patterns of cell determination in many of the included groups are so similar (see Siewing 1969, 1980), and functional reasons for this so non-apparent, that shared ancestry seems rather likely.

This still says no more than that many animal phyla were descendants of an early metazoan in which these patterns were already established. As this early metazoan 'proto-platyhelminth' is very close to the postulated archemetazoan in design, some of the archemetazoans may have acquired spiral determinate cleavage very early on and given rise to what we now call the Spiralia. In that case the term would be valid and useful, though other groups could have acquired less precisely patterned spirality independently, to confuse the issue. Alternatively, the archemetazoan itself may have had some or all of this developmental pattern originally, passing it on to all its descendants, other phyla merely losing it as egg membranes were developed, or as cleavage proceeded faster and determination processes of the egg slowed; there is an increasing appreciation of the possibility that deuterostomy could have developed from protostomy (e.g. Salvini-Plawen 1980*a*; and earlier chapters). In that case many groups may have lost their highly organised spiralian features independently; so phyla that are now included in the designation

Spiralia could well be at least as closely related to classic deuterostomes as to each other. We have no way of deciding between these possibilities, so the term should only be retained with reservations, accepting that we do not know how far it tells us of shared ancestry.

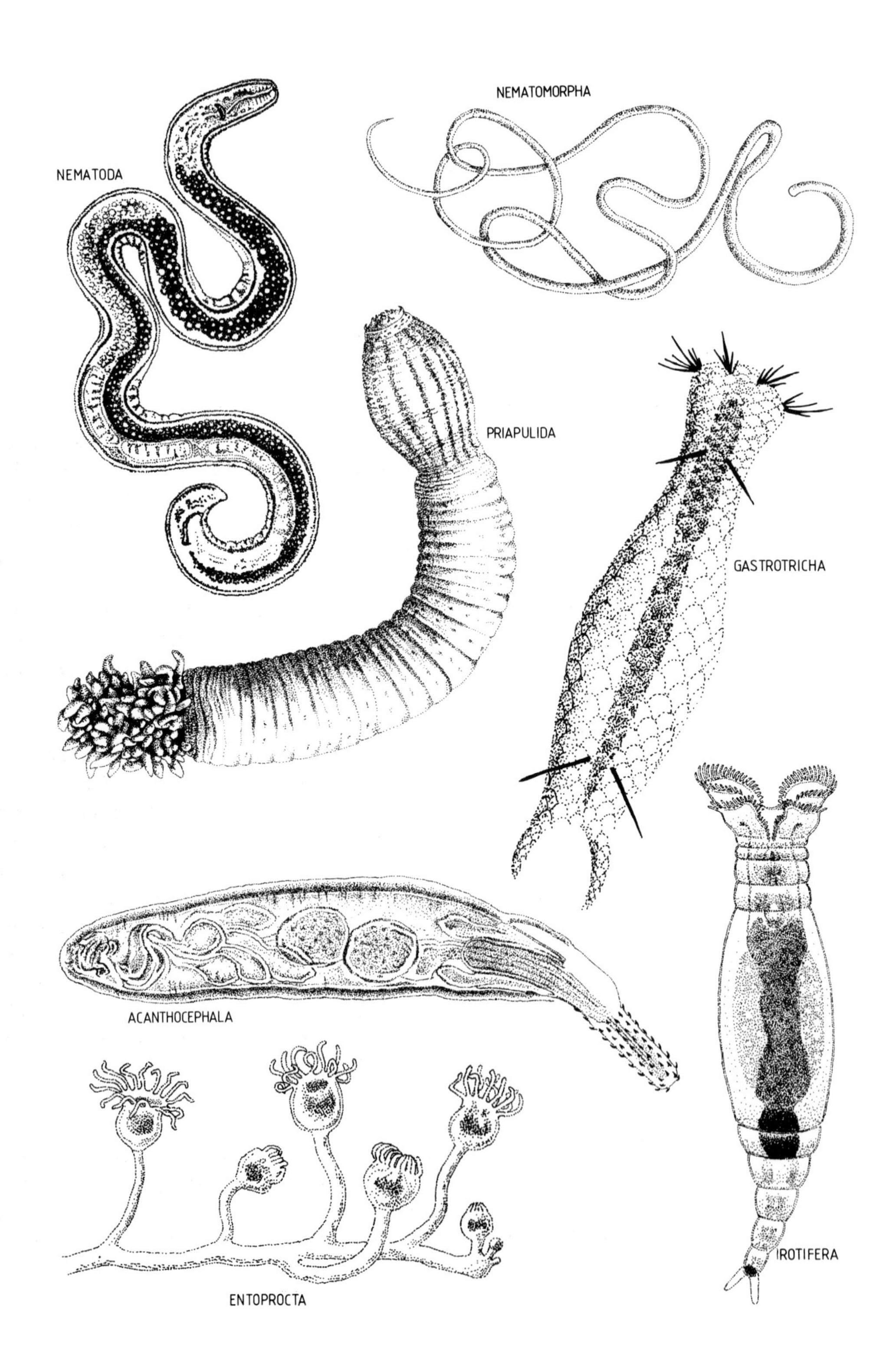
NEMATOMORPHA
NEMATODA
PRIAPULIDA
GASTROTRICHA
ACANTHOCEPHALA
ROTIFERA
ENTOPROCTA

9 The pseudocoelomates

9.1 Introduction

The pseudocoelomates have always been the most trying group of invertebrates as far as phylogeneticists are concerned, and are certainly destined to retain this status for the foreseeable future. All the animals concerned are small (in diameter at least, though some are very elongate), and all are more or less soft-bodied, so there is little possibility of obtaining evidence from the fossil record. About eight distinct groups of animals can be included in this assemblage, and at some time or other almost every possible permutation of affinities between them has been proposed. There is not even a consensus about their rank - do they collectively constitute one phylum, should some of the animals be set apart as one phylum and the rest placed elsewhere, or is each group of phyletic status in its own right? Are any of them related to each other at all, in fact?

Classically, a sub-set of these phyla have often been united as the phylum Nemathelminthes, or more recently Aschelminthes. Hyman (1951*b*) included six groups in her aschelminth assemblage: the gastrotrichs, rotifers, kinorhynchs, nematodes, nematomorphs, and priapulids. She set the acanthocephalans and entoprocts apart as separate groups. Other authors and many subsequent text-books have followed similar schemes, though the priapulids have sometimes been taken out since there has been controversy about the nature of their body cavity. However, there have also been recent attempts to ally the priapulids with the acanthocephalans (Conway Morris & Crompton 1982), or the rotifers with the acanthocephalans (Lorenzen 1985), the latter author reinstating priapulids with a kinorhynch/gastrotrich/nematode assemblage. Meanwhile Nielsen (1971, 1977*a,b*) has reiterated an older view that entoprocts are closely allied with ectoproct bryozoans; and Rieger & Mainitz (1977) argue for a link between gastrotrichs and the acoelomate gnathostomulids, since at best the gastrotrichs have a pseudocoelom that is all but obliterated. Inglis (1985) believes that all the groups are independently evolved from a flatworm stage and have no links to each other. Even this brief summary indicates the enormous scope for phylogenetic speculation offered by these curious animals.

9.2 Pseudocoelomate characters

Introductory text-books of zoology generally give a set of features some or all of which are displayed by the pseudocoelomate animals. These include small size, vermiform shape with mouth and anus, limited cephalisation often with elements of radial symmetry (see fig 2.1*C*), reduced surface ciliation, and a pronounced cuticle that often shows a degree of superficial segmentation. Internally there may be a lack of circular musculature, simple tubular gut with a specialised triradiate pharynx, simple protonephridia but no respiratory or circulatory adaptation, a constancy of cell number (eutely), and a pseudocoelomic cavity with mesoderm located only to the outside of the cavity and organs lying freely within it. The animals generally have determinate cleavage, sometimes with elements of spirality, and a dioecious habit.

A simplified review of the occurrence of many of these features is shown in table 9.1, which amply serves to underline the variety of possible relationships that could be proposed according to which features are chosen as important. In this section each major feature is dealt with in turn, considering both its phylogenetic value and its applicability to the animals in question.

The pseudocoelom

The nature and origin of the body cavity known as a pseudocoelom were discussed at some length in chapter 2. Essentially it can be considered as a persistent blastocoel, occupying the space between ectoderm plus mesoderm on its outside, and endoderm on its inside, so that the gut has no musculature of its own and organs lie free in the cavity. In these ways a pseudocoelom differs from the coelomic cavities, which occur within the mesoderm and are entirely bounded by it, and which have their own complete cellular lining forming a peritoneum that also gives rise to mesenteries holding the organs in place.

In practice, a pseudocoelom is very hard to define precisely, and much of the argument covered in chapter 2 centred on two major issues. Firstly, there are problems in deciding if a cavity is continuous in space or time with the blastocoel, and in many pseudocoelomates it is not. The developmental stages are commonly very small and never develop an obvious blastocoel; this is particularly clear in Acanthocephala, where the pseudocoelomic cavity appears quite late, as a split in the tissues of a previously solid body (so could technically be called a schizocoel). Therefore it is often necessary to rely on adult features to decide the nature of a cavity. This leads to the second issue, the question of whether or not a cavity has a specific organised cellular lining; if not, it is usually called a pseudocoelom. Detecting such linings depends on

Table 9.1 *Pseudocoelomate characteristics*

	Pseudo-coelom	Eutely	No cilia	No circular muscle	Cuticle	Proto-nephridia	Tube gut + pharynx	Dioecious	Introvert
Aschelminthes									
Gastrotrichs	no?	no?	NO	NO	+	+	+	NO	NO
Rotifers	+	+	NO	NO	+int	+	+	+	NO
Kinorhynchs	+	no?	+	part	+	+	+	+	+
Nematodes	+	+	+	+	+	NO	+	+	no?
Nematomorphs	+	+	+	+	+	NO	NO	+	+
Other pseudocoelomates									
Acanthocephalans	+	part	+	NO	+int	+	NO	+	no?
Entoprocts	+	NO	NO	+	+	+	no pharynx	+	NO
Priapulids	no??	NO	+	NO	+	+	+	+	+
Possible relatives									
Gnathostomulids	NO	no?	NO	NO	??	no?	often no anus, no pharynx	NO	NO

the interpretation of light and electron micrographs, and leaves some room for disagreement, so that, as was seen in table 2.3, there is still debate about the status of several major groups.

Hence within the groups generally accepted as 'pseudocoelomates', a number of different conditions prevail (see Maggenti 1976). Several have much reduced or almost completely obliterated body cavities concomitant with reduced size. Gastrotrichs in particular have a tissue-filled body, often divided longitudinally into three compartments none of which appears to have a lining (Rieger *et al.* 1974); and a rather similar state is characteristic of nematomorphs. In both groups, the lateral compartments contain gonads, so by some people's definitions are designated as coeloms (Remane 1963*a*,*c*; Teuchert 1977; Siewing 1976). Other groups of pseudocoelomates have cavities that are apparently largely cell vacuoles, as in the nematodes where a few large 'pseudocoelocytes' in fixed positions occupy a considerable volume (Hyman 1951*b*); and there are also numerous amoeboid cells present in the cavities of entoprocts, rotifers and kinorhynchs. Yet others have apparently lined cavities, that may more properly be called coeloms; but the lining of rotifer body cavities though substantial is anucleate, so the term pseudocoelom is retained by Remane (1963*c*), whereas the priapulids have changed their position repeatedly as new analyses of their cavity lining have been published (see section 9.3). Even the nematodes show something very like a proper mesothelial cavity lining in the way their muscle cells are arranged (Remane 1963*c*; Clark 1979), in many ways more convincing as a peritoneum than that of some classic coelomates!

It has already been suggested that body cavities are extremely polyphyletic in origin. The pseudocoelomates in fact represent a continuum with the acoelomates in many ways (Salvini-Plawen 1980*a*), showing varying degrees of retention or re-establishment of the primary body cavity according to their size and functional needs. It seems likely that these phyla represent an unholy alliance of many different ancestries, as far as this character goes.

Eutely

A numerical constancy of cells composing particular organs, or in some cases the whole animal, is a peculiar feature of many pseudocoelomates. Mitosis appears to cease following embryonic development, and further growth is by cell enlargement. The numbers of cells making up the tissues are then characteristic for each species, and may be very low - only a few hundred or few thousand in total. Nematodes, whose embryology is more thoroughly studied than that of any other phylum, are particularly well known in this regard (in *Caenorhabditis*, for example, there are only about 600 cells, of which around half form the nervous system). Eutely also seems to be characteristic of some other phyla within the old group Aschelminthes (rotifers and nematomorphs), though perhaps absent in gastrotrichs and kinorhynchs (Lang 1963). Partial eutely, of the nervous system and epidermis, is found in acanthocephalans.

Eutely is not restricted to pseudocoelomates, though, and may be a general phenomenon associated with very small size. Tardigrades for example have been said to show cell number constancy, at least in the epidermis (though this is disputed, and certainly does not apply to other tissues); they too are generally tiny, less than 0.2 mm in diameter. In any case this character is almost certainly a continuum, rather than an absolute phenomenon; it may have more use at a lower taxonomic level (for example certain rotifer families all having the same number of cells in particular organs) than in defining phyla.

Musculature

Pseudocoelomates are often credited with rather peculiar musculature in several respects. Some or all of them are supposed to lack circular muscles, to have no muscle around the gut, and to have an aberrant innervation system in which muscle processes run to the nerves rather than nerves sending branches to the muscle. None of these three generalisations stands up very well to close examination.

Nematodes and nematomorphs clearly lack circular muscle in the body wall. This lack causes some limitation of movement, but together with their high internal hydrostatic pressures and tough cuticles the longitudinal muscles of these worms give an integrated and reasonably effective locomotory system (Clark 1964; Inglis 1983), usually involving flexing of the body. Entoprocts also lack circular muscle; being sessile and stalked, their only quick action is a retraction, and circulars are not necessary to give the subsequent slow re-extension. All the other groups covered here have at least some circular muscle, though in kinorhynchs it is largely restricted to the first 'segment' of the body (Lang 1963). But only in the priapulids are the body wall muscles substantial and arranged in neat concentric longitudinal and circular layers.

It is unclear whether a lack of circular muscles is a primary trait or represents a loss. For the mobile worms, the lack of circulars does seem to impose restraints on movement and cause a degree of inefficiency, such that it would be hard to envisage their loss being favoured by selection. It is easier in functional terms to see the lack of circulars as a primary design feature, that has necessitated and directed certain other peculiar specialisations in the nematodes. However, recent reports that some circular fibres do exist in at least one genus of nematodes (*Eudorylaimus*; see Lorenzen 1985), the same genus also possessing cilia in its gut, does suggest that these worms might be descended from more typical ciliated ancestors with standard musculature; this scenario is discussed in more detail by Inglis (1985). The lack of circular muscles in entoprocts can readily be accepted as a derived feature on functional grounds. Perhaps reservations should therefore be set aside and the whole phenomenon regarded as a secondary effect. This would make it of little phylogenetic use, and even a link between nematodes and nematomorphs could not then be convincingly indicated as they might have lost circulars in response to similar mechanical constraints resulting from similar size, shape and habitats.

The lack of gut musculature is a direct consequence of a true pseudocoelomic state in which mesoderm only occurs around the outer margin of the body cavity. While most of these animals do have muscular pharynxes and muscle sphincters around the hindgut, the true endodermal part of the gut usually does lack muscle. This inevitably affects the feeding mechanisms profoundly; since peristaltic churning of the gut contents is prohibited, most of the pseudocoelomate animals have to be liquid or small-particle feeders. In the nematodes particularly, where there is also a high internal pressure to contend with, a complex pharyngeal anatomy is required to get food into the gut at all. In nematomorphs the problem is circumvented by degeneration of the gut, and in the parasitic stages food is absorbed through the body wall. The acanthocephalans have no gut at all and use this absorptive technique throughout their life. Both groups therefore inevitably lack gut musculature and for them this feature cannot reasonably be counted as a shared similarity with other pseudocoelomates. In the case of priapulids, there is a striking difference from other groups in that their gut is highly muscularised and achieves normal peristalsis.

The apparent oddity of pseudocoelomate muscles in having a muscle-to-nerve innervation pattern is also unacceptable as a shared derived character of pseudocoelomates, for reasons discussed in chapter 6. Many invertebrates have been shown to use this same innervation system via muscle processes, and though the phenomenon is often more striking in some pseudocoelomates this may be largely because of their constant cell number so that in the larger species the cells and their processes are unusually large and visible.

A final muscular peculiarity sometimes cited for several of the pseudocoelomate groups is that the highly muscular suctorial pharynx has a lumen that in cross-section is Y-shaped. This is particularly noticeable in gastrotrichs, nematodes, rotifers and kinorhynchs. However, several other groups of invertebrates show a similar pattern (some platyhelminths, tardigrades, and bryozoans), suggesting that it is simply an optimum way to pack and operate muscle around an extensible lumen. It is very unlikely to be homologous in the pseudocoelomate groups that show it, not least because in some the lumen is an upright Y, and in others it is inverted (Inglis 1985).

In terms of musculature, then, there are no obvious homologies that can unite the pseudocoelomate groups, nor can any sub-set of the groups be convincingly allied.

Cuticle and ciliation

Pseudocoelomate groups are generally well-endowed with cuticles (frequently tough enough to require moulting for growth), and lack external ciliation. This is particularly true of the most familiar group, the nematodes, but as with other characteristics it looks less convincing when one turns to the more esoteric animals. Most obviously, gastrotrichs and rotifers are both very easily diagnosed by their characteristic ciliations, on the ventral surfaces and on the corona or wheel organ respectively. Entoprocts also have obvious ciliation, especially on the feeding tentacles. All these groups also have cuticles; in gastrotrichs this has the curious property of actually covering even the individual cilia (Rieger & Rieger 1977), whilst in rotifers it may be thickened up in parts into a box-like lorica.

The cuticles of the pseudocoelomate groups do not bear any marked overall resemblances one to another (Storch 1984; Inglis 1985), and since most animal phyla have some form of cuticular secretion (chapter 6) the mere presence of an epidermal covering cannot be of much phylogenetic value. If the trend from simple highly microvillar cuticles to denser structures without epidermal surface projections is a real one (Rieger - chapter 6), then groups with the former variety of cuticle (gastrotrichs, and perhaps rotifers) should be more primitive than the nematodes, nematomorphs and kinorhynchs with strong and structureless or fibrous coverings. Such generalisations perhaps do not achieve very much. However, there are a few specific cases of similarity that have been interpreted as significant. One is the presence of an intra-epidermal cuticular material in the two groups Rotifera and Acanthocephala (Welsch & Storch 1976), an arrangement that is sufficiently unusual to suggest common ancestry (see Lorenzen 1985), and which may also occur in Chaetognatha (van der Land & Nørrevang 1985). Another example is the similarity of the nematomorph cuticle to that of one order of nematodes, the mermithids (Eakin & Brandenburger 1974; Batson 1979). Rieger & Rieger (1977) and Ruppert (1982) also suggest that the nematode cuticle shows some similarity to that of gastrotrichs, particularly in the lamellation of the outer layers; though Inglis (1985) disputes this and suggests (on rather limited evidence) that the nematode cuticle could best be seen as intra-epidermal.

Some authors have also drawn attention to the tendency of pseudocoelomates to show superficial segmentation of the cuticle; this is particularly striking in kinorhynchs (where underlying epidermis and muscles, and even the nervous system, are also affected), but is also evident in some rotifers, nematodes and acanthocephalans. However, given the probable multiple origins of different forms of segmentation, as was discussed in chapter 2, these similarities are most unlikely to be reliable indications of close affinity between the groups concerned.

Reproduction and developmental patterns

Most groups of pseudocoelomates are dioecious and use internal fertilisation mechanisms. However, at one extreme the gastrotrichs are unusual in their hermaphroditism, and at the other the priapulids stand out as having the so-called 'primitive' sperm type (chapter 6) and external fertilisation, perhaps associated with their larger size and hence an ability to store gametes.

Some of these animals show an apparent affinity with the 'protostome' (or 'spiralian') assemblage in respect of their embryology (see chapter 5). Entoprocts in particular have fairly classic spiral determinate cleavage, with 4d-derived mesoderm; and modified forms of these processes can be traced in gastrotrichs and rotifers. The latter group shows some resemblance to the protostome trochophore larva (see Remane (1963*c*) for a discussion of this point), though they have no real gastrulation process and limited mesoderm formation. On these grounds, some authors have linked the pseudocoelomates to the protostome groups. But there is a lack of conviction about the protostome affinities of pseudocoelomates; the resemblances are few, and the embryology of groups such as the nematodes is notable for its *lack* of similarity to any other group (chapter 5). Any suggested link with protostomes seems largely to rely on the unsupported and indeed improbable view that all animals *must* be either protostome or deuterostome in essence. Authors who adopt these ideas hold the pseudocoelomates to be either early protostome offshoots of flatworm-like creatures (e.g. Ruttner-Kolisko 1963), or regressive stages from protostome coelomates such as the early annelids (Remane 1958; Lorenzen 1985; see also chapters 2 and 5).

However priapulids have classic radial cleavage, and the embryo hatches as an unciliated stereogastrula, with no affinity to a trochophore larva. Since the authors that see pseudocoelomates as regressed spiralian coelomates tend also to regard priapulids as the most primitive of the assemblage (having larger sizes, cavities nearer to true coeloms, and 'primitive' reproductive traits), this apparently non-spiralian development comes as something of an embarrassment.

So, in common with virtually every preceding review of the pseudocoelomates (see Hyman 1951*b*; Remane 1963*c*; Clark 1979; Inglis 1985), no single homology emerges to unite the group. Is there any justification for regarding sub-sets of the groups as being related, and if so which are the most likely affinities?

9.3 The pseudocoelomate groups and possible affinities

Nematodes

The nematodes are by far the most important of the pseudocoelomate groups in terms of familiarity and ecological impact, and they constitute one of the largest of all animal phyla. They are also very readily characterised, with a standard elongate cylindrical shape and limited size range. In most respects the nematodes are classic pseudocoelomates, as table 9.1 indicates, lacking only the single supposedly definitive character of protonephridia (they have one or two elongate gland cells for osmoregulatory purposes instead). Their interrelated features of thick collagenous cuticle, high pressure body cavity, and four blocks of strong longitudinal musculature (fig 9.1) form an unusual locomotory system suited to efficient movement in viscous media and surface films (conditions found in many substrates and within other organisms), though less effective in free swimming or crawling. There is little cephalisation

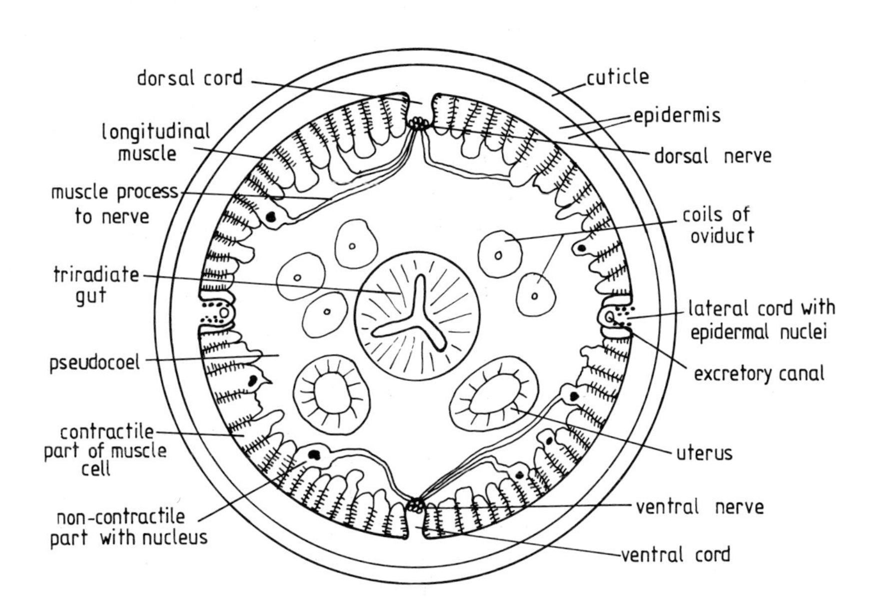

Fig 9.1. Transverse section of a nematode, (roundworm), with various characteristic 'pseudocoelomate' features, (actual diameter <0.1mm-5mm).

and an array of rather simple sensillae; the epidermis is thickened into cords at four points between the muscle blocks (the lateral cords containing the excretory cells); and feeding involves a rather elaborate triradiate pharynx often endowed with spines and teeth. Much of the internal space is given over to complex dioecious reproductive apparatus.

Most recent authors regard the nematodes as having a fairly close relationship with two other groups, the nematomorphs and the gastrotrichs. They share with the former group a lack of circular muscles and structurally similar longitudinal muscles and cuticles, though it is as yet unknown whether the nematomorph cuticle contains the same very aberrant high molecular weight collagen that has been isolated from nematodes (see chapter 4). Mermithid nematodes also have life-cycles very like those of nematomorphs (Lorenzen 1985), and both larval nematomorphs and one genus of nematodes (*Kinonchulus* - Riemann, quoted in Lorenzen 1985) have introvert-like front ends. Gastrotrichs resemble nematodes in their nervous systems, developmental patterns and pharyngeal structures (Teuchert 1968, 1977; Ruppert 1982); though an often quoted similarity of having 'unusual' myoepithelial cells in the pharynx wall should probably be discounted in view of the discussion of muscle ultrastructures in chapter 6 (and see Inglis 1985). One group of gastrotrichs in particular have similar cuticular structure (Rieger & Rieger 1977). The similarity of the body cavity in nematomorphs and some gastrotrichs has already been noted.

Thus, while in many ways nematodes seem set apart from other major phyla altogether, and were the source of many anomalies when details of structural and chemical evidence were considered in earlier chapters, there do seem to be some reasonably convincing grounds for placing the nematodes together with the nematomorphs and gastrotrichs as a coherent phylogenetic unit. However, it is worth recalling that Hyman (1951*b*) used quite different sets of similarities (admittedly some of them now rather dubious, and a few of them decisively disproved) to argue that the nematodes were most closely allied to kinorhynchs and priapulids! And recently Inglis (1985) has rejected most of the features discussed here as poor indicators of relatedness. It is unfortunate that so many of the techniques that have been used in recent years to study phylogeny - electron microscopy, and chemical and genetic analyses - have not yet been applied to the more abstruse pseudocoelomates to settle these matters.

Nematomorphs

These horsehair worms or 'gordian worms' are rather unfamiliar creatures, exclusively comprising highly specialised parasites whose juvenile life is spent in the body cavities of arthropods. Similar in external form to nematodes (though usually thinner and much longer), they have nevertheless been recognised as distinct for nearly one-and-a-half centuries. They have similar body walls and cuticles (although the ultrastructure and chemistry of the latter may be different), but have a greater infilling of the body cavity by mesoderm, giving a rather solid acoelomate appearance. The body has a longitudinally-partitioned effect (Eakin & Brandenburger 1974), and the animals lack a functional gut for most of their life (fig 9.2). They have only one or two thickened epidermal cords, which do not normally break the longitudinal muscles up into blocks. They apparently lack any excretory system. During development their eggs usually cleave symmetrically, but the peculiar rhomboid or 'T' configuration of nematodes has sometimes been observed (Hyman 1951*b*). The same author notes a striking similarity of the resulting larva of gordioid nematomorphs (fig 9.3) with the priapulids.

For reasons set out above, a close relationship with the nematodes seems quite likely, and a slightly more distant link to gastrotrichs can be tentatively accepted.

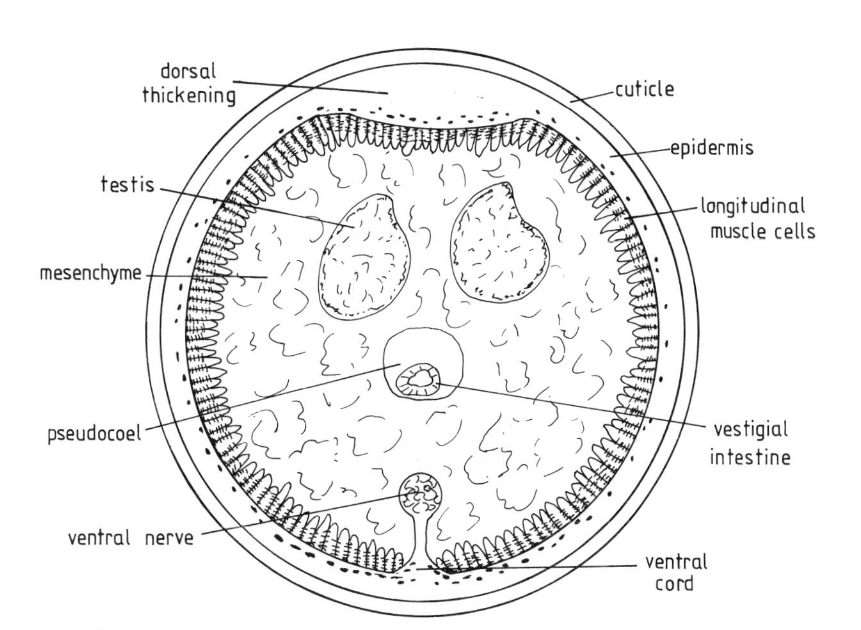

Fig 9.2. Transverse section of a nematomorph or horsehair worm (actual diameter <1mm)

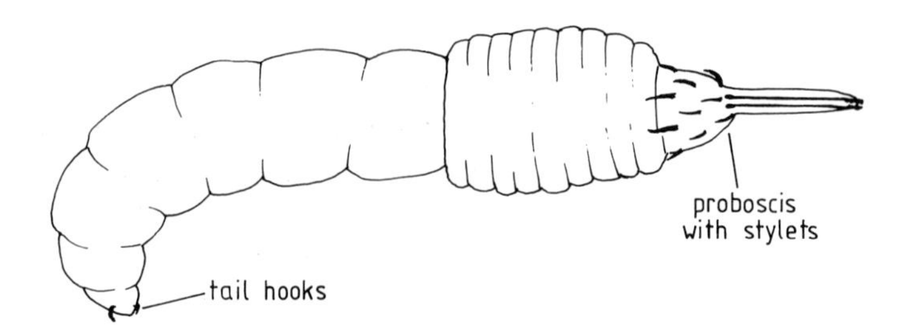

Fig 9.3. The larva of a gordioid nematomorph, supposedly resembling an adult priapulid.

Gastrotrichs

These minute aquatic animals glide along on ventral cilia in a fashion reminiscent of turbellarians, and thus contravene immediately one of the supposedly pseudocoelomate features in having external ciliation. They slightly rectify this situation by also having an external cuticle, which is thin enough to cover separately each individual cilium (fig 6.5), and which can be linked to the nematode condition (Rieger 1976). Table 9.1 confirms, though, that in many ways they do not quite fit as ideal pseudocoelomates. Their pseudocoelom is almost totally obliterated by mesoderm (hence they look remarkably like acoelomates in transverse section - fig 9.4); they possess circular muscle, though neither this nor the longitudinals are extensive; and they are nearly all hermaphrodite.

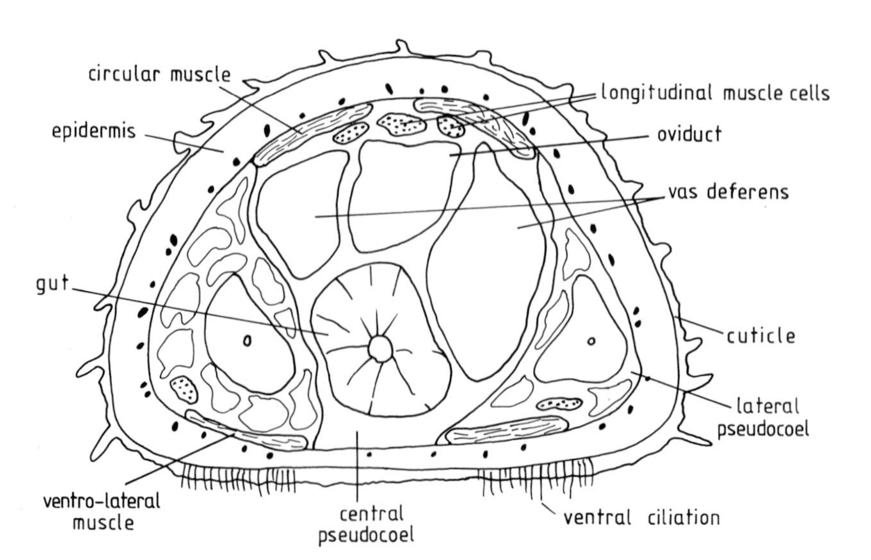

Fig 9.4. Transverse section of a gastrotrich (20-50μm diameter). The body cavity is very limited and the group is nearly 'acoelomate'.

Gastrotrichs were considered to be close relatives of the rotifers almost from the moment of their discovery, and the two groups were often united in the taxon 'Trochelminthes'. Both possess cilia (primarily from multiciliated cells) and usually thin cuticles, and both have limited amounts of longitudinal and circular muscle. However, recent authors, particularly since the studies of Remane and of Rieger, have come to regard them as much closer relatives of the nematodes (see also Hyman 1951*b*). They share with the nematodes the various features set out above, and the link between these groups is now reasonably uncontroversial.

Some similarities between gastrotrichs and the acoelomate platyhelminths should also be noted, though, perhaps giving clues to the origins of the pseudocoelomate condition. In particular, Rieger (1976; and Rieger & Mainitz 1977; see chapter 8) has stressed the probable links between these two groups and a third type of 'worm', the gnathostomulids. The monociliate gnathostomulids provide a possible connection between monociliate metazoan ancestors, the similar but multiciliate turbellarians, and the also similar but perhaps pseudocoelomate gastrotrichs, possessing both types of ciliation and in particular retaining the primitive monociliate condition in the flame cells. Gastrotrichs could then be something like the starting point from which both nematodes and nematomorphs on the one hand, and perhaps rotifers on the other, could have evolved. Boaden (1985) has recently gone so far as to describe the gastrotrichs as 'archecoelomates', carrying the implication that they represent a starting point for many evolutionary lines.

Rotifers (& Loricifera)

Amongst the pseudocoelomates, these 'wheel animals' are second in abundance only to nematodes. Their size range is similar to that of ciliate protozoans, though, and this and their predominantly fresh-water habit make them unique amongst metazoans and probably account for many of their peculiarities. The front part, forming the feeding corona, is strongly ciliated (entirely multiciliate), and the rest of the body is encased in an obvious cuticle (fig 6.5), sometimes elaborated into a thick lorica. Posteriorly there is usually a foot, bearing adhesive toes. Rotifers have a simple epidermis, and underlying this are numerous small individual muscles (fig 9.5), roughly corresponding to circular and longitudinal fibres but probably only very distantly derived from continuous muscle sheets. There is a fairly spacious body cavity, between the body wall and the viscera, which is in most respects clearly a pseudocoelom, and is ontogenetically continuous with the blastocoel; but it does possess a thin anucleate lining. Sensory organs, and in particular eyes (Clément 1985), are abundant and diverse.

Embryologically the rotifers are peculiar, partly because of their very small size and eutely, and partly because they so often develop parthenogenetically. There are elements of spirality and also of bilaterality in the cleavage, but the rotifers cannot readily be related to any other lower invertebrates; the nearest relationships seem to be to acoel turbellarians and to acanthocephalans (Hyman 1951*b*).

In terms of adult morphology, close relationships to gastrotrichs have been largely discounted (see above), and early suggestions of a link with arthropods (on the basis of the cuticle and its apparent segmentation in some forms) are even more improbable. Hatschek's slightly later view (1891) that rotifers are persistent trochophores, from annelid-like ancestors, is also open to question - the prototroch cilia of a trochophore are not homologous with the coronal cilia of normal rotifers, and one species that does appear to have a ciliary girdle (*Trochosphaera*) is a very aberrant and specialised rotifer. Rotifer embryology does not in any way indicate that they are regressions from 'higher' forms.

However, an origin from something like a creeping flatworm seems quite probable to most recent authors; the coronal cilia then represent a specialisation

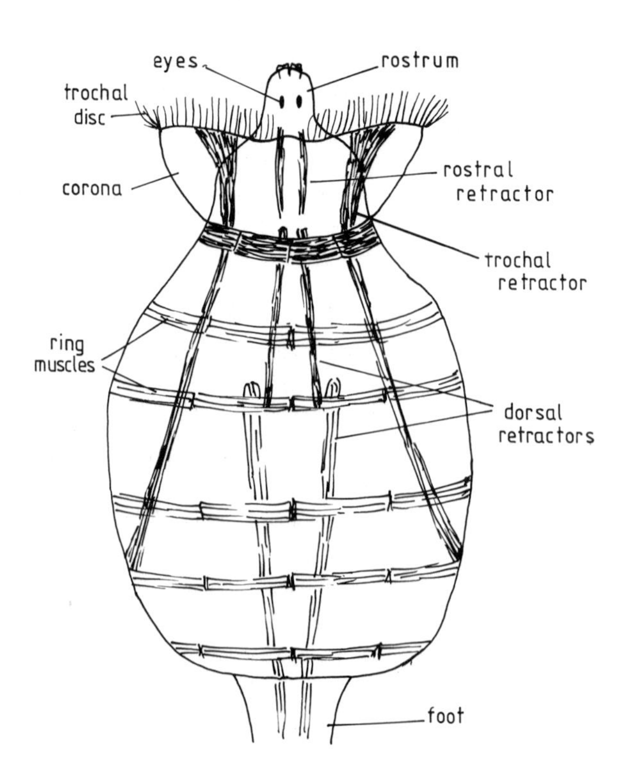

Fig 9.5. The general morphology and musculature of a rotifer (*Rotatoria*), with limited circular and longitudinal bands; (modified from Hyman 1951*b*).

of the antero-ventral cilia of a turbellarian, so that a resemblance to a ciliated trochophore is not totally surprising. This view is supported by the similarity of rotifer and rhabdocoel excretory systems, and of the female reproductive systems in the two groups. A similarity of rotifer and gnathostomulid pharynxes can also be noted. However, rotifers have a clear body cavity, a through gut and much-reduced muscle sheaths, so the links to either of the worm-like acoelomate groups cannot be too close.

Lorenzen (1985) suggests that within the pseudocoelomates there is a close link between rotifers and acanthocephalans, on the bases of similarly intra-epidermal cuticles, similar retractable proboscis structures, and a pair of similar invaginations called lemnisci in the neck region occurring in both acanthocephalans and in bdelloid rotifers. To this list could be added the embryological similarities already referred to, and the cloacal similarities discussed by Remane (1963*c*). An analysis by Clément (1985) strongly denies that rotifers are possibly a regression from a coelomate state, and largely agrees with their affinity to acanthocephalans (both groups being quite close to the turbellarians). This link therefore seems both popular and moderately plausible, the two groups Rotifera and Acanthocephala being set somewhat apart from the triumvirate of nematodes, nematomorphs and gastrotrichs and each set being derived from acoelomate ancestors.

The recently discovered new phylum Loricifera (Kristensen 1983) may belong close to the rotifers; they are somewhat similar in general form, and probably pseudocoelomic. Nielsen (1985) puts them close to kinorhynchs and priapulids, but there is too little detailed information on them as yet to speculate about their final phylogenetic resting place.

Kinorhynchs

The kinorhynchs (or Echinodera in some texts) are the least familiar and hardest to find of the pseudocoelomates. All are marine, and mostly interstitial, with a spiny, cuticularised, and markedly segmented body of 13 'zonites', devoid of cilia. Reference to table 9.1 shows that they probably best exemplify the pseudocoelomate condition, and since they effectively show all the requisite characters they can be (and have been) allied with every other group with about equal confidence!

Beneath the plates of thickened cuticle, the epidermis of kinorhynchs is partially syncytial and is condensed into cords like those of nematodes and nematomorphs mid-dorsally and laterally (fig 9.6). Longitudinal muscle bands occur, but circulars are only present in the first one or two zonites, perhaps being modified into the dorso-ventral muscle strands more posteriorly. There

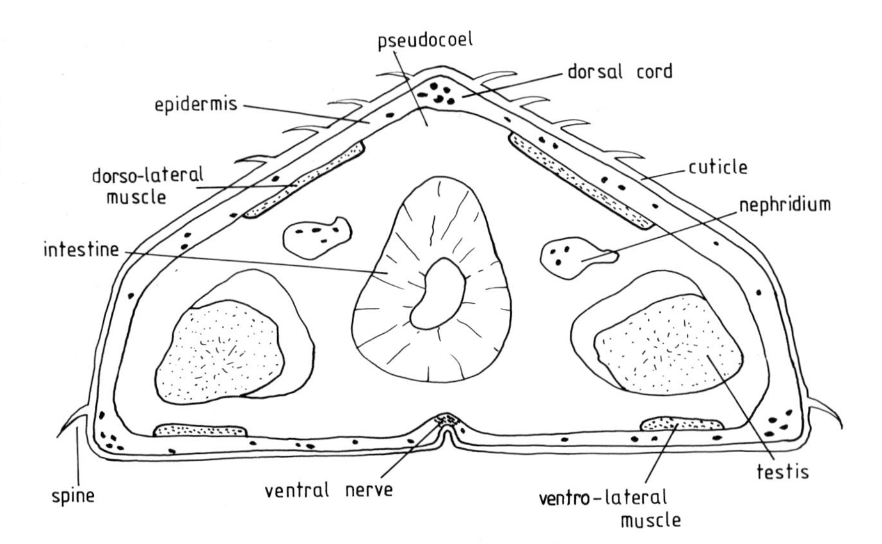

Fig 9.6. A kinorhynch, or echinoderid, in transverse section (actual diameter 0.1-0.5mm). Circular muscle is absent from most of the body.

is a spacious pseudocoel, with many active amoebocytes, and the gut is very similar to that of gastrotrichs and nematodes, lacking proper musculature.

Kinorhynchs were initially likened to arthropods, with whom they share the segmented, cuticularised state. But apart from other obvious differences, their segmentation is fundamentally superficial, starting in the cuticle and spreading inwards, whereas that of the arthropods is basically a mesodermal phenomenon. Their affinities probably do lie instead with other pseudocoelomate groups, but they seem about equidistant from each of the other types described and may be an independent offshoot from some common stem. Lorenzen (1985) makes much of their eversible first zonite, forming an introvert, and so links them closely to priapulids and more distantly to nematomorphs (larval introverts) and nematodes (introvert in one genus).

Acanthocephalans

Hyman (1951*b*) excluded the acanthocephalans from her phylum Aschelminthes, and thus contributed new vigour to the long-standing controversy over their relationships. She conceded that they were pseudocoelomate, and granted aschelminth-like characters of an eversible proboscis, superficial segmentation, and a degree of eutely. But against their status as aschelminths she cited the pronounced circular musculature (fig 9.7), the peculiar mode of formation of the pseudocoel (not contiguous with the blastocoel, but arising as a split in the inner cell mass), the absence of a gut, a

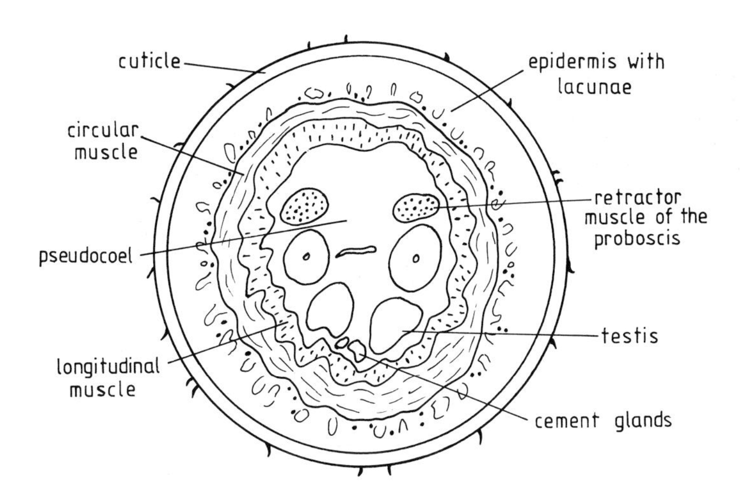

Fig 9.7. Transverse section of a parasitic acanthocephalan worm, lacking a gut, (actual diameter 0.5-1mm).

rather flatworm-like reproductive system, and developmental patterns resembling those of cestodes. A position independent of both platyhelminths and aschelminths was proposed. A somewhat later review by Nicholas & Hynes (1963), largely based on embryology, was similarly indecisive about the status of acanthocephalans.

There is little doubt that the endoparasitic nature of these worms has profound effects on their morphology and helps to obscure their phylogenetic relationships; similarities to cestodes are probably convergences due to similar lifestyles. Recently, authors have tended to re-unite the acanthocephalans with the rest of the pseudocoelomates, (see Whitfield 1971) and features allying them with the rotifers have already been stressed. A recent review by Conway Morris & Crompton (1982) dissented from this view and revived a tradition linking the acanthocephalans with priapulids instead, on the basis of 'overall appearance', similar muscle ultrastructure, and protonephridial resemblances. In addition they regard the Burgess Shale fossils *Ancalagon* and *Ottoia* (see chapter 3) as evidence for the antiquity of acanthocephalans, and of their link with priapulids, respectively. However, Lorenzen criticises these views on the grounds that the acanthocephalans do not really have an introvert, their proboscis being non-homologous with those of priapulids and kinorhynchs; he argues strongly for a link with rotifers (Lorentzen 1985).

Entoprocts

Entoprocts (or Kamptozoa, in much of the German literature) have had an erratic past, often tied in with the ectoprocts in the phylum Bryozoa or even Polyzoa (see chapter 13). These two groups of animals certainly bear a strong superficial resemblance, but for most authors the ectoprocts have a true coelomic status, and have taken the name Bryozoa all to themselves, whilst the simpler and less diverse entoprocts have been taken 'down' to a more appropriate pseudocoelomate designation (but see Nielsen 1977*a*,*b*, and chapter 13, for a dissenting view). The entoprocts have ciliated tentacles in a ring that surrounds both mouth and anus, whilst ectoprocts have (as the name suggests) the anus outside the tentacular ring, and this ring is organised on a ridge called the lophophore, the whole structure being retractable. Ciliary beat patterns in the two groups are almost diametrically opposite, water flowing up into the tentacular ring in the entoprocts and down through the lophophore in the ectoprocts/bryozoans. Furthermore, only the entoprocts have protonephridia, characteristic of 'lower' animals and found commonly in the pseudocoelomates.

If the entoprocts are indeed pseudocoelomate, there is little or no consensus about their position in that assemblage. Hyman (1951*b*) excluded them from the Aschelminthes because the gut is U-shaped (unlike any other pseudocoelomate admittedly, yet a normal concomitant of sessile lifestyles), and lacks a pharynx. They possess both cilia (on the tentacles) and a cuticle (on the rest of the body), and they have sparse and purely longitudinal muscle fibres, with no gut-wall muscle at all (fig 9.8). The 'pseudocoel' is quite spacious, but is filled with a gelatinous material and some mesodermal cells, so that some authors prefer to call the group acoelomate (e.g. Salvini-Plawen 1980*a*). Developmentally the entoprocts are in many respects rather classical protostomes, with spiral determinate cleavage and a 4d-derived mesoderm; though they show enough differences to suggest separate derivation from a

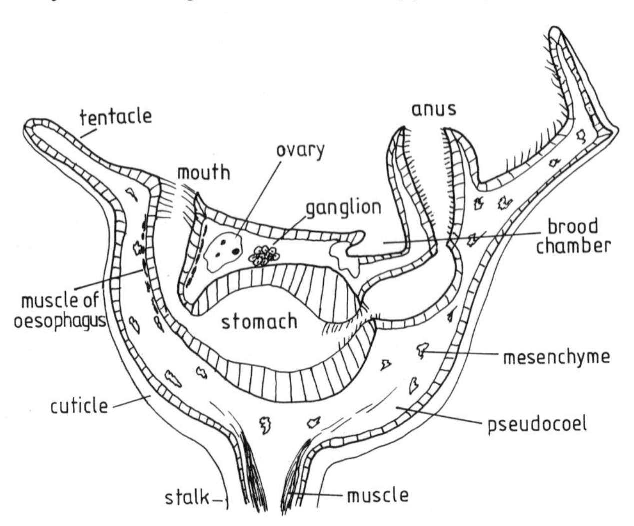

Fig 9.8. Sagittal section of an entoproct, modified from Hyman (1951*b*), (actual size 0.5-2mm).

flatworm stage rather than common ancestry with molluscs or annelids. The larva has often been described as a modified trochophore, but even Hyman is sceptical about this (see the discussion in chapter 5 on this point). At least some entoprocts (e.g. *Loxosoma*) have larvae more nearly resembling rotifers.

Because of this, Hyman follows earlier authors and suggests a link between entoprocts and rotifers, particularly pointing out similarities with collethecacean rotifers (gut structure, protonephridia, eyes); and she links the pre-oral organs of entoprocts to the lateral antennae of rotifers.

Priapulids

Priapulids, with only about 15 living species, are perhaps the most controversial of all the groups covered in this chapter, and many authors have precluded them from the pseudocoelomate assemblage altogether. In the 1950s Hyman allowed them an uncertain association with aschelminths, the nature of their body cavity being then unknown; but a report by Shapeero (1961) that this cavity had a true coelomic lining and that mesenteries suspended the organs seemed to decisively sever them from such links. The priapulids therefore spent nearly twenty years as coelomates, before Malakhov (1980) challenged the observations of Shapeero and reported no such cavity lining. More recently careful studies by Nørrevang (see van der Land & Nørrevang 1985) have disclosed that the appearance of a cellular lining results from the peculiar position of muscle cell nuclei; there is no separate cellular peritoneum, the 'mesenteries' are formed aberrantly from extracellular material, and the priapulids are not conventionally coelomate. They should therefore be dealt with here.

These worms are larger than most other pseudocoelomates, with a more conspicuous body cavity and well-organised circular and longitudinal muscle layers round gut and body wall (fig 9.9). They have a pronounced introvert with teeth often in a pentamerous array, used for burrowing locomotion and for feeding, but are otherwise unsegmented. They are clothed in a chitinous cuticle, relatively thick and structureless, perhaps most similar to that of some nematodes and kinorhynchs; the nervous system also resembles that of kinorhynchs. They have monociliate flame cells, a possible primitive feature (chapter 6) shared with gastrotrichs and the acoelomate gnathostomulids. They are peculiar amongst pseudocoelomates in having spermatozoa of the 'primitive' type (chapter 6), and most species use external fertilisation. Their embryology is distinctly non-spiralian, with radial cleavage and gastrulation by a form of polar ingression.

This does not give many clues as to where the priapulids belong. Many authors have recently come to regard them as a rather primitive group, partly because of their reproductive behaviour and their long-standing fossil history (chapter 3). If the view that large size is primitive is accepted, as some studies of life-history patterns discussed in chapter 5 suggest, then the priapulids might make a convenient starting point for the pseudocoelomates as a whole, representing the first step in a regression from the larger coelomate state, perhaps from a starting point near the sipunculans (see Candia Carnevali & Ferraguti 1979). Later groups of pseudocoelomates would plausibly show smaller sizes, loss of muscle layers and increasing eutely as responses to miniaturisation. However this story does not accord well with the relationships between other pseudocoelomate groups presented above, mostly suggesting an origin from simple acoelomate forms. Nor does it fit well with the non-spiralian embryology of priapulids and the more definite spiral traces in other groups, as most authors who believe the large coelomate condition to be primitive seek to derive 'regressive' pseudocoelomates and acoelomates from spiralian annelid-type ancestors.

The alternatives would be to take the priapulids as independent descendants of an acoelomate precursor, or to view them as offshoots of some particular pseudocoelomate group. Joffe (1979) suggested their embryology allied them to nematomorphs; Malakhov (1980) linked them to the nematomorphs and kinorhynchs; Lang (1963) preferred just a kinorhynch link, as do van der Land & Nørrevang (1985) and Lorenzen (1985), relying on introvert similarities; Por and Bromley (1974) see links to the rotifers; and Conway Morris & Crompton (1982), as discussed above, believe the acanthocephalans are close derived relatives. The association with kinorhynchs on the basis of cuticle similarities, and comparable introvert operation and structure, seems

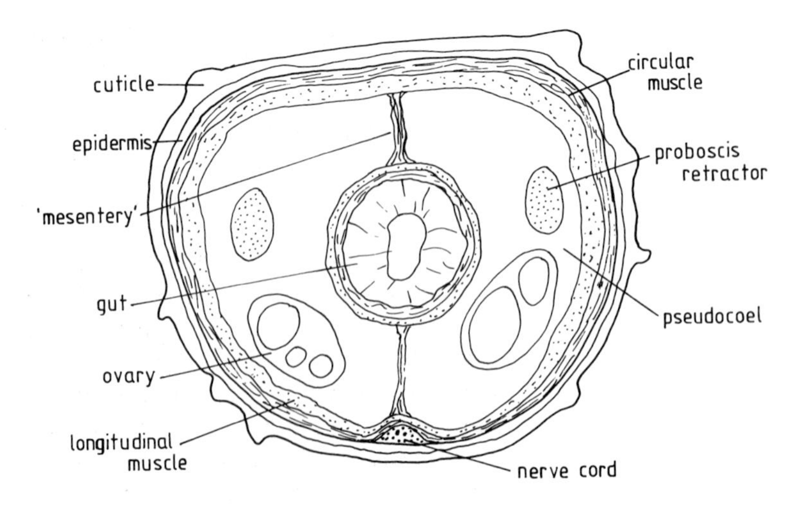

Fig 9.9. Transverse section of a priapulid, with spacious body cavity and mesenteries, (actual diameter up to 15mm).

most plausible, and the possibility that rotifers and acanthocephalans are descendants from a similar ancestor cannot be ignored (Nielsen 1985).

9.4 Conclusions

There are no clearcut and easy messages emerging as yet about the phylogeny of pseudocoelomates. But from the quite extensive recent literature, there is perhaps a consensus view to be found that can give some framework to our views on these peculiar groups. The following points would not find too many dissenting views.

(1) None of the pseudocoelomate groups shows clear affinities to the classic 'protostome' groups discussed in the last chapter, though elements of spiralian development, for example in entoprocts and rotifers, may indicate descent from proto-platyhelminths that had already acquired these developmental patterns.

(2) Gastrotrichs, nematodes and nematomorphs may be related to each other. The latter two groups probably derived from gastrotrich-like ancestors, adding more complex cuticles and pronounced eutely, and losing the circular muscle; with gut loss in the nematomorphs associated with obligate endoparasitism. Gastrotrichs may provide a link via gnathostomulids with acoelomate ancestors.

(3) Rotifers and acanthocephalans are probably related, the bdelloid rotifers having the most marked similarity to the parasitic gutless acanthocephalans. Old ideas that rotifers (and by inference other pseudocoelomates) are persistent neotenous derivatives of trochophore larval forms are implausible.

(4) Kinorhynchs are classically pseudocoelomate but of uncertain affinity; they may be most closely related to priapulids. However priapulids in some respects are the most distinct of all the groups and may be a quite separate development from acoelomates.

(5) Entoprocts show no clear affinities to the other pseudocoelomate groups. They are possibly closest to rotifers, but again could equally represent a separate lineage from platyhelminth-like ancestors.

If these points are accepted (perhaps in decreasing order of confidence!), a set of relationships such as those shown in fig 9.10 could be proposed. This suggests real phyletic units within the old pseudocoelomate/aschelminth assemblage, each of which might be linked to the others but most probably represent independent derivations from acoelomate ancestors. The separate and independent establishments of primary body cavities, giving potential for increased sizes, could account for the elements of parallel character acquisition by these groups (see Inglis 1985); whilst the different habitats and lifestyles adopted clearly contribute to the differences within groups and the convergences between them. It is also worth recalling that at least two other groups of animals share some of the pseudocoelomate characters and may merit inclusion in this scheme: the tardigrades, with cuticles resembling gastrotrichs and a degree of eutely, and the chaetognaths with a pseudocoelomic adult cavity and no circular muscle. Nematodes, rotifers and tardigrades also share a capacity for anhydrobiosis, withstanding extreme desiccation using very similar mechanisms (Crowe *et al.* 1984). All these points serve to underline the caution needed in giving weight to any particular character.

Whether the scheme in fig 9.10 bears any relationship to actual evolutionary pathways is of course open to furious debate. It should serve to emphasise that pseudocoelomates form no part of a dichotomous protostome/deuterostome scheme of the animal kingdom; and it does at least give a plausible structure to otherwise chaotic areas of speculation, based on relatively recent reviews and analyses of the problems. In particular it stresses the almost certainly polyphyletic nature of the pseudocoelomate assemblage, though retaining a degree of ambivalence on whether this polyphyly involves two or three main branches or should be extended to make almost every phylum a quite separate shoot of the evolutionary 'tree'. The absence of any really convincing homologies between groups of phyla and even between any individual phyla may indicate that more, rather than less, polyphyly is to be invoked. At the moment this kind of scheme is probably the best we can do.

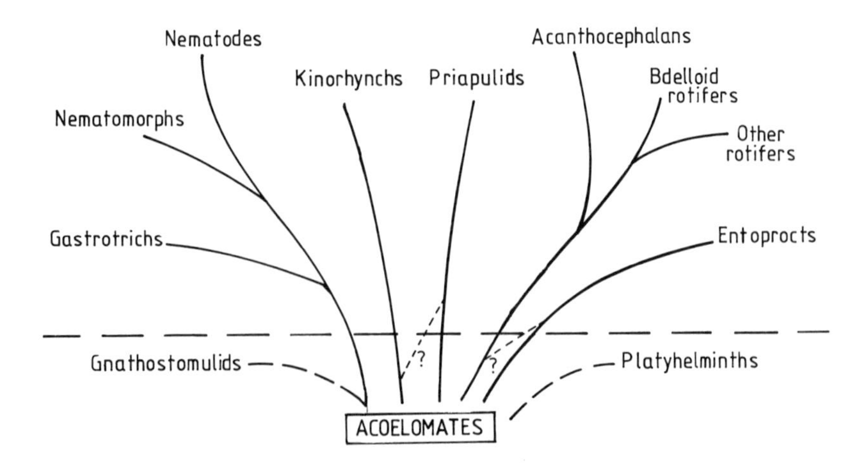

Fig 9.10. A tentative scheme for relationships between the pseudocoelomate groups (see text). Polyphyletic origins seem certain, though sub-sets of the phyla may have some association beyond the acoelomate ancestral stage.

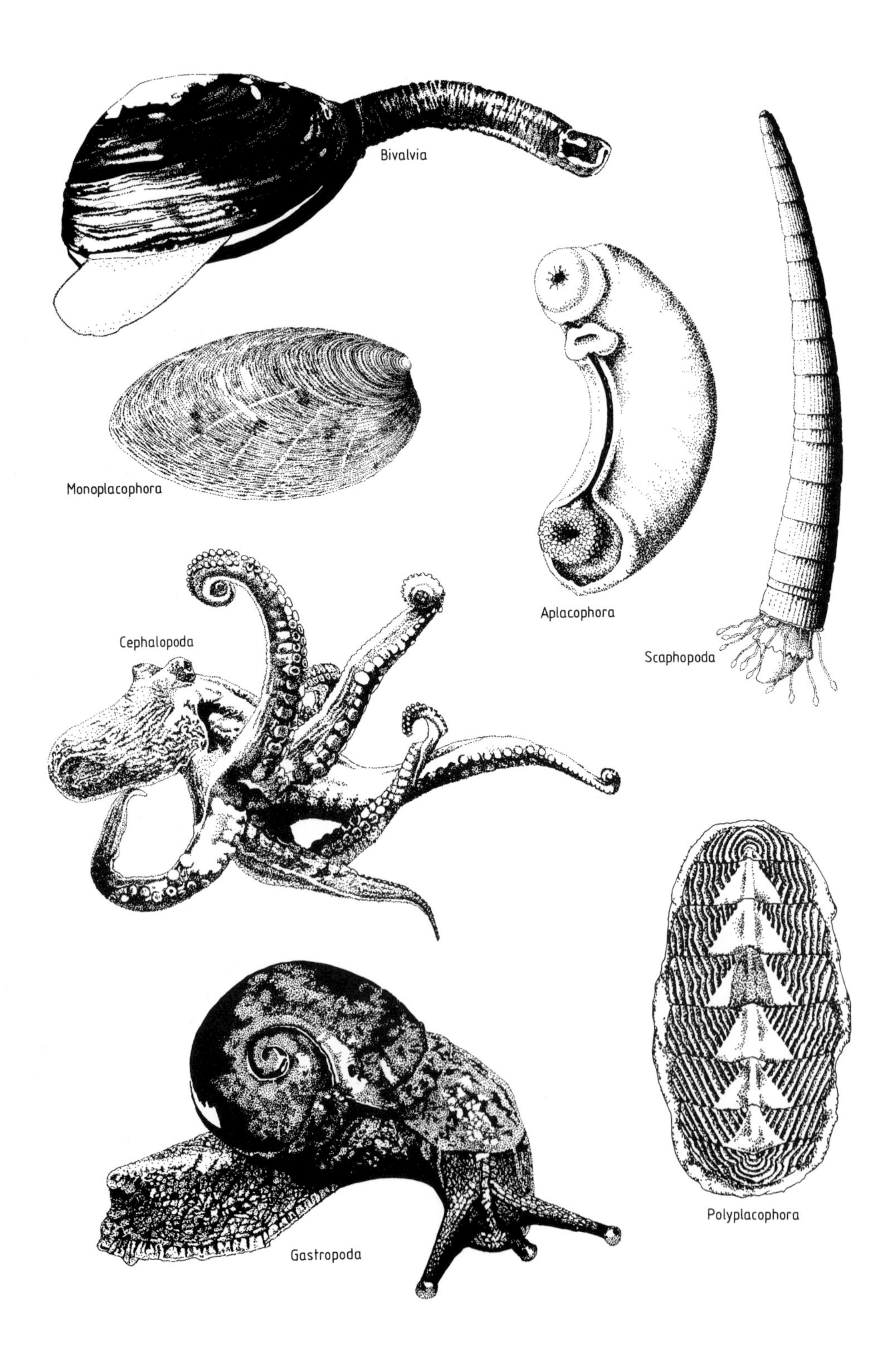
Bivalvia
Monoplacophora
Aplacophora
Scaphopoda
Cephalopoda
Polyplacophora
Gastropoda

10 The phylogenetic position of molluscs

10.1 Introduction

The phylum Mollusca is structurally and ecologically enormously diverse, and highly speciose. Inevitably this makes the search for ancestral forms and possible origins more difficult, and there have been endless controversies over the nature of the molluscan 'archetype', and over relations between the various sub-phyla or classes that it gave rise to. However, there is perhaps a stronger current consensus over the status of the molluscs, and their position relative to other phyla, than there has been for many decades; consequently this chapter presents reasonably uncontroversial material, with which most modern treatments would concur.

Nevertheless many text-books and the simplified phylogenetic trees commonly presented therein still treat the molluscs in a rather outdated manner and insist on their close affinities to annelids and arthropods, endowing them with coeloms (and sometimes even segmentation) of a homologous if reduced nature. Such traditional views usually combine these three groups as the major spiralian or protostome invertebrates, at the apex of the left-hand 'branch' of a neatly dichotomous 'tree'. It is therefore necessary to consider the evidence that gave them this position for the best part of a century, before turning to the insights from morphology and embryology that have, within the last two decades, led to their relocation as derivatives of a turbellarian-like form (a view actually initiated in the 1890s).

The traditional 'annelid theory' for the origins of molluscs is fast losing ground to the newer views, but there remains some dispute over the relative affinities of the annelids, molluscs and flatworms. There are several possibilities: the former two groups may have evolved together from flatworms and then split before segmentation was acquired; or they may have separated a little earlier, before a coelomic cavity was established; or they may have had quite separate origins from amongst a turbellariform assemblage. These alternative phylogenies are summarised in fig 10.1.

Fig 10.1. The various possible origins of the molluscs, in relation to other important spirally cleaving groups of invertebrates. Each scheme is discussed in the text.

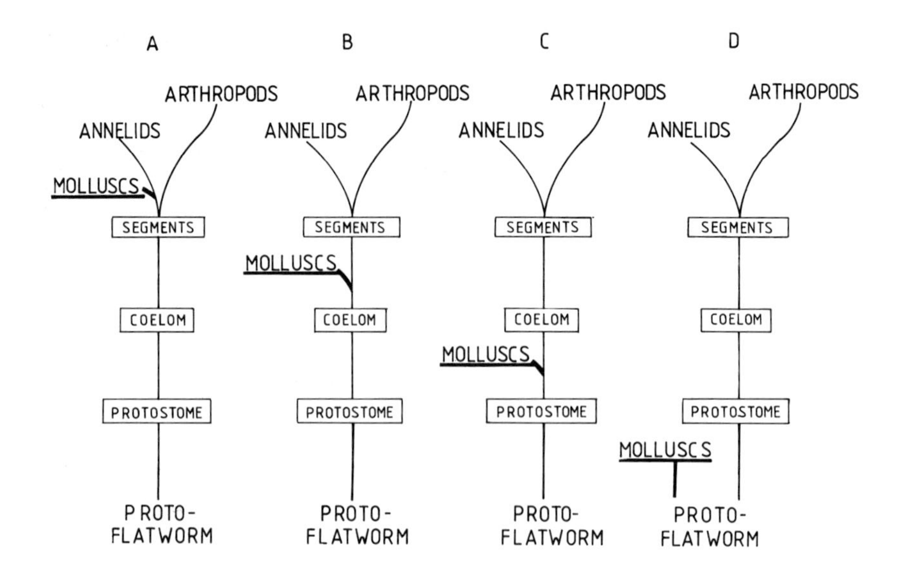

Clearly these differing views of evolutionary sequences have distinct consequences for the likely ancestral form of molluscs, and for relations between the molluscan classes. Therefore a review of current phylogenies must finally be tested against the palaeontological record of molluscs, with which this chapter concludes.

10.2 Basic molluscan anatomy: the 'archetype'

Essentially similar hypothetical mollusc ancestors, incorporating in a simple form all the features that unite modern molluscs, can be found in any textbook; a typical example is given in fig 10.2. Such diagrams serve to underline the basic similarities of members of the phylum, given that molluscs are in practice quite hard to characterise. The foot differentiated from a visceral mass, the protective calcareous shell, a mantle cavity housing gills, a condition of tetraneury (four main longitudinal connectives) in the nervous system, and a chitinous radula are all useful pointers, but no one feature is on its own diagnostic as each has been lost (or modified almost beyond recognition) repeatedly. Hence an archetype incorporating all these characters, with in addition a haemocoelic body cavity, a through-gut, and a pericardial cavity ('coelom'), is at least a useful reminder of the phyletic unit being considered. And it has the (perhaps dubious) merit of resembling a simple gastropod without the complication of torsion, or a monoplacophoran such as *Neopilina* without any compounding serial repetitions of structures.

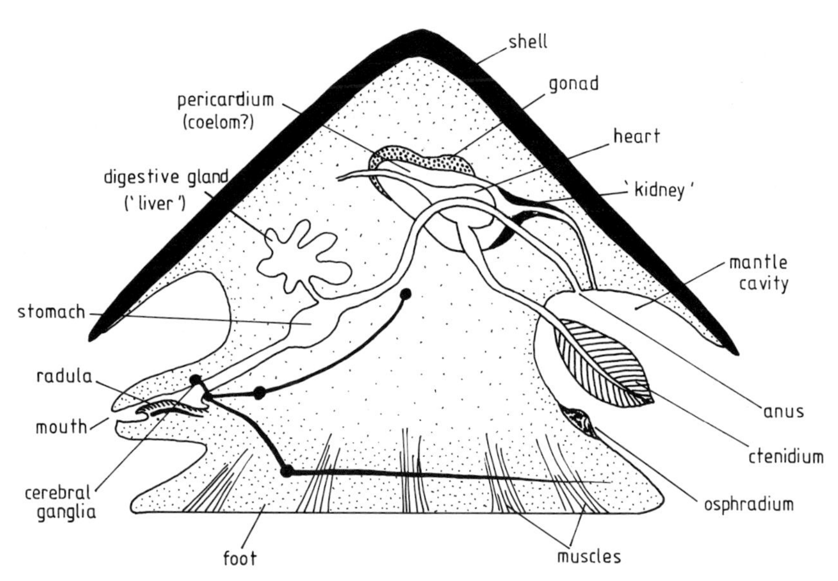

Fig 10.2. A traditional molluscan archetype, with all the classic features present in simple form; (modified from various sources).

Unfortunately such a 'paper animal' has its drawbacks, and has perhaps proved a positive phylogenetic hindrance. Each molluscan class is usually represented as emerging from this construction, by loss or particular emphasis of certain characters, giving no indication of relationships between classes. Yet the archetype has no actual role in phylogeny, and never actually existed; it is based upon the end-products of molluscan evolution, as they have been stabilised over time, and not upon primitive features. Like any archetype it may tend to constrain rather than revitalise conceptions of molluscan ancestry, and it should therefore be summarily dismissed as an evolutionary construct. This archetype is not an ancestor, merely the post hoc attempt at constructing or imagining one; and it may bear no resemblance to any actual early mollusc. Some features of this 'classic' archetype are almost certainly invalid, as will be seen. So it is used here merely as a useful reminder of general body plan, as we move on to consider the comparative morphology of the group.

10.3 The status of molluscs

Molluscs as a segmented group?

The classic view of the animal kingdom places molluscs near to annelids and arthropods, and rather distant from the 'lower' spiralians (fig 10.1*A*). This view perhaps reached its zenith with Lemche (1959*a*,*b*), whose work on recently discovered monoplacophorans led him to revive older views that molluscs were indeed primitively segmented. All the protostome phyla were thus united by their embryology, and there was a convenient hierarchy of comparative anatomies; flatworms provided the raw materials, nemerteans first achieved a through-gut and improved circulation, then molluscs and annelids achieved true coeloms and segmentation, before arthropods added the exoskeleton and began to lose their ancestral characters as a consequence. Lemche maintained that there were even closer relationships between annelids and molluscs than between annelids and arthropods - only the former groups have spiral cleavage and trochophore larvae. He further believed that polyclads, nemerteans and perhaps cestodes all bore witness to the incipient segmentation in the protostome line, thus requiring its presence in molluscan stock; in fact for him and some of his contemporaries the segmentation of *Neopilina* merely confirmed what must self-evidently have been the case in primitive molluscs, and chitons with their multiple shell plates also reflected a past when segmentation was a definitive molluscan character.

A first step in establishing the actual status of molluscs must therefore be to disentangle them from the segmented groups. This is not difficult; few modern zoologists retain Lemche's belief that the apparent metamerism of *Neopilina* is a result of reduction from a fully segmented ancestor. Some of the problems of defining segmentation and metamerism were considered in chapter 2; here it should suffice to point out some of the problems with *Neopilina* itself, matters that have already been comprehensively reviewed by Vagvolgyi (1967), Salvini-Plawen (1969, 1981) and Stasek (1972).

Some molluscs do show a serial repetition of certain organs (fig 10.3), as do many vermiform invertebrates. But segmentation in the form found in the annelids and arthropods, and in chordates, involves a strict mesodermal and coelomic repetition, with each pair of organs or muscles recurring in the same fashion in every segment and with an organised developmental origin for this repetition. No mollusc shows this kind of eumetamerism, and the group should therefore at best be regarded as pseudometameric (see Clark 1979). For example, *Neopilina galatheae* (fig 10.3*A*) has two pairs of atria, five pairs of gills, six pairs of excretory (osmoregulatory) organs, eight pairs of pedal retractor muscles and ten paired nerve commissures. Other species of monoplacophorans have somewhat different repetitive sequences. Chitons (fig 10.3*B*) always have eight shell plates and hence eight associated muscle pairs, but their gills though repeated are not even strictly paired, and certainly not segmentally arranged, while the nerve commissures repeat out of phase with the shell plates and muscles. In all these cases, the rudiments of the various repeating organs are developmentally uncoordinated as well as spatially irregular, for they are initiated at different times and develop at different rates. It seems clear that repetition of organs in molluscs arose quite independently of the ordered segmentation in annelids, by multiplying or breaking up existing organs as the need arose, with a limited subsequent regularisation of the repeating sequences.

The non-segmented nature of molluscs is also likely on functional grounds. Locomotion in all the more primitive groups of molluscs is by a combined ciliary and mucus gliding action over hard substrates, reminiscent of the locomotion of flatworms; segmentation of a hydraulic body cavity would not greatly aid such an action (see Clark 1964), and splitting the musculature into blocks rather than continuous sheets would also be counter-productive in the ancestral form assuming it to be very small. To suppose that molluscs have evolved 'back' to such a system from a highly organised and economical septate body cavity with peristaltic action is illogical, despite Johansson's (1952) proposal that this might have happened in some annelids as a response to life on rocky surfaces (and notwithstanding its partial occurrence in the special case of leeches). Indeed the recorded history of molluscs has if anything been away from the ciliary-mucus glide and towards more elaborate hydraulic systems, as in burrowing bivalves and in cephalopods (see Vagvolgyi 1967).

Evidence for the view that molluscs evolved their pseudometamerism from the eumetameric/segmented annelids, by reduction, is therefore extremely thin. Such a view cannot be supported by comparative morphology or embryology, and is unlikely on functional grounds. The molluscs are not descended from a segmented ancestor; and their origins can therefore be set further 'down' the

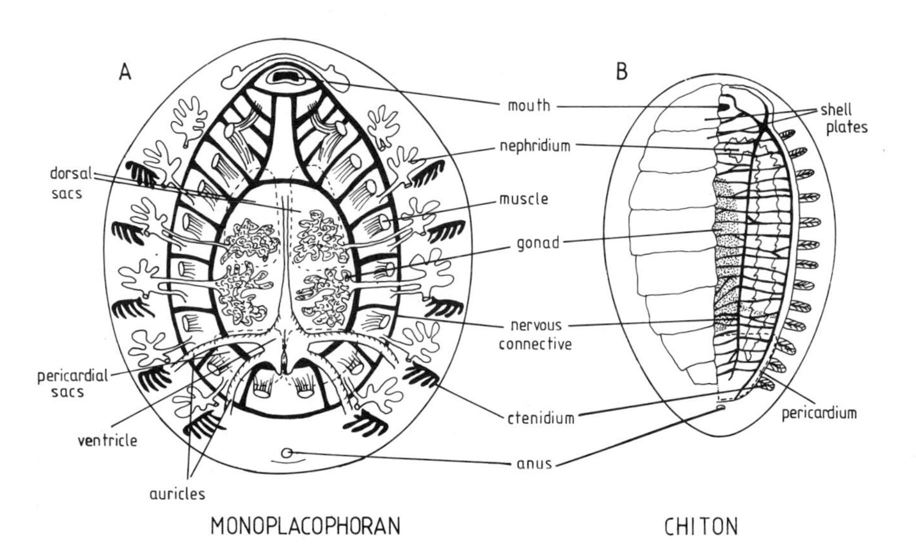

Fig 10.3. The serial repetition of structures in *(A)* a monoplacophoran, *Neopilina galathaea*, and *(B)* a polyplacophoran chiton.

protostomian lineage, as in fig 10.1*B* -*D*, away from the very close association with annelids that Lemche advocated.

Molluscs as coelomates?

Conventionally, the molluscs have a spacious haemocoel as their main body cavity, and a coelom of small size around the heart, forming a pericardial cavity and generally housing the gonads and excretory organs. (see fig 10.2). But many recent authors have denied them the status of 'coelomate' (Clark 1979; Salvini-Plawen 1981; Runnegar & Pojeta 1985). Stasek (1972) for example regards the pericardial coelomic space as an incidental and non-homologous development as compared to the annelid coelom, and states that molluscs are effectively an acoelomate phylum having 'escaped that designation on a technicality'.

The definitions of a coelom, and its probable polyphyletic origins, were considered in chapter 2. There is little doubt that the pericardial cavity of molluscs is theoretically a 'coelom' as conventionally and minimally defined, for it is entirely bounded by mesoderm: this is the technicality to which Stasek referred. But the molluscan 'coelom' does not develop within the mesodermal bands, as occurs in annelid embryology. These bands of cells remain solid and unsegmented in the larval mollusc, developing later into the various mesodermal organs, of which the pericardium is just one. The single 'coelomic' cavity which later arises therein, and which is functionally necessary to permit the heart to beat independently, may become partially sub-divided into pericardial and nephridial areas, but it never enlarges into a major hydraulic cavity pertaining to the whole body. The cavity is moderately large in the monoplacophorans, housing the heart and gonads; but only in modern cephalopods is a really spacious expanded 'coelom' to be found, and there it functions principally as a flotation device, often filled with ammoniacal fluids. The molluscan 'coelom' therefore cannot be regarded as homologous with the annelid coelom, and it seems pointless in practice to regard it as a coelom at all. Adopting the view discussed in chapter 2, the word coelom should at best be reserved as a non-specific covering term for all spacious hydraulic secondary body cavities, and it is thus certainly an inappropriate term for molluscs. Its use in relation to this phylum seems partly a throw-back to gonocoel theories of coelomic origin ('it must be a coelom because it contains the gonads': see Clark 1979); and partly to the view that molluscs and annelids must be close relatives ('it must be a coelom because annelids have a coelom').

The conclusion, then, is that molluscs are not a coelomate group according to any useful definition of that term; and the mesodermal cavity they possess is not homologous with the annelid-arthropod coelom. (One of the few recent dissenting views may be found in Götting, 1980*a,b*.) They can best be regarded as acoelomate, or perhaps as 'pseudocoelomate' following Clark (1979), since a relatively spacious primary body cavity is present. Even the term 'haemocoel' should probably be avoided, as it implies homology with the arthropod condition. Salvini-Plawen (1980*a*, 1981) includes the molluscs in a super-phyletic grouping termed the 'Mesenchymata', together with platyhelminths, nemerteans and entoprocts. With this view of their status, molluscs can be moved yet another step 'down' the hierarchy of spiralian phyla (fig 10.1*C*); they are neither segmented nor coelomate, and certainly cannot be designated as part of a triumvirate of advanced protostomes with annelids and arthropods.

At this point it is worth recalling briefly the 'archecoelomate' arguments discussed elsewhere (see particularly chapter 2), suggesting that relatively large coelomate animals actually preceded the acoelomate flatworm stage. If that were the case molluscs might reasonably represent a half-way stage in the *loss* of a spacious body cavity from something like a sipunculan/priapulid. Most of the arguments advanced above imply that this is functionally unlikely; and taken together with arguments against the general principle that this was how the Metazoa evolved, as given in the earlier chapters, perhaps this view can be dismissed from further serious consideration, retaining the view that molluscs are derivatives of acoelomates like flatworms rather than the reverse being true.

Molluscs as 'protostomes'?

A further step in establishing molluscan status must be to consider their embryology and its implications. Conventionally they show classic protostome development (see chapter 5), with spiral determinate cleavage, a primary mouth formed from the blastopore, 4d-derived mesoderm containing a schizocoelic coelom, and a trochophore larva. These features should indicate a common ancestry with annelids and certain other worm-like phyla, and set the molluscs some way 'up' the protostome lineage as in fig 10.1*C*; a view strongly supported by Hammarsten & Runnström (1925) and more recently by Vagvolgyi (1967). The general validity of these embryological features was discussed in chapter 5; here matters need only be reviewed briefly, in relation to the expression of such features in both annelids and in flatworms. Good reviews of the details of molluscan embryology are given by Raven (1966) and Verdonk & van den Biggelaar (1983).

Molluscs do have spiral cleavage, except where (as in cephalopods and some gastropods) the embryo is meroblastic and excessive yolk obscures and distorts the planes of cleavage. Exactly the same can be said of annelids, flatworms, nemerteans and many other groups. The occurrence of spiral cleavage cannot therefore be used as evidence of a particularly close relationship between annelids and molluscs, or of descent of one group from the other; both groups could reasonably have acquired it independently from flatworm ancestors. Furthermore, the cleavage of molluscs has its own idiosyncracies (Vagvolgyi 1967; Götting 1980*a*,*b*), for example in the formation of a 'molluscan cross' (fig 10.4*A*). The cells that will give rise to the ciliary band at the equator of the larva assume a cross shape in the early embryonic stages, centred on the animal pole; and whereas the arms of the cross are inter-radial in annelids and their close relatives, they are distinctly radial in many molluscs (and also in sipunculans - see chapter 8). These cells are also derived from different micromeres in the two phyla Annelida and Mollusca. Again, no clear link between molluscs and annelids seems to exist here, though some believe a link between molluscs and sipunculans is indicated (Inglis 1985), while others discount all the differences of embryology as trivial (Götting 1980*a*,*b*) and keep annelids and molluscs fairly close together phylogenetically. But the flatworms and nemerteans, usually having no differentiation of a cross, may be the lowest common denominator for this feature, which cannot be given great phylogenetic weight (Salvini-Plawen 1985).

Many species of mollusc are also unusual in having 'polar lobes' (fig 10.4*B*) during their early cleavage, causing an apparent three-cell phase (the

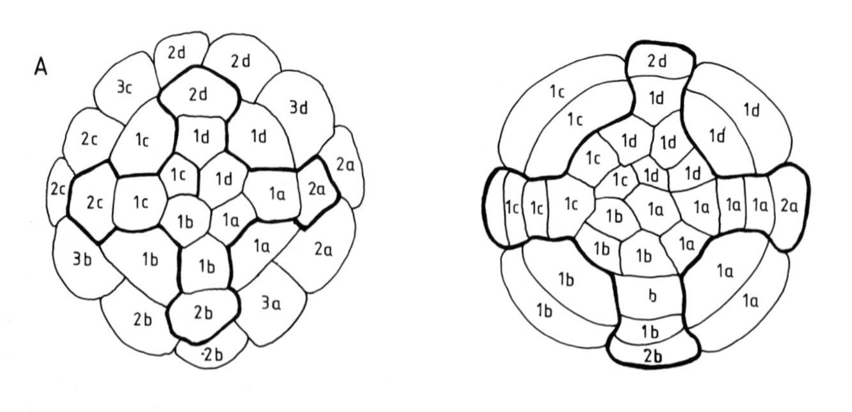

Fig 10.4. The embryology of a mollusc showing *(A)* the appearance and development of the 'molluscan cross', with the prototroch cells as a radially disposed cross centred on the animal pole; and *(B)* the presence of a polar lobe, giving an apparent 'three-cell' stage.

'trefoil' conformation). This can be found in most molluscan classes, but its absence in archegastropods and increasing apparency in meso- and neogastropods may argue for a fairly late acquisition in the phylum; thus the polar lobes, though setting the molluscs somewhat apart, cannot help in revealing the direction of evolution between the main phyla here considered.

Molluscan gastrulation begins relatively early in development, when there are 70-200 cells, and varies according to the size of the blastocoel. Where this is spacious, the gastrula forms by invagination and the blastopore is distinct. In other cases, gastrulation is epibolic, the micromeres growing over and around the macromeres. However, in both types of gastrulation the molluscs are characterised by a very wide blastopore (though whether or not this can be regarded as a homologous structure in the two cases is a moot point considered in chapter 5), which later narrows into a slit. Almost invariably this slit then closes from back to front, leaving a small aperture to become the mouth. Hence, by normal definitions, the mouth is primary and is formed from the blastopore: and the molluscs are protostomes. There are a few exceptions to this behaviour though; in some molluscs the entire blastopore remains open, and in others it completely closes (giving rise directly to neither mouth nor anus), whilst in at least a few cases the blastopore clearly becomes the anus.

The molluscan mesoderm is formed from the 4d cell (with the presumably secondary exception of *Viviparus*), as also is the hind–gut. This 'endomesoderm' cell normally divides into two, and then starts to sink into the blastocoel. Some enteroblasts are split off, and the remaining 4d derivatives then become the two true mesodermal teloblasts, budding in either direction to produce the mesodermal bands. However, as already mentioned, these bands do not cavitate to produce a spacious schizocoelic coelom, and the 'coelomic' cavity in the pericardium arises later in just one piece of the mesodermal tissue. Hence, though technically 'schizocoelic', the molluscs are not homologous with the annelids in these aspects of their embryology.

Finally, the 'trochophore' larva does indeed occur in molluscs, where it normally precedes a unique veliger larval stage (fig 10.5); and this has perhaps been regarded as the strongest feature linking them to annelids. The essential features defining this larval form are discussed in chapter 5; for present purposes, the important point to remember is that the term has been much abused and many reportedly trochophoral types are in fact dissimilar in important details (Salvini-Plawen 1973, 1980*b*). For example supposedly rather similar larvae also occur in polyclads (Müller's larva), in nemerteans (the pilidium larva) and in sipunculans, but none of these should really be called trochophores according to Salvini-Plawen. The molluscan 'trochophore' lacks a metatroch, lacks apical ocelli, and lacks protonephridia; these differences are enough for Salvini-Plawen to allow it at best the term 'pseudotrochophore'. In any case larval similarities bear a high risk of being convergent even between widely separated phyla due to the similar selective pressures on all small and ciliated marine dispersive phases. Annelid and mollusc trochophores could be similar due to common descent, but a separate derivation from flatworm ancestors under similar constraints seems very much more probable. At best, all these groups may be united by the more primitive 'pericalymma' larval type (Salvini-Plawen 1973, 1985), from which all 'trochophores' may be convergently derived.

Overall, embryological evidence leaves it possible that molluscs belong on a common line with the annelids (fig 10.1*C*), but much more likely that they are a quite separate evolutionary offshoot from a small, acoelomate, ciliated ancestor (fig 10.1*D*). Given that there are relatively few 'options' for each of the embryological features concerned, it is entirely plausible that annelids and molluscs should show a degree of similarity even if they have been evolved separately from a proto-flatworm stage; and most recent authors have concurred in seeing the two groups as only minimally related. Salvini-Plawen (1985) allows molluscs to be part of a Spiralia assemblage, but of a mesenchymatous grade closer to flatworms and nemerteans. An analysis of adult form in the molluscs, and of the manner in which such form could be derived from flatworms, may give further support to the view that molluscs are phylogenetically distant from the annelids and other coelomate worms, and no part of any reasonable 'protostome' assemblage.

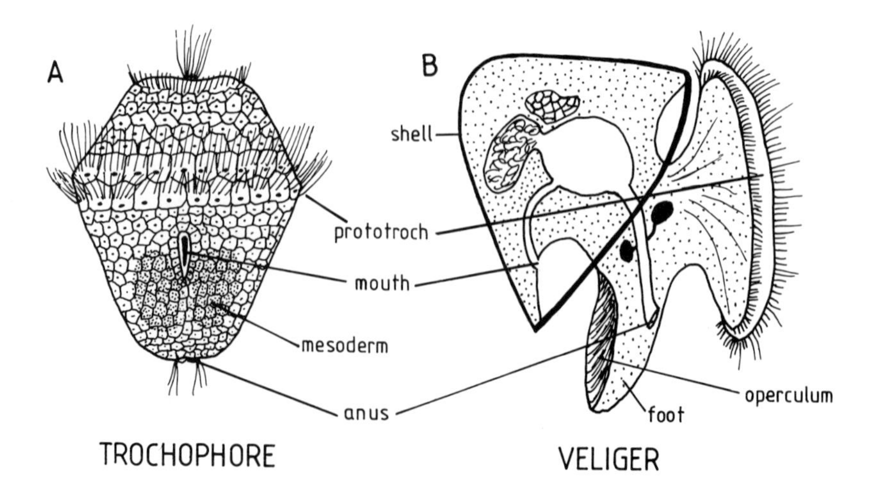

Fig 10.5. The trochophore and veliger larvae of a typical marine mollusc with non-yolky embryology.

10.4 A flatworm origin for molluscs - possible mechanisms

The preceding section examined evidence for the traditional status of molluscs, and concluded that they are neither primitively segmented nor coelomate, and are not necessarily part of a common lineage with the annelids. This leaves them as a separate and independent offshoot from the early acoelomate metazoans (fig 10.1*D*), exemplified by extant free-living flatworms (turbellarians); simple small, ciliated, acoelomate forms, creeping over substrates with a typically worm-like ciliary-mucus glide, and with a degree of repetition in various organ systems. Such a view of their ancestry is now the consensus amongst zoologists (Stasek 1972; Salvini-Plawen 1981, 1985; Runnegar & Pojeta 1985). It is strongly supported by a comparison of adult forms; flatworms and molluscs share rather similar nervous systems, musculatures, and modes of intracellular digestion; and there may be homologous features in their epithelia (see chapter 6). But what selective pressures, and what initial changes of form, could have plausibly led from a flatworm ancestor to the evolution of the first true molluscs? Specifically, is there a likely sequence for the acquisition of the main molluscan traits that are distinct from flatworms: body cavity, through-gut, shell, mantle cavity, gills, visceral mass, condensation of organ systems, and radula?

Acquisition of a through-gut

Any small early metazoan acquiring an anus would apparently be at an advantage over its competitors (see chapter 6) in being able to pass food continuously through its gut and achieve local specialisation of gut function to enhance its efficiency in dealing with particular nutrients; overall, such an animal should have potentially higher growth rates. Thus Yochelson (1979) has argued that this may have been the first and vital step in the evolution of molluscs from flatworms. With its through-gut the animal could grow faster and larger, which should have a number of consequences. Perhaps it would require a shell for protection as it became increasingly rewarding as food for others; or, with improved digestion of more exotic foodstuffs, would be faced with waste products that could best be disposed of as an inert shell. (The reasons for shell development are considered further in the next section.) Once the shell was present, for whatever reason, the animal would require separation of the locomotory surfaces (foot) from the increasing mass of viscera constrained by the shell; and it would require increased surfaces for respiration (gills), seeking to house these protectively under the eaves of the shell, in a mantle cavity.

The problem with this view is that a through-gut is ineffective, and possibly even counter-productive, in an animal lacking some form of circulation - minimally a very fluid parenchyma or incipient body cavity. In a flatworm the gut itself is diverticulate and serves to distribute nutrients; but in a flatworm with an anus, tubular gut and through-flow of food, many parts of the body would receive no digested foodstuffs. As the through-gut achieved its potential by specialising in particular types of absorption in particular areas this problem would be exacerbated - some portions of the body tissues would be adjacent to gut surfaces absorbing peptides, others might receive only fatty acids, and others again would be effectively starved. Clearly a through-gut must be accompanied by a body cavity with some circulation of distributive fluids, as the nemerteans clearly show us (chapter 8). Such a gut is only likely to be evolved, then, in animals above a certain size, notwithstanding its occurrence in some modern minute metazoans of secondarily reduced size. The acquisition of a body cavity by the acoelomate molluscan ancestor would therefore have to precede acquisition of an anus, or at least to accompany such a change; and improvements to the gut alone are unlikely to have been the starting point for molluscan evolution.

Acquisition of a shell

The shell is, at least superficially, the most characteristic element of the molluscan diagnosis, and as such a plausible starting point for their evolution. The shell may have been organic and mucoid initially (Stasek 1972), as the conchiolin matrix precedes the calcium salts in modern molluscan ontogeny; but it must have been mineralised very early in molluscan history. Perhaps initially this involved spicule formation (Runnegar & Pojeta 1985), as still occurs in the probably primitive aplacophorans (Scheltema 1978; Salvini-Plawen 1981, 1985), and in some chitons.

The potential for mineralisation is probably inherent in the eucaryote cell, and deposition of calcium carbonate has occurred repeatedly in diverse groups through evolution; yet there is no real consensus as to the prime reasons for such depositions (chapter 3). The assumption that shells are mainly selected for to provide protection from predators is held by Glaessner (1984), in a comprehensive discussion of these problems, to be 'naive'. He cites a number of alternative functions: energy conservation in locomotion, giving fixed muscle attachments (see Clark 1964; Vogel & Gutmann 1981); storage of excess calcium, given the vital role of calcium ion regulation in cell metabolism (Lowenstam & Margulis 1980); a waste product sink; support, to raise the body above the sea floor and facilitate food gathering; and built-in channelling of water currents. It is clear that biomineralisation became widespread at the

Pre-Cambrian/Cambrian transition (again Glaessner gives a review of possible geophysical reasons for this; and see chapter 3), so that at least for some animals a major spur for shell development may have been that other phyla became similarly endowed with hard and potentially offensive structures. Early molluscs would have gained several advantages from their shells, in any case; perhaps further debate on the question of which advantage was the dominant selective force is superfluous here.

Stasek (1972) presented a convincing argument for the role of shell development in entraining the essential molluscan framework (fig 10.6) (though it should be noted that his starting point was a flatworm that had already acquired an anus, and that this may not be a necessary or helpful assumption in view of the scenarios discussed above). Once present (fig 10.6*A*), in whatever form and for whatever reason, a dorsal covering or shell would severely reduce the free respiratory surface of a proto-platyhelminth, especially if atmospheric oxygen levels were lower than at present (chapter 3), and would necessitate some elaboration of the animal's surface as gills (fig 10.6*B*). These would ideally be placed postero-laterally to take advantage of anterior-posterior water currents, and perhaps several pairs would be developed initially to provide as much oxygen uptake as possible, giving rise to a degree of pseudometamerism. However gills alone would not keep the central part of the animal's tissues (beneath the shell) oxygenated, and the addition of circulating fluids would be an almost immediate necessity - the animal must have acquired a body cavity, probably via cavitation of the mesoderm to produce 'haemocoelic' sinuses (fig 10.6*C*). Once this step was taken, the animal could get larger; the shell could provide eaves to protect the gills, and foot and viscera could be separated to give the characteristic simple

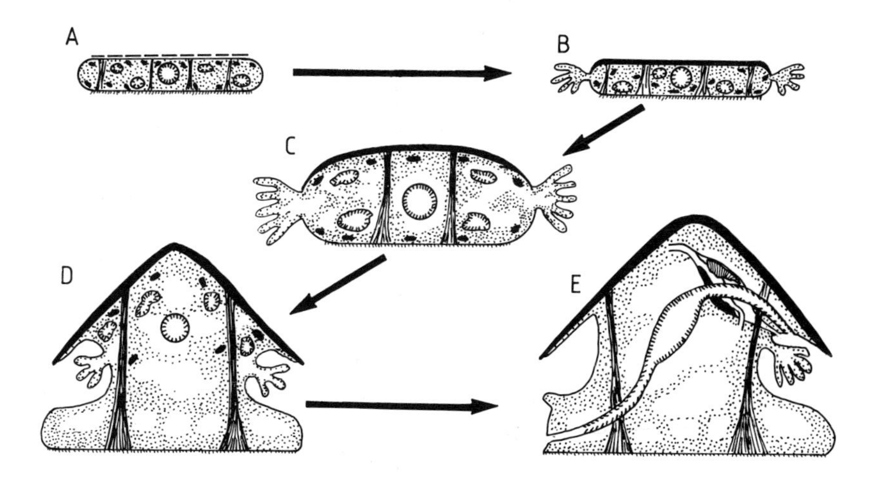

Fig 10.6. A scheme for the evolution of the molluscan Bauplan from a flatworm, largely brought about by the addition of a shell. Further explanation is given in the text.

molluscan shape (fig 10.6*D*). With locomotory activity thus restricted to the still flexible foot, typical tetraneury with only ventral and ventro-lateral connectives persisting would be appropriate (see fig 2.9). All these changes could have taken place in concert and rather rapidly (Runnegar & Pojeta 1985).

Two further changes, not made explicit by Stasek, might also be predicted now. This larger creature, with its body cavity, could at last make use of a through-gut, and indeed positively required one to support its size and metabolic requirements. And its circulation would also allow it to condense all the scattered organ systems that were so essential to its acoelomate ancestor - for example, instead of numerous flame cells throughout the body tissues, a single pair of excretory organs would suffice to clear all waste products deposited in the circulating fluids. The animal would thus have acquired virtually all the features that define a mollusc (fig 10.6*E*) as a simple consequence of adding a shell to a flatworm. The only exception is the radula, which must be viewed as a separate and unique acquisition of the ancestral molluscs.

Clearly, however, the shell had immediate disadvantages for the ancestor that could only be avoided if a body cavity was acquired, and this argument therefore rests on a complex of features being developed more or less synchronously. But it does, in Stasek's terminology, explain how the 'molluscan framework' could be 'unfolded' from a flatworm origin.

Acquisition of a body cavity

Each of the above scenarios required the addition of a body cavity at or very soon after the inception of the primary feature considered. It seems possible, therefore, that acquisition of that cavity was itself the starting point for the evolution of molluscs. But there are clearly difficulties with such a view, in that a body cavity and the increase in size it allows cannot explain why the molluscs took on the form they now show. Flatworms with retained blastocoels plausibly evolve into larger 'pseudocoelomate' worms, whereas with added secondary cavities they might give rise to unsegmented 'coelomate' worms; either way the immediate outcome should be worms resembling modern sipunculans and priapulids. It clearly requires some additional feature to direct changes towards a typically molluscan plan; and the most logical explanation, as was shown above, is the shell, producing the characteristic 'incomplete exoskeleton' design discussed in chapter 2, with all the advantages that entails. It is relatively unimportant whether this change preceded, accompanied or immediately followed the acquisition of an internal cavity; the shell was the critical development.

It may be concluded, then, that derivation of molluscs from flatworms is feasible in terms of functional morphology, and that the evolution of some form of shell was probably the essential feature in separating this phylum from early worms. It is much more difficult to devise any satisfactory scheme deriving molluscs from more advanced worms already possessing body cavities or septae, while maintaining functional continuity. From a flatworm ancestor, the shell itself is enough to entrain virtually all the other features that characterise the basic mollusc, and changes in (or even losses of) the shell have been the major effector in subsequent separation of the classes and subclasses of molluscs. By its very nature a shell clearly confers some specific and wide-ranging advantages on its possessor, and it is quite likely to have been acquired several times by different groups of early metazoans (as is fairly clear from the fossil record, discussed in chapter 3). So the shell cannot in itself define a mollusc. Accordingly the radula must be recalled to the forefront of the argument, as a separate and necessary step for a hypothetical flatworm ancestor to take in order to give rise to the true molluscs; and certainly in some classes of the molluscs, notably gastropods (Graham 1979), the radula has been a major spur to adaptive radiation. In essence, then, a mollusc is the end-product of a proto-platyhelminth that acquired not one but two independent features, shell and radula. The presence of chitinous teeth in the pharynx of the gnathostomulids, which in other respects closely resemble extant platyhelminths, suggests that independent acquisition of a chitinous tooth-bearing structure by flatworms is not implausible (Riedl 1969; Stasek 1972). There were certainly other lineages of primitive shelled forms in the Cambrian fauna (quite apart from the shelled brachiopods, discussed in chapter 13); but of these, only the molluscan line, perhaps precisely because of the advantages of the separately acquired radula, survived and diversified.

10.5 Molluscan palaeontology

Having developed a view of how molluscs may have evolved, as an independent lineage, from early flatworm-like creatures, it is possible to test this hypothesis as far as possible against the known fossil record, for the molluscs are one of the very few groups where this exercise is feasible (chapter 3). Good reviews of fossil molluscs and molluscan-like forms are given by Runnegar & Pojeta (1974, 1985), Runnegar (1980), Pojeta (1980) and Glaessner (1984).

A first point that can be made is that there were indeed a number of shelled animals in the early Cambrian that most authors do not regard as molluscs - several independent 'experiments' with adding shells to worms do seem to have been attempted in the early evolution of Metazoa. It is in fact very difficult to separate true molluscs from fossils which are merely mollusc-like (Yochelson 1979); with only the shell to go on, and no preserved radula associated, the best clues may be had from shape and microstructure of the remains, but a conclusive answer is rarely achieved. Much disagreement remains over these early shells, therefore, but most authorities now agree that some of them, notably tentaculites, cornulitids and the extremely abundant hyolithids, probably belong to extinct non-molluscan phyla. Dissenting views and useful discussions may be obtained from Yochelson (1978, 1979).

The earliest Cambrian (Tommotian) rocks contain an array of convincing true molluscs, though, and most of the extant groups were well-established by this period (fig 10.7). These early fossils were generally very small limpet-like creatures, slightly planospiral or with a helical coil to the single valve. Such 'micromollusc' forms, of which *Helcionella*, *Scenella* and *Aldanella* are characteristic, have been assigned to various classes, but Runnegar & Pojeta (1974) make out a compelling case for regarding them as early monoplacophorans, and deem them to be ancestral to all molluscan groups except the Aplacophora, making descent of molluscs from large coelomates seem most unlikely. Molluscs remained small and of low diversity throughout the Cambrian, including a few archegastropods, rostroconchs and bivalves amongst the more numerous monoplacophorans. There was then further radiation and considerable size increase in the Ordovician (Runnegar & Pojeta 1985).

Thus in size and shell form the very earliest shelled molluscs seem to be in accord with our expectations of molluscan evolution, being readily derivable from flatworms by the route described above; and perhaps the changes in fossil molluscs at the start of the Ordovician correspond with their proposed acquisition of larger internal body spaces, gills and mantle cavities, representing the point at which the full molluscan body plan was realised, and resulting in explosive radiation. In fact Glaessner (1984, page 163) concludes that, in relation to a turbellarian ancestry, 'Stasek's reconstruction of stages leading through a stage of a covering cuticle to a calcified shell has been confirmed by the study of numerous Tommotian Monoplacophora and their descendants'. Conveniently for our theories, the first known fossil radulas are also from the earliest Cambrian rocks (Firby & Durham 1974).

From such roots, gastropods (with distinct heads and torsion) were probably derived also in the Early Cambrian; though the arguments persist as to whether torsion precedes shell coiling (in which case all coiled shells are gastropods) or is caused by it, in which case some of these early coiled shells need not be gastropodan. The perennial arguments about the causes and consequences of torsion are not relevant here, but good discussions can be found in Stasek (1972), Russell-Hunter (1979), and Pennington & Chia (1985).

Chitons appeared in the Late Cambrian, with forms such as *Preacanthochiton* and perhaps *Matthevia*, probably by an anomalous acquisition of several centres of calcification in a monoplacophoran; (though again Yochelson (1979) interprets matters differently, placing these polyplacophorans much nearer to the root of the monoplacophoran/ cephalopodan assemblage, and gastropods more distant).

Cephalopods make their first appearance as fossils in the Late Cambrian, from septate shells such as *Knightoconus*, by acquisition of the diagnostic siphuncle to give forms like *Plectronoceras* (Holland 1979).

Bivalved shells appear in the very earliest Cambrian rocks, though most of them are referred to the extinct class Rostroconchia, lacking ligaments and paired adductors and having a univalve protoconch stage; *Heraultipegma* (formerly *Heraultia)* is probably the oldest of these. After diversifying through the mid-Palaeozoic era the rostroconchs went extinct in the Permian. Somewhere within the early Cambrian they had given rise to a true lamellibranch/pelecypod offshoot (Morris 1979), *Fordilla* being the first clear example, with bivalve calcification throughout life. Rostroconchs may also have given rise to scaphopods (where the shell originates dorsally and grows out to either side before joining up as a tube), already well-developed when first found as fossils in Ordovician rocks. These views of molluscan ancestry are summarised in figs 10.7 and 10.8; and further details are given in Runnegar & Pojeta (1985).

The problem of the aplacophorans remains, as they have no real fossil record. Since Salvini-Plawen (1969) drew attention to their phylogenetic significance, having only thin chitinous coverings in which calcareous spicules are embedded, most authors have accorded them unique status in a subphylum Aculifera, well separated from the other 'Placophora' or 'Conchifera' molluscs (see Stasek 1972; Runnegar & Pojeta 1974, 1985; Scheltema 1978; Salvini-Plawen 1985). They are almost certainly very 'primitive' molluscs, and geologically very old. Of the burrowing Caudofoveata and the browsing Solenogastres (better kept separate as two high-ranking taxa) only the latter retain a vestigial foot in the form of the ventral groove, moving in a manner very similar to flatworms; and both groups have very simple radulas. Some late Pre-Cambrian trace fossils of locomotion trails do indicate a body form of this type, rather than that of a simple worm (Glaessner 1984), but there is no other fossil record for this essentially soft-bodied group. Almost certainly the

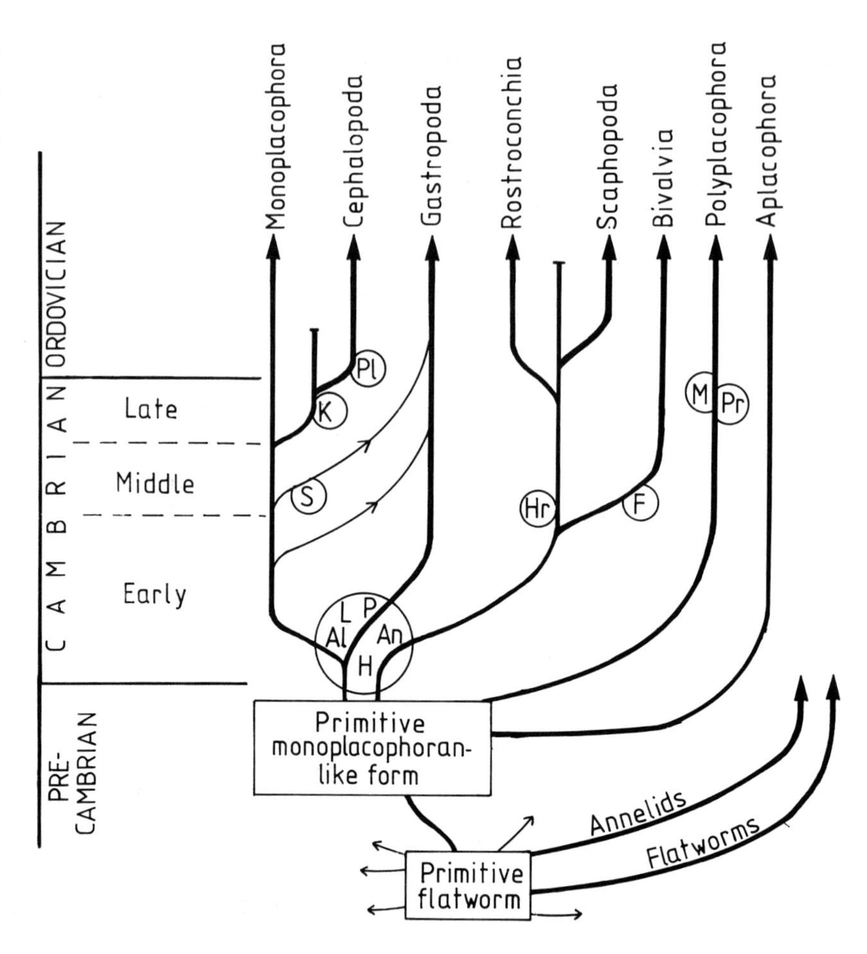

Fig 10.7. The geological pattern of fossil molluscs and suggested relationships between them, based on data in Runnegar & Pojeta (1974). Al,*Aldanella*; An,*Anabaraella*; F,*Fordilla*; H,*Helcionella*; Hr, *Heraultipegma*; K,*Knightoconus*; L,*Latouchella*; M,*Matthevia*; P, *Pelagiella*; Pl, *Plectronoceras*; Pr, *Preacanthochiton*; S,*Scenella*.

nearly shell-less molluscs did precede the more classic forms though (so that aplacophorans are not, as some texts imply, a later degenerate group). The early aculiferans probably had a reasonable endowment of dorsal spicules, so that the proposed sequence of evolutionary events caused by the restrictions of a dorsal covering should still make sense. Indeed, as Rieger & Sterrer (1975) indicate that the integument of some turbellarians bears calcareous spicules, there may be a homology here that would reinforce the derivation of aculiferans from flatworms, with other molluscs as a slightly later invention. A complete picture of the relationships between molluscan groups incorporating these views is given by Salvini-Plawen (1985), and a simplified version of this scheme is shown in fig 10.9.

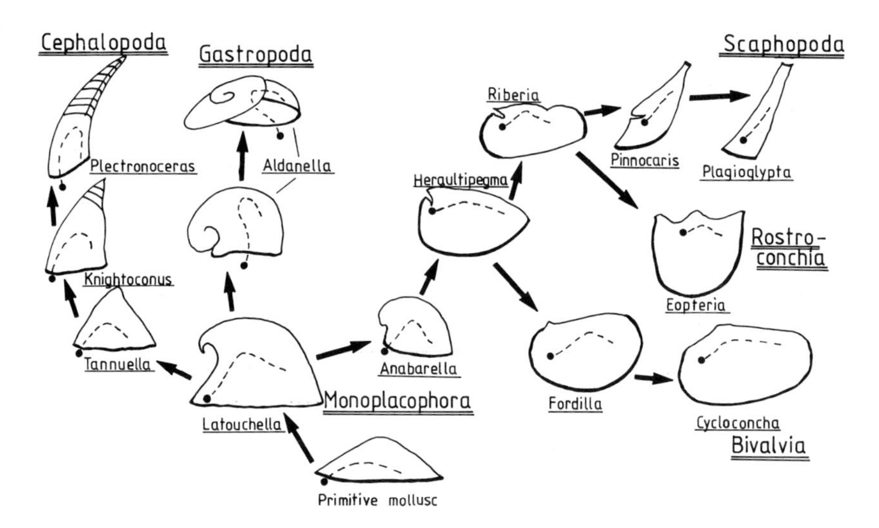

Fig 10.8. The origins of monovalved and bivalved molluscan groups from a monoplacophoran-type ancestor, modified from Runnegar & Pojeta (1974).

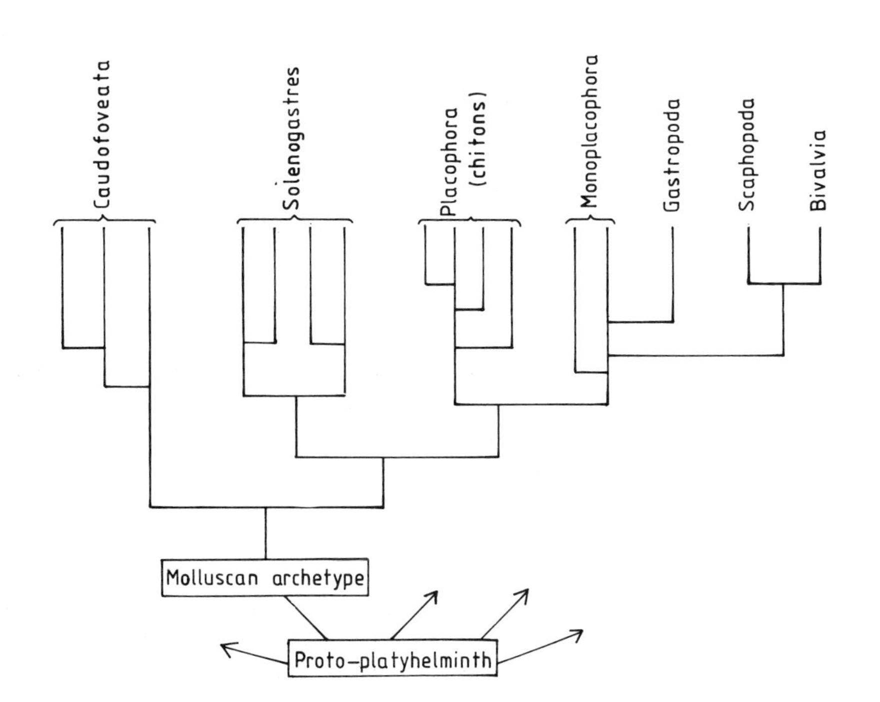

Fig 10.9. A phylogenetic scheme of molluscan radiation, including the aplacophorans; (based on Salvini-Plawen 1985).

10.6 Conclusions

On balance, the fossil record does not conflict with other evidence indicative of molluscan phylogenies set out earlier, and in some respects palaeontological findings substantially support the proposed flatworm origins. The best we can do with available knowledge is to allow the molluscs a unique status as shelled, acoelomate/pseudocoelomate derivatives from the ancestral flatworm-like metazoans, having little or no relationship with the classical major sub-divisions of the animal kingdom and a fundamental design showing little in common with other phyla.

The argument that molluscan morphology and evolution has been largely determined by the acquisition of a shell closely resembles one we shall meet shortly in considering arthropods (chapter 11) - in that case, the presence of a complete exoskeleton necessitates all the other characteristics that are associated with the insects, crustaceans and arachnids. For that group, such an argument can be used to support a polyphyletic viewpoint - that the arthropod assemblage in fact contains several independent lineages which have come to resemble each other through constrained convergent evolution. Could the same apply to the molluscs? Are we in fact dealing with separate groups, only resembling each other because the shell, separately acquired for obvious defensive purposes by various early flatworms, dictates their final form? This is clearly a possibility, but several points militate against a polyphyletic view of the molluscs. Firstly, the shell is a much less restrictive construction than the arthropod cuticle, and leaves the final design of its owner more open-ended as was discussed in chapter 2; convergent evolution is therefore less likely. Secondly, the one feature of molluscs that cannot be attributed to shell-acquisition is the radula, yet this occurs in some members, at least, of every molluscan class. Its specialised nature makes multiple independent origins unlikely; more probably it was acquired in an early acoelomate shelled form. Another possible homology throughout the molluscan classes is expressed in the gills (as specialised 'ctenidia') and pallial complex, discussed by Russell-Hunter (1979), and showing an apparent constancy of organisation that owes nothing to flatworm origins.

Since we are given two or more clearcut and apparently independent features defining the molluscan phylum, polyphyly at this level becomes a non-parsimonious explanation (though the possibility that several of the conventional classes of molluscs are in fact polyphyletic remains strong, as Stasek 1972 points out). It is more reasonable to accept that the Mollusca, despite their extraordinary present diversity, are a natural and acceptable grouping, set apart from all other phyla within the animal kingdom but strongly influenced by their flatworm ancestry.

UNIRAMIA - Myriapoda
UNIRAMIA - Insecta
CHELICERATA
CRUSTACEA

11 Arthropod phylogeny

11.1 Introduction

All the animals that are included within the arthropod assemblage show a degree of similarity frequently lacking in other groups that are nervertheless recognised without dispute as single phyla. For example it is quite easy to accept a relationship between all the leg-bearing spiders and insects, crabs and woodlice; but difficult to appreciate the common ancestry of the snails, clams and squids that together constitute the molluscs. It is therefore odd, at first sight, that there should be particular disagreement about the status of arthropods as a single phylum.

Historically, theories involving monophyly certainly prevailed, and the occurrence of a major dispute about polyphyly is fairly recent. All the arthropods share a cuticle made of chitin and proteins, all show segmentation with at least some segments bearing paired articulated limbs, and all show rather similar patterns of cephalisation and pre-oral segments. Thus early authors found few problems in uniting them as a taxon, and the term 'arthropod' itself dates back to von Siebold in the mid-nineteenth century, coined as testimony to the jointed legs of all these animals. Through the nineteenth and early twentieth centuries the group retained its identity, either as a phylum or as a sub-group (with the annelids) of an even larger phylum called Articulata. The main features of the four principal types that constitute the arthropods (Crustacea, Insecta, Myriapoda and Chelicerata) are shown in table 11.1. Different phylogenetic schemes that have been proposed to link these four sub-groups together are usefully reviewed by Tiegs & Manton (1958); it is evident that essentially monophyletic schemes can differ enormously in detail, as virtually every possible pairing and hierarchy of groups has been attempted. Hence taxa such as 'Mandibulata' or 'Antennata' (insects, myriapods and crustaceans), or 'Tracheata' (insects, myriapods and chelicerates - as opposed to the crustacean 'Branchiata'), or 'Schizoramia' (trilobites and chelicerates, sometimes with crustaceans) have been canvassed. Current advocates of monophyly still show no accord about the status of groups within their phylum, though a few points of agreement have emerged to (if anything) complicate the issue. For example the myriapods are now often split into four separate groups, each of equal status to the insects, and frequently allied with them to give three main assemblages of extant 'arthropods'. In addition the numerous extinct trilobites have come to be regarded as a further distinct major taxon, lacking very clear affiliations with the extant groups. Despite these rather limited agreements, and some persistent problem areas, many authors retain some version of the classical view that all arthropods are related more closely to each other than to members of other phyla, that their nearest common ancestor was itself definable as an arthropod, and that their similarities are too great to be accounted for by anything but shared ancestry.

However, since the exhaustive works of Sidnie Manton were published between the 1940s and 1979 there is now an alternative view, with evidence amassed to show that three major groups - Crustacea, Chelicerata, and Uniramia (essentially insects plus 'myriapods') - should be recognised as phyla in their own right, having no common ancestor that was itself an arthropod, and thus representing separate end-products of the process of 'arthropodisation'. The similarities clearly visible between groups then have to be regarded as extensive convergence, creating new explanatory problems. The evidence for and against these perhaps surprising and still controversial ideas forms the subject of this chapter.

Table 11.1 *The main extant arthropod groups - a simple diagnosis*

	Appendages			Tagmosis	Other features
	Antennae	Legs	Mouthparts		
Crustacea	2 pairs	*n* pairs, biramous	mandibles maxillae 1 maxillae 2	variable	nauplius larva
Insecta	1 pair	3 pairs, appear uniramous	mandibles maxillae 1 maxillae 2	3: head thorax abdomen	2 pairs wings (usually)
Myriapoda	1 pair	*n* pairs, appear uniramous	mandibles maxillae 1 (+ variable)	2: head trunk	
Chelicerata	0	4 pairs, appear uniramous	chelicerae (+ pedipalps)	2: prosoma opisthosoma (=cephalothorax, abdomen)	

11.2 The case for monophyly

The general argument

In one sense, the case for a monophyletic grouping of all arthropods hardly needs to be made: they are self-evidently all rather similar animals. It is by no means difficult to assemble lists of common features; a typical example is given in table 11.2, modified from Snodgrass (1938), Sharov (1966) and

Table 11.2 *The shared features of arthropods*

1) Cuticle, secreted by epidermis, with chitin and protein predominating.
2) Localised sclerotisation of cuticle.
3) Metameric segmentation of trunk, with articular membranes between sclerites.
4) Pre-oral segments present.
5) Tagmosis, especially cephalisation.
6) Periodic moulting of cuticle, controlled by ecdysones.
7) Segmental jointed appendages.
8) Similar inter-segmental tendon systems.
9) Dorsal and ventral longitudinal muscle metamerically arranged.
10) All muscle striated.
11) Muscle tonofibrillae penetrate cuticle.
12) Lack of cilia.
13) Brain with at least ocular protocerebrum and usually antennal deuterocerebrum; and paired ventral nerve cords.
14) Presence of compound eyes.
15) Cuticular lining of fore- and hindguts.
16) Haemocoel as principal body cavity, characteristic blood distribution and ostiate heart. Coelom limited, perhaps due to reduction.

Boudreaux (1979). In the face of such a broad spectrum of similarities, covering many aspects of external morphology and internal anatomy, it rests with proponents of polyphyletic views to present strong counter-arguments and to explain away the resemblances as far as they can.

Nevertheless, it is also instructive to examine some of the detailed arguments used by other authors in support of their monophyletic interpretations.

Cuticle and moulting

Perhaps the most obvious similarity between all extant arthropods is their possession of a continuous multi-layered exoskeleton, secreted by the epidermal cells and constituted principally from fibrils of chitin and a matrix of proteins. Three layers are generally present, with a thin epicuticle overlying hardened exocuticle and softer endocuticular material that can be resorbed and recycled. Not only is the basic chemistry of the cuticle similar in all groups, but the microstructural patterns of the exo- and endocuticle are also the same in insects, arachnids and crustaceans. All show the helicoidal arrangement of chitin fibrils described by Bouligand (1972), with successive horizontal layers of the microfibrils oriented at a slight angle to the adjacent layers; good reviews of this and other aspects of cuticles in the groups can be found in Bereiter-Hahn *et al.* (1984).

Besides this constancy of chemistry and structure, all arthropod cuticles of course also share a common pattern of moulting and growth, controlled by a range of hormones in which ecdysone-type steroids play a central role.

Head structure

The segmentation and appendages of the head in different arthropods have been a long-standing source of controversy, and have been interpreted so differently as to provide either a principal support for a monophyletic scheme (Sharov 1966; Weygoldt 1979) or a back-up to polyphyly (Manton 1977; Manton & Anderson 1979). A summary of head segmentation patterns, with some of the controversies, is given in fig 11.1; clearly different groups seem to have antennae, mouths and jaws on quite different segments.

Snodgrass (1938) gives a careful analysis of arthropod heads, maintaining that a simple condition of the presegmental acron plus one true segment, as currently found in anostracan crustaceans, was 'unquestionably the primitive head of all the mandibulate arthropods'. Weygoldt (1979) presents a similar cohesive picture of homologies in insect and crustacean heads, retaining their designation as Mandibulata, and he believes that some homologies with chelicerates can also be identified. He proposes a primitive arthropod with pre-oral antennae, four pairs of post-oral biramous appendages, and a large labrum. Sharov (1966) has presented a more controversial tabulation of head homologies, invoking both labral and ocular segments in front of the antennae to try and fit all arthropods and annelids into one scheme. Each of these schemes has also differed about the relation of external segmentation to internal divisions of the brain. However, whilst no single view of head segmentation has achieved wide acceptance, it is clear from the literature that head morphology is commonly regarded as good evidence for shared ancestry in the arthropods.

Whatever the homologies involved, arthropod heads do show very marked dissimilarities in extant adults. Crustaceans are essentially trignathan (mandibles, maxillae 1 and maxillae 2), though the second maxillae are absent in some primitive forms. Onychophorans such as *Peripatus* have only one pair of mouthparts, described as 'feeding claws'. Myriapods are basically dignathan, but with extra mouthparts derived from the trunk in some of the classes, notably the centipede poison claws; and there is apparently full trignathy in the class Symphyla. Insects are also trignathan. Chelicerates have fewer pre-oral segments than other groups, with an associated lack of deuterocerebrum in the brain, and often only the chelicerae are used in feeding. It is not easy to derive all these patterns from any one ancestor, yet given the variation of mouthparts even within small and clearly related groups it is no easier to discount monophyly of the whole assemblage either.

One further aspect of head structure, the problem of differing mouth positions in chelicerates (segments 2-3) and other groups (segments 3-4), is perhaps less controversial now. All arthropods differ from annelids in having

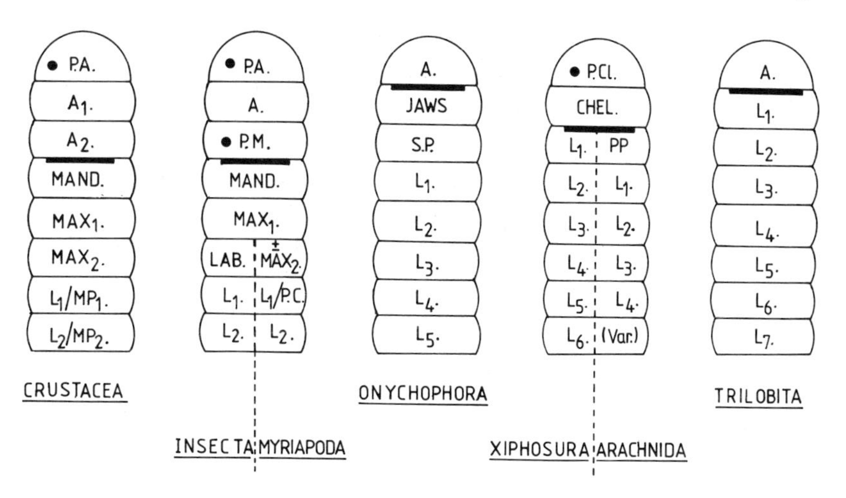

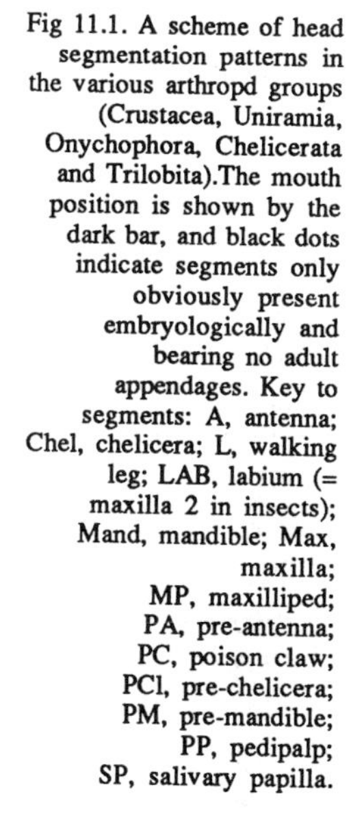

Fig 11.1. A scheme of head segmentation patterns in the various arthropd groups (Crustacea, Uniramia, Onychophora, Chelicerata and Trilobita).The mouth position is shown by the dark bar, and black dots indicate segments only obviously present embryologically and bearing no adult appendages. Key to segments: A, antenna; Chel, chelicera; L, walking leg; LAB, labium (= maxilla 2 in insects); Mand, mandible; Max, maxilla; MP, maxilliped; PA, pre-antenna; PC, poison claw; PCl, pre-chelicera; PM, pre-mandible; PP, pedipalp; SP, salivary papilla.

a fundamentally ventral and backwardly-directed (as opposed to terminal) mouth, suitable for particulate feeding; this change had already occurred by the Cambrian, and was accompanied by a pre-oral shifting of one or several segments, notably an antennal somite. In some groups, particularly those adopting a terrestrial lifestyle and eating bulkier foodstuffs, the mouth may then have returned to a somewhat more terminal position by the anterior segments bending upwards, with one or more of them being lost. Because of these movements, determined by functional needs, the mouth could reasonably come to lie between different segments in modern arthropods and still be fully compatible with monophyletic origins.

Eye structure

Both compound (multi-faceted) and simple ocellar eyes are widespread in the arthropods. Trilobites often had both types, as does *Limulus*; modern arachnids have only simple eyes, and most insects and crustaceans have both types. It has frequently been claimed that as compound eyes do occur in all three major branches of the arthropods, they must be a monophyletic group. Occurrence alone is inconclusive, of course; other authors have maintained with equal conviction that as compound eyes contain only the same components as ocelli they could have evolved convergently and are only superficially similar. Recently, however, Paulus (1979) has reviewed the detailed structure of eyes more carefully, and concludes that their similarities are too great to be accounted for by convergence; rather, the primitive arthropod head must have borne compound eyes right from the beginning, and there has been only limited subsequent modification in different groups. Eye structure is taken as an important support for monophyly.

In detail, all compound eyes are constructed from ommatidia, each with a central rhabdom formed from the microvillous borders of individual photoreceptor (retinula) cells. This rhabdom shows marked variation within and between groups, and Paulus recognises three basic types which reflect functional considerations; the rhabdom therefore does not help when considering phylogeny. However, the overall ommatidium and the cellular constituents thereof are less functionally constrained, yet are nearly identical in most insects and crustaceans, supporting the 'Mandibulata' assemblage. The constituent cells can be matched precisely, with four cells primitively contributing to the cone and seven (or perhaps eight) to the rhabdom. Myriapods have a different eye structure, though (simple in all but the centipede *Scutigera*); and chelicerate eyes are very different again, lacking a crystalline cone between lens and rhabdom and with an anomalous disposition of the retinula cells.

Paulus' monophyletic interpretation of arthropod eyes clearly relies heavily on the insect-crustacean similarities, and therefore requires myriapods, the insect's sister-group, to have secondarily modified eyes rather than a primitively simple form. It also requires homology of the mandibulate ommatidium with the ommatidium of *Limulus*, though this is less convincing and Paulus retains the possibility of separate evolution of the two types from a precursor. In either case, arachnids subsequently abandoned the faceted eye. They now have a basic set of four simple median eyes; and as this may also be the normal quota of median eyes in mandibulates, the chelicerates should in Paulus' view be set well apart from the insects and crustaceans but within the same phylum.

Reproductive and visceral features

The reproductive biology of arthropods is exceptionally variable; whether in relation to sperm structure, mode of sperm transfer, or female reproductive specialisations (see Baccetti 1979; Schaller 1979), they appear to show greater variation than any single accepted phylum. Much of this is of course related to the habitat diversity, necessitating varying degrees of motility of sperm and of protection of sperm, eggs or embryos. However the single pair of gonads are rather simple, and have a relatively unspecialised embryology. Hence few conclusions can be drawn from most aspects of arthropod reproduction, and if anything it is the degree of convergence according to habitat that is striking. However, Baccetti (1979) has conducted a careful review of sperm morphology that supersedes earlier less critical studies. This reveals a basic aquatic sperm structure in *Limulus*, only slightly modified in the more primitive members of the major groups Arachnida, Myriapoda, Crustacea and Insecta. He believes that, after a common origin, there has been independent radiation in each main line, with some similar evolutionary pathways being taken where circumstances dictate. The issue of sperm morphology in relation to fertilisation pattern, as discussed in chapter 6, should be borne in mind here though; it may be that the common aquatic sperm Baccetti detects for all arthropod groups is no more than the common primitive type for all metazoans. A shared primitive feature is of course unhelpful, as it could have been independently inherited in the various lineages.

Visceral morphology has also been taken as indicative of monophyly in arthropods; a review is given by Clarke (1979). All groups share the early breakdown of coelomic cavities and their substitution by an open haemocoel, and this has affected other aspects of internal organisation substantially. In particular, an open internal 'box' remains, in which tissues and organs are very free to move; this leads to some problems for positional interpretation, as it has liberated internal structure from the marked physical constraints that limit the external morphology of arthropods. But it also means that similarities of internal form are perhaps less likely to be convergent, and may indicate real common ancestry.

Arthropod gut morphologies differ at a gross level in that chelicerates and crustaceans possess caeca (digestive glands), perhaps of different origin in the two groups, whereas uniramian guts are nearly always a simple tube. Cutting across this classification, insects and chelicerates have peritrophic membranes but crustaceans generally do not; the same is true for Malpighian tubules, common in insects and occasional in some orders of arachnids. However, at the level of fine structure clear similarities seem to emerge, especially in the disposition of gut muscle fibres: all groups have longitudinal muscle inside a circular layer in fore- and hindguts, with a reversal of these layers in the non-cuticularised midgut.

Respiratory apparatus has long been the classic case of convergence in arthropods. On almost any phylogenetic scheme, tracheae, gills, and lamellate book-lung systems have all had to be evolved several times over; thus little significance can be drawn from their occurrence, but monophyly is certainly not precluded.

Turning to the circulatory systems matters are clearer. Between all groups there are fundamental similarities. These can be found at the level of embryological origin; of adult form, with a tubular ostiate heart, main pericardial septum, and other distributive specialisations; and of content, in the form of haemocytes (see Gupta 1979). A common evolutionary derivation of the haemocoel in all the arthropods therefore seems likely, and this is taken by Clarke (1979) as a major support for monophyly.

Taken together with the obvious external resemblances of all arthropods already referred to, these detailed features of both external and internal anatomy seem to constitute a formidable battery of evidence in favour of a monophyletic origin of the arthropod groups, and a retention of the Arthropoda as a valid phlum. The case for polyphyly must not only counter all this evidence, but must present solid and incontrovertible proofs of separate origins of the groups, and a plausible explanation for the observed similarities. All of this it has been able to do, with a remarkable degree of success, and the remainder of this chapter examines the manner in which these aims are achieved and the grounds on which polyphyletic views are founded.

11.3 The case for polyphyly

The case for polyphyletic origins of the various arthropod groups rests largely on a belief that they are too different from each other for any conceivable ancestor to have been functional unless one goes right back to a soft-bodied form. The arthropods would then have to have evolved separately from one or more groups of worms, probably already endowed with segmentation; this source is often described as proto-annelids, but was not necessarily made up of close relatives of the annelids as we now know them.

Three complementary approaches to establish polyphyly can be, and indeed have to be, taken; as outlined above, there must be specific evidence for separate ancestry, a counter argument for each of the observed similarities held to be indicative of monophyly, and a general case for accepting marked and repeated convergence between independently evolved cuticularised groups.

The specific arguments for polyphyly will be considered first; clearly direct evidence of separate and incompatible origins for the various groups is needed. Here the stress on the particular condition that any proposed common ancestor must have been a fully functional form in its own right is especially important, and stems largely from Manton's work. Arguments invoking functional morphology have centred on the structure and mechanism of limbs and of jaws above all. Interpretations that imply incompatible evolutionary routes have also been drawn from studies of comparative embryology of the groups. Additionally, recent fossil studies can be taken to support polyphyletic origins and multiple arthropodisations of soft ancestors. These four main lines of evidence therefore need separate analysis.

Functional morphology of limbs

All arthropods share a common limb structure, based on serial tubular elements of sclerotised cuticle (podomeres) with intervening softer cuticle forming joints. There is, after all, no other way to design a limb in an animal with a continuous hard exoskeleton, so the design tells us nothing useful in a phylogenetic sense. Nor can the precise number or nature of the podomeres and joints give useful information about relationships, since there is at least as much variation within any one group as between major groups, due to the enormously diverse functions of the limbs. There is, however, a fundamental variation in limb form that has been largely conserved through evolution, and is now manifested as the complex biramous (two-branched) limbs of crustaceans in contrast to the simpler uniramous limbs of modern insects, myriapods and arachnids (fig 11.2).

In crustaceans the outer branch or ramus of the leg (exopodite) is borne on the second segment, or basipodite (fig 11.2*A*); though the coxa and basis are often fused and termed the protopodite. The exopodite is commonly modified as a filter-feeding structure bearing long setae (fig 11.2*B*), or as an aid to swimming. In the familiar decapod crabs and crayfish (fig 11.2*C*) the walking legs (actually the endopods) appear to be uniramous, though they do bear gills, inserting on the coxa and tucked upwards within the carapace; but even here the remaining abdominal appendages, or pleopods, remain visibly two-branched. The biramous rule persists throughout the diverse orders of crustaceans, and even where limbs of uniramous form occur traces of their biramous ancestry usually appear in embryology. In fact the biramous leg has been present in crustaceans throughout their known fossil history with little essential modification.

By contrast the groups now designated as Uniramia (the insects or hexapods, and the four groups of myriapodous arthropods) have and apparently always have had a simple unbranched limb (fig 11.2*D*,*E*). Manton's analysis of the mechanics of the various animal's limb actions (1974, 1977) underlines a basic similarity and stresses the common gaits of uniramians not found in other arthropods. It also indicates a very early divergence of the uniramian groups, their limb operations each being designed for a particular lifestyle; the slow herbivorous millipedes, for example, require a quite different action to force the stout body through humus than do the fast

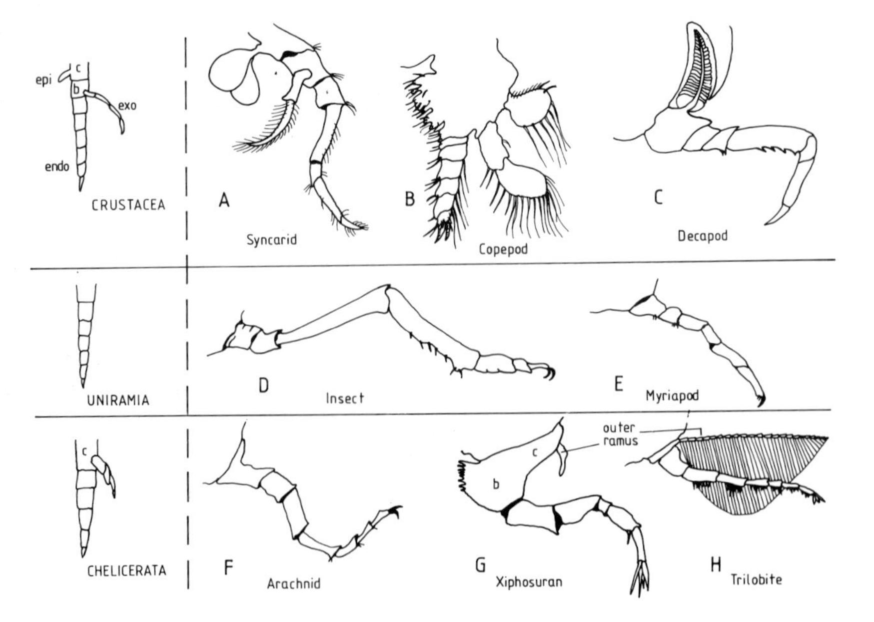

Fig 11.2. Limb structure in adult arthropods. The leg is biramous in crustaceans, but uniramous in insects and myriapods; and in chelicerates and trilobites a primitively biramous condition, with the exite/outer ramus more proximally placed than in crustaceans, is likely; (b, basipodite; c, coxopodite).

and sinuous carnivorous centipedes. The group Uniramia may also include the enigmatic onychophorans allied to *Peripatus*, which share the uniramous leg structure and gaits but perhaps show in their relatively soft 'lobopod' version of limb the starting point for the variety of actions now found in this assemblage of essentially terrestrial animals.

The third major group, Chelicerata, also show a uniramous limb form in all the familiar spiders, scorpions and related creatures that we know as arachnids (fig 11.2*F*). But the marine members of this group - *Limulus*, and its extinct relatives in the merostome group - bear distinct outer rami on the protopodite of the small opisthosomal limbs, and on the leg of the sixth prosomal segment (fig 11.2*G*). The trilobites also had this outer ramus (fig 11.2*H*), and many schemes place the group Trilobitomorpha closer to the chelicerates than to other extant groups. Thus the chelicerates may themselves have been primitively biramous, the modern terrestrial forms retaining only the endopod necessary for walking. But, critically, the outer ramus in the chelicerate line was borne not distally on the 'basipodite' portion of the limb, but usually more proximally on the region corresponding to the coxopodite. Crustaceans do sometimes have an extension in this region, generally a small flattened plate used for respiratory exchange and termed an exite or epipodite (see fig 11.2); there have been attempts to ally this to the merostome and trilobite outer ramus, but as the latter itself often bears complex lamellate structures this seems an unlikely homology.

Hence there are two quite different biramous designs; the crustaceans bearing the major outer branch on the basipodite, and the chelicerates and trilobites bearing a similar but non-homologous branch on the coxopodite. Neither pattern can conveniently be derived from the other without invoking an intermediate either lacking an outer branch and so deprived of apparatus vital for aquatic life (whether for swimming, feeding or breathing), or of improbable triramous design with a superfluity of appendages never yet found in the fossil record. Both biramous forms could of course be derived from a uniramous ancestor, but that does not tally with the fossil record either, as biramous legs occur prolifically long before truly uniramous appendages become common. Indeed monophyletic theorists have often had great difficulty in deciding on the common ancestral leg they require, and seem about equally split between biramous and uniramous precursors. It should be accepted, then, that there are apparently three basic types of arthropod limb, neatly corresponding to the three major extant groups, and the most economical hypothesis may be that they are independently evolved.

Functional morphology of jaws

In all arthropods the feeding apparatus is formed from modified segmental limbs, and usually several pairs of these become mouthparts of some kind as shown in fig 11.1. But in all modern types one pair are particularly modified as the main biting structures or jaws. Controversy still reigns over precisely which segment has contributed the appendages from which these formed, as head segmentation (see section 11.2) has been much modified through evolution and segments have been lost or fused several times in different lines. But at least the nature and action of the jaws, according to Manton's (1964, 1974, 1977) interpretation, show distinct and irreconcilable patterns that clearly imply a polyphyletic origin.

Essentially, jaws could be formed either from the whole limb, with fusion of the podomeres, such that the tip of the original limb actually performed the bite; or from the base of the limb only, the distal segments being lost, giving a 'gnathobase' jaw. In Manton's view, the whole-limb jaw occurs in all uniramians, while gnathobasic jaws are found in both crustaceans and chelicerates (see table 11.3). This of course cuts right across the traditional description of the mouthparts as 'chelicerae' or as 'mandibles', with the old category Mandibulata embracing crustaceans and insects. If Manton's interpretation is accepted, the jaws of these latter groups are very far from homologous.

Table 11.3 *Types and mechanisms of jaws in arthropods*

	Type	Rolling jaw	Biting jaw
	Action	Promotor-remotor swing	Transverse depressor-levator movement
Apparatus			
Gnathobase		CRUSTACEA (May be secondarily modified to a transverse bite)	CHELICERATA
Whole limb		INSECTA (May be secondarily modified to a transverse bite)	MYRIAPODA

The use of the term mandibulate arose from a similarity of action in the bite of insects and many crustaceans, involving a rolling action that gives a grinding effect unlike the biting action of chelicerae; and an apparent similarity of position just behind the mouth on the fourth segment of the head (fig 11.1). These resemblances may be readily accounted for in the Mantonian polyphyletic sheme, though; for while the crustacean gnathobasic mandibles are indeed quite unlike the simpler gnathobases of chelicerates and most extinct arthropod forms, they are also dissimilar in origin to the insect whole-limb mandibles having subsequently converged in relation to feeding habit to produce the modern similarity.

The crustacean gnathobasic jaws move on an axis ventro-lateral to the head, with an action derived from the promotor-remotor swing of the original limb (a clear explanation of the limb mechanisms involved is given by Manton 1977). The jaws therefore primitively rolled across each other while chewing. In more advanced forms the rear corner of the mandible becomes modified as a cutting edge and the whole jaw comes to lie obliquely, so that its action is effectively horizontal and resembles a depressor-levator system instead. This derived condition, which may appear very like the insect jaw mechanism, is especially obvious in isopods where predation and scavenging has replaced the primitive crustacean dependence on fine particle feeding; but the mandibles of such animals cannot be homologised with the condition found in insects where the whole-limb jaw effects the bite.

In uniramians as a whole, the capacity for secondary modification of the jaw mechanism is again in evidence, lending support to the proposed basic evolutionary convergence of bites in insects and crustaceans. Though all uniramians are said by Manton to have a whole-limb jaw, and to bite with its tip, the jaw only retains its joints in some myriapods, and appears unjointed in the insects and in *Peripatus*. Jointed jaws seem to have been primitively endowed with a transverse bite, retained in millipedes but now somewhat modified in the carnivorous centipedes. But the unjointed jaws of hexapods probably had a rolling action initially, using the promotor-remotor musculature of the limb, now secondarily modified to a transverse bite.

Boudreaux (1979) believes that the multiple actions of jaws in these modern groups can be reconciled with a monophyletic scheme by proposing an ancestor with dorsal sclerites but a soft arthrodial membrane undersurface; this would permit acondylic limbs and jaws, operated by haemocoelic pressure, from which all the current morphologies and actions could be derived. Whether this ancestor could in itself be considered an 'arthropod' is open to question; nor is it clear that it would be a very sound functional design.

It should also be noted that the 'whole-limb jaw' of uniramians remains a highly controversial notion (see Lauterbach 1972; Boudreaux 1979; Weygoldt

1979). Manton defends her views by pointing to the visible segmentation of certain myriapod jaws, which might seem more likely to be a vestige of the ancestral condition than a secondary development. But Weygoldt retaliates by drawing attention to the secondary articulations of some isopod jaws that give an appearance of segmentation, and to the peculiar anatomy and musculature of the myriapod jaw's podomeres, which might indicate that it is merely a secondarily segmented coxa. Boudreaux meanwhile maintains that in terms of musculature and tendons *all* arthropod jaws are of a coxal type.

While the jaws of uniramians and crustaceans do show similarities that perhaps justify the common term 'mandibles', even if the state has been arrived at repeatedly and convergently, the jaws of chelicerates are indisputably different enough to merit their separate designation as 'chelicerae'. The primitive jaws here were derived from gnathobases, with a clear depressor-levator action in the transverse plane of the body. In *Limulus* this condition persists in a particularly simple form, all five of the post-oral limbs contributing their gnathobases to a chewing role whilst the rest of the limb remains locomotory. Here strong adductors (depressors) pull the gnathobases together, giving a grinding action suited to dealing with small particles from the substrate moving forwards towards the mouth. All arachnids retain a similar system, usually with just one pair of gnathobases operative on the pedipalps, though the chelicerae themselves have become adapted for piercing and sucking.

Thus analysis of jaw mechanisms suggests at least three quite separate origins of modern arthropods from an earlier stock; and since no animal could effectively bite with both the tip and the base of a limb, or survive a transition stage when neither part of the limb was functional in feeding, it is proposed that the common ancestor must have been soft-bodied, before cuticularised mouthparts were evolved. (Giving it only a cuticularised dorsum, as Boudreaux suggested, seems merely to avoid the critical issue.) This view is supported by the fact that many of the earliest fossil arthropods had no defined feeding limbs at all; and by the jaws of trilobites, which are of yet another gnathobasic form. Hence, and most conveniently, the groups defined by Manton's work on mouthparts are in exact agreement with the groupings already derived from a consideration of leg structure, and none of the groups can be successfully derived from each other. On these grounds a polyphyletic view of the arthropods seems unavoidable.

Embryological evidence

The traditional analyses of comparative embryology, outlined in chapter 5, have little contribution to make to the arthropod debate. Virtually all arthropods have very yolky eggs and incomplete cleavage of the blastomeres; development takes place very superficially, nuclei being distributed over the external surface of the yolk. Hence the arthropod embryo cannot readily be compared with the spirally cleaving, determinate embryos of annelids and other possible ancestral groups by conventional means. An alternative and unified approach was pioneered by Anderson (1973, 1979), using 'fate maps' to describe developmental patterns, and once again demanding that functional constraints should be appreciated. This technique remains controversial (e.g. Weygoldt 1979), but stands as one of the rare attempts to compare embryology in groups where patterns are otherwise clouded by the presence of yolk.

The crustacean fate map (fig 11.3*B*) shows a pattern derived from spiral cleavage, but with a profound reversal of the fate of certain blastomere derivatives as compared to annelids. This fundamental difference also occurs in crustaceans with little yolk and thus amenable to more conventional interpretation, and therefore seems indisputable (see Weygoldt 1979).

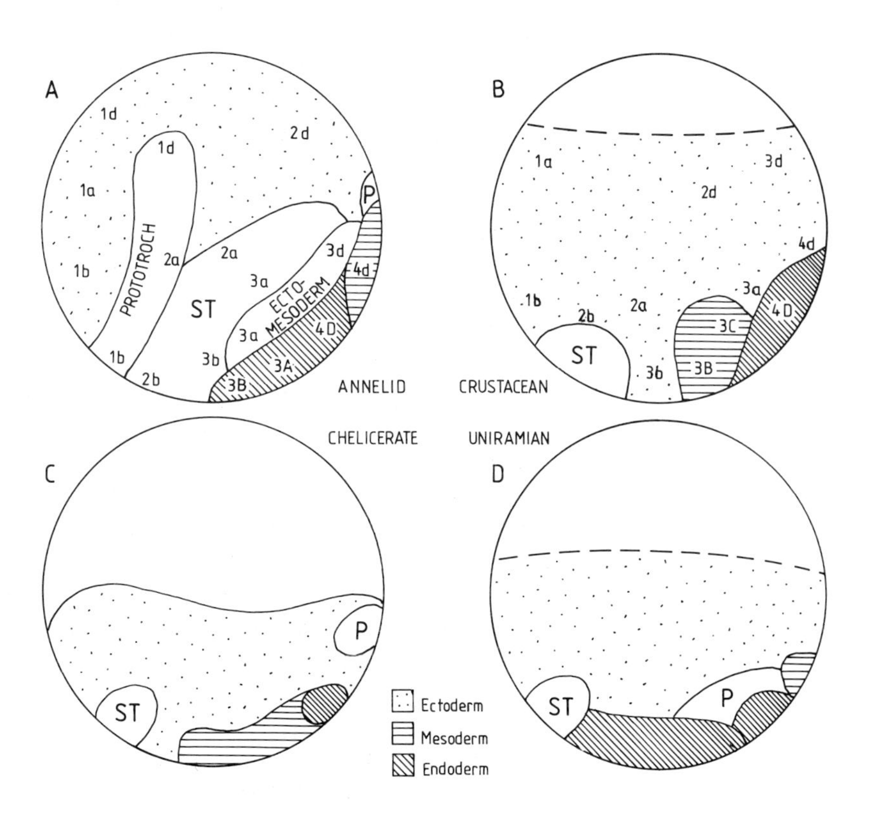

Fig 11.3. Fate maps of the embryos of arthropods, based on the works of Anderson. The uniramian fate map may be derivable from classic annelid embryology, but that of crustaceans is very different; (P, proctodaeum; ST, stomodaeum).

Whereas in annelids and many other groups the mesoderm arises from blastomere 4d (see chapter 5) and lies behind the presumptive midgut (endoderm) tissues, in a crustacean embryo mesoderm arises from the 3A, 3B and 3C blastomere equivalents and lies in front of the midgut, with the 4d cell forming ectoderm. Anderson concluded that crustaceans were so different from, and incompatible with, the annelid pattern of embryology, that whilst they did evolve from some form of worm within the spirally cleaving assemblage their origin must have lain at some distance from those groups with 4d mesoderm and was perhaps close to the ancestral platyhelminths.

Similarly, the fate map analysis of uniramians (fig 11.3*D*) reveals a constant pattern of development in Myriapoda, Hexapoda and Onychophora; though Weygoldt (1979) points out the complications caused by unusually marked movements of tissue rudiments around the insect embryo, and thus dislikes Anderson's interpretations. Anderson believes the fate of presumptive blastomeres in insects is closer to that of annelids and molluscs than to the crustaceans, with mesoderm arising behind the midgut. He backs up his view by demonstrating that onychophoran embryology shows particularly clear similarities to that of the clitellate annelids, where yolk is also present. These findings imply that metamerism in the uniramians may be homologous with that of annelids, whereas it may have had a separate origin in the crustaceans; such a view is by no means unlikely given the known multiple origins of segmentation (chapter 2), and according to Anderson it is substantiated by detailed analysis of the differing mechanisms of somite development in the various groups.

Within the uniramians Anderson describes separate patterns of development for the myriapods and hexapods, neither derivable from the other, with Chilopoda (centipedes) being the most generalised of the myriapod groups and showing links to the onychophorans. Insects are held to be unrelated to any existing myriapods, and use of fate maps suggests a close link between thysanuran apterygotes and the true Pterygota, in accordance with the classical view.

The use of fate map techniques rather falls down in relation to chelicerates, however. Some of these start with total cleavage, but further development is generally superficial. Many examples have been described, and the basic fate map of these is distinctive (fig 11.3*C*), but none have revealed an acceptable 'generalised' pattern, and even *Limulus* embryos give no clues to ancestral links. The relation of midgut and mesoderm, critical to Anderson's descriptions of uniramian and crustacean ancestry, is totally unhelpful for chelicerates as the midgut tissues remain internal throughout the blastula stage. Anderson has ruled out a relation with uniramians, but has as yet failed to separate chelicerates from crustaceans convincingly, on embryological grounds alone; the chelicerates are probably part of a loose non-4d-mesoderm assemblage, and there even remains some suggestion that the phylum may itself be polyphyletic (Anderson 1982).

There is no overall agreement about the validity of Anderson's conclusions. A good critique of his techniques is presented by Weygoldt (1979), though this gives no clear alternative for the interpretation of early arthropod development. Some recent authors (e.g. Zilch 1979) have reasserted a more traditional view about crustacean embryology and cleavage patterns that would put this group back as descendants of annelids and close to uniramians. But if the fate map analysis is to be taken as a useful tool, then its conclusions are in very satisfying agreement with the independently derived views from comparative adult limb and jaw morphology. That is, there are three main lines of extant arthropods, with no common ancestor beyond a distant, soft-bodied, non-arthropod stage. Chelicerates and crustaceans probably stand apart from each other, and distinct from annelids. Both are distant from the uniramians which, while perhaps more nearly derived from annelid ancestors, are themselves readily split into three major lines diverging at a very early stage.

Fossil evidence

Most recent surveys of fossil arthropods appear to show that the distinction between major groups is very deep-seated (see Bergström 1979; Manton & Anderson 1979; Whittington 1979). Earlier reviews (e.g. Sharov 1966; Cisne 1974) that seemed to reconcile known fossils with a monophyletic scheme have been effectively superseded by the improved description of numerous Cambrian forms, notably from the Burgess Shale beds (see chapter 3). Though available remains do not give a clear picture of arthropod evolution, as major groups were already established by the time these early fossilisations occurred, it does now seem likely that many 'arthropodised' groups existed then that cannot be placed in, or en route to, the extant assemblages. Nor can they be readily derived from any one common ancestor, though convincing remains pre-dating the mid-Cambrian are not available to prove this point. Thus the fossil record attests to the past occurrence of multiple experiments in arthropodisation, and any scepticism that modern groups could have arisen by similarly diverse and multiple routes must surely lose its foundation.

Fossils that can be assigned to modern groups are remarkably rare. There are a few undoubted mid-Cambrian crustaceans; two genera, *Perspicaris* and *Canadaspis* (fig 11.4*A*), are assigned to the group Phyllocarida and resemble modern leptostracans such as *Nebalia*. Another form, *Waptia*, is also now granted crustacean affinities. Apart from these ancestral malacostracans, the class Ostracoda also seems to have had a very early origin as many ostracodlike carapaces appear in the early Ordovician, and there were also possible branchiopods in this period. None of these early crustaceans show links to their contemporary chelicerae-bearing or trilobitomorph forms.

Cambrian uniramians are still represented only by the marine *Aysheaia* (fig 11.4*B*), originally accepted as an early onychophoran form (Tiegs & Manton 1958) but now of unclear affinities (see Bergström 1979) since its 'antennae' have been redescribed as displaced front legs. The status of the possibly uniramian form *Xenusion* is also now in doubt. Fossils therefore do not help to tie in onychophorans with the myriapod-insect groups in quite the manner required by Manton's scheme. Indeed, with the exception of a few possible Silurian myriapodous types, true uniramians do not appear until the lower Devonian with the multi-legged Arthropleurida, and hexapods are first represented even later by a Devonian collembolan.

Chelicerate xiphosurans first appeared in the Cambrian. Most of the early fossil forms are represented only by the prosoma, and the status of their legs and chelicerae are therefore doubtful; hence arachnid evolution remains controversial (see Bergström 1975, 1979; Whittington 1979). The aglaspidids (fig 11.4*C*), formerly regarded as chelicerates, are now of particularly dubious status (Whittington 1979), undermining the supposed link between trilobites and xiphosurans somewhat. The Cambrian fossil known as *Cheloniellon* (fig 11.4*D*) has also been interpreted as indicative of such a link, though its two pairs of antennae are more reminiscent of crustaceans (Manton & Anderson

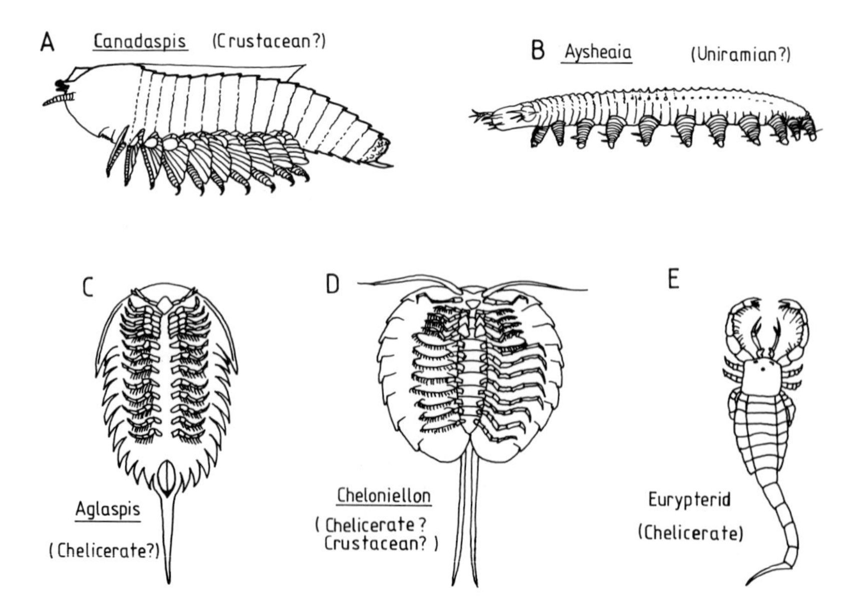

Fig 11.4. Arthropod fossils that have been assigned to extant groups, with varying degrees of confidence, (redrawn from several sources, not to scale).

1979) and it probably deserves quite separate status. Only much later do matters of chelicerate evolution become clearer; eurypterids (fig 11.4*E*) are present from the Ordovician period, and the true Scorpionidea date back to the middle Silurian, when they may have been aquatic. Terrestrial arachnids of other types appear in the early Devonian, possibly on a quite separate evolutionary line from the scorpions.

Besides these fossils that can be assigned a status with reasonable confidence, and which make up only a small percentage of the arthropod remains in early deposits, there are a host of other forms of extraordinary diversity. In the highly localised Burgess Shales alone there are at least thirty other genera of non-mineralised arthropods (discounting seventeen types of trilobites), often with unique suites of characteristics. Three examples, *Mimetaster, Yohoia* and *Marella,* are shown in fig 11.5. From recent reconstructions, of their appendages in particular, it is quite clear that they cannot be united in any natural group, as Størmer (1944) attempted by calling them all Trilobitoidea (indeed some of Størmer's original group, including the remarkable *Opabinia* (fig 11.5*D*) are now not regarded as arthropods at all, having no legs). These Cambrian fossils do not provide real evidence of links between chelicerates, trilobites or crustaceans. The trilobite-crustacean link is in fact usually discounted now (e.g. Schram 1982), but a few authors still retain it (Hessler & Newman 1975; Cisne 1982). Bergström (1979), amongst others, prefers a connection between chelicerates and trilobites. Most recent

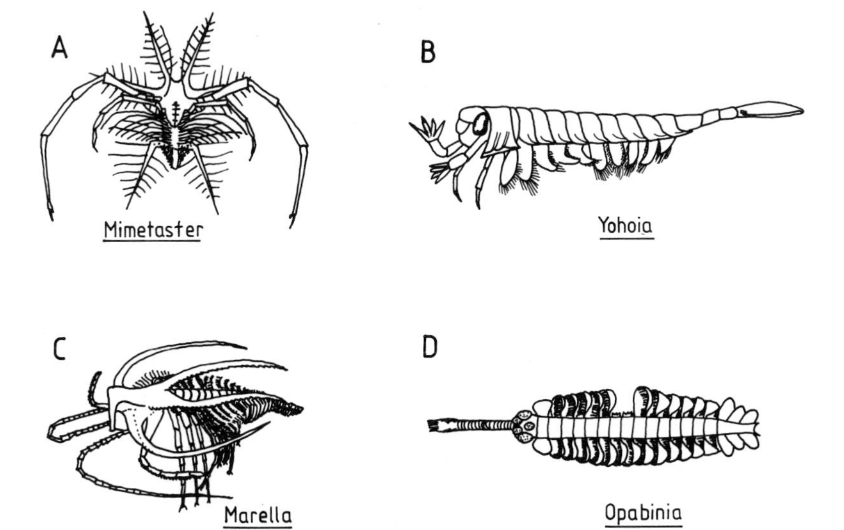

Fig 11.5. Fossils that are arthropodised but clearly belong to no known group; *Opabinia* should not even be termed an arthropod at all, as it has no articulated limbs, (redrawn from various sources, not to scale).

authors reject even this link, and many recognise trilobites themselves as composed of at least three separate and at best only distantly related lineages.

In summary, the most reasonable conclusion to be drawn from careful analysis of the available arthropod fossil material is that arthropods are not just di- or triphyletic but have arisen repeatedly, with the modern forms representing just a few of the more successful experiments in applying a continuous, partially stiffened cuticle to a soft-bodied worm. This is the conclusion reached by Bergström (1979), by Whittington (1979) and his various co-workers, by Manton & Anderson (1979) and by most other palaeontologists familiar with the Cambrian material. All of their studies lead to the type of phylogenetic pattern shown in fig 11.6. Most of the Cambrian 'arthropods' died out by the end of the Palaeozoic, including the dominant trilobites, and thereafter only the three most successful lines continued to diversify. Unless further major finds from the early Cambrian or the immediate pre-Cambrian periods are made, we are unlikely ever to discover just how each of the major arthropodisations occurred, and from which worm-like ancestors they came.

Further information on phyletic relations at lower taxonomic levels within the various groups of arthropods is given by Bergström (1979). Though not directly relevant to the present discussion, such studies do confirm the view that major groups were already highly diversified and distinct from each other at their first appearance, though often the gaps between 'phyla' seem to be not much greater than the gaps within them (for example between merostomes and

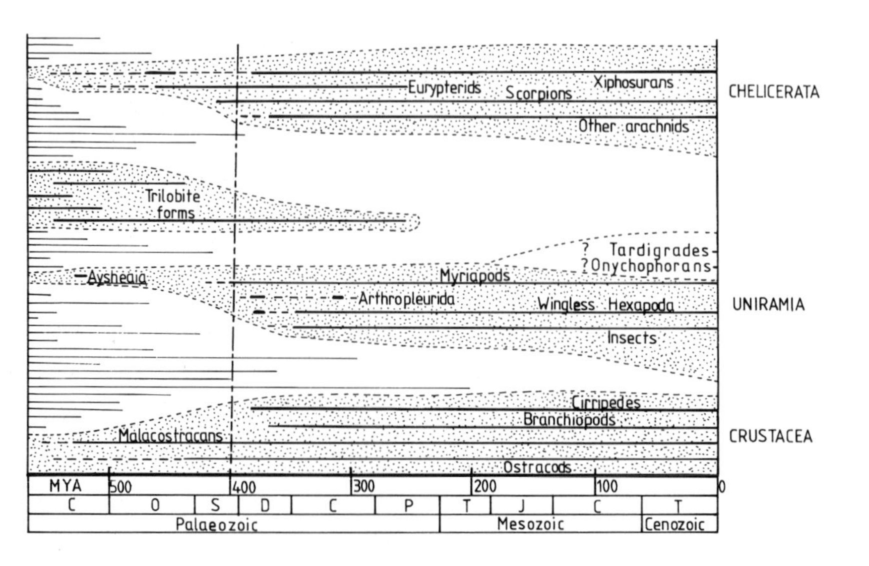

Fig 11.6. The fossil record of arthropods, including many groups now extinct and apparently unrelated to extant groups. Modified from data in Whittington (1979) and Bergström (1979).

arachnids). Most certainly, these studies reveal the prevalence of repeated and precise convergent evolution, over and over again in every group.

Countering the evidence for monophyly

So far in this section, specific evidence for polyphyletic origins of arthropods has been considered, and this evidence is clearly rather strong. The second major approach required to undermine conventional views of monophyly is a critical consideration of those other pieces of specific evidence already advanced in its support. There are several possible reasons why the various similarities we discussed in section 11.2 may be invalid or less than convincing. Either the features themselves could be suspect and not common to all groups as has been claimed; or those same features might occur quite independently elsewhere (preferably in distant phyla), indicating at least a strong possibility of their convergent nature in arthropods; or they may be inevitable consequences of some other feature, so should not be counted independently. This last point relates to the important question of the effects of the arthropod cuticle itself, and the constraints it imposes that may force convergence on any animal possessing it, a matter to which we will return.

The first indisputable similarity given before was the actual cuticle structure and chemistry. This is certainly one of the linch-pins of many monophyletic views; but recent evidence has left it open to severe attack. Firstly the helicoidal deposition of chitin fibrils is no longer known exclusively in 'arthropods', as it has been demonstrated in the tubes of pogonophorans (Gupta & Little 1975). It may in fact occur wherever chitin is laid down in a protein matrix, as a self-assembling liquid crystal system (Bouligand 1972), in which case the similar construction of cuticles in differing arthropods is probably fairly irrelevant as a phylogenetic character. Secondly, and following neatly from this point, it turns out that in so far as the cuticles of arthropods can be variable they frequently are. Chitin and proteins are the commonest epidermal secretions in many invertebrates (chapter 4), and thus make natural building materials for stiffened cuticles; but further hardening requires extra chemical reactions, not usually found in soft-bodied ancestors, and here variation could occur. The three main groups of modern arthropods have often achieved quite different forms of exocuticle hardening (see Bereiter-Hahn *et al.* 1984). Insects are generally said to use a quinone-type tanning between cuticular proteins (as do many other invertebrates in localised structures such as bristles - see chapter 4). Arachnids, or at least *Limulus* and scorpions, commonly use disulphide bonding. Crustaceans incorporate inorganic carbonate and phosphate salts in the exocuticle to achieve extra strength, in addition to partial chemical tanning. Despite subsequent convergences and divergences between groups, these essential differences largely persist, suggesting separate initial evolution. Finally, the control of cuticular moulting by similar ecdysone hormones in all arthropods is not convincing evidence for monophyly either, as ecdysone turns out to occur in other groups too - for example in nematodes where it again controls the moult cycle. It thus seems to be a standard biological accompaniment to the necessity of shedding a cuticle. The fact that the source of ecdysone is dissimilar in arthropods may confirm its separate acquisiton by the various groups; chelicerates do not possess any ecdysial glands that can be homologised with the discrete endocrine glands familiar in insects and crustaceans (Tombes 1979). Thus several aspects of the cuticular biology of arthropods are all consistent with separate origins, and the similarities they do exhibit are likely to be convergent features.

As regards head structure, committed 'monophyletic' authors still do not agree among themselves on any one interpretation of head segmentation homologies. There is certainly a close similarity between insect and crustacean heads, with corresponding positions of mouth and mouthparts (fig 11.1) and common possession of mandibles; the group 'Mandibulata' is generally supported by such features, though the Mantonian view on the non-homology of the mandibles themselves should be remembered. Attempts to incorporate the chelicerates into a common scheme are much less successful, and the existence of a variety of 'reduced' pre-antennary or even pre-cheliceral segments, required to allow homologies to be drawn, remains controversial. Even Sharov (1966), whose attempt to force all groups into a common plan was notoriously extreme, had to concede (page 170) that 'cephalisation occurred ... independently in each of the groups of arthropods'.

Eye structure again seems to support a close relation of insects and crustaceans. But there are enough relatively gross differences between the eyes of mandibulates, for example in the distribution of pigment cells, and the control of pigment by light directly in insects but by hormones in crustaceans, to raise doubt about their monophyletic status (see Horridge G.A., in Manton 1977). Furthermore, Odselius & Elofsson (1981) point out that there are considerable ultrastructural differences in compound eyes, consistent with broad taxonomic groupings, and suggest parallel development of insect and crustacean eyes. There are even more difficulties in allying both myriapods and chelicerates with the classic mandibulate eye. It is thus not at all certain that there was one common ancestor of all modern eye forms that was itself arthropodised; probably the differing forms should be derived instead from the eyes of separate soft-bodied invertebrate prototypes. A few polychaetes such as *Branchiomma* do have a variety of composite eye, after all; and the compound eyes of *Scutigera* are also likely to represent convergence arising secondarily as a special adaptation in this fast hunter. The presence of a cuticle would so constrain the nature of eyes and their optical properties that convergence from separate ancestries might be very profound; this is by no means implausible if one recalls the extraordinary convergence of cephalopod and vertebrate eyes, where there is not even a great similarity of materials to impose restraints.

Finally, various features of internal anatomy have been taken as evidence of monophyly, and here the constraints imposed by cuticle cannot be readily invoked. Nevertheless, there are at least as many features that are different between groups as there are similarities, and it is impossible to tell which are more fundamental. Very few features turn out to be common to all three (or four) major groups. Those which are, like the gut musculature or the granulocyte haemocytes, may be simply due to the common embryology of the tissues involved; they may even be present in platyhelminth or annelidan groups as well, in which case all the arthropod groups could easily have inherited them separately. More frequently, similarities occur between two groups and not the third, yet it is often crustaceans and chelicerates, or chelicerates and insects that are in accord; whereas for non-visceral features, as we have seen, the closest agreement commonly lies within the mandibulate (insect-crustacean) assemblage. Matters are, at best, confusing.

Overall, then, the evidence for monophyly is patchy and no single feature is incontrovertible. One is left with the simple fact that all arthropods appear rather similar; but where specific homologies do seem to exist, they rarely encompass all the major extant groups, let alone known fossils. It is not even possible to agree on which of the modern forms are sister-groups, as some of the features used will always cut across any particular scheme that could be proposed.

The general arguments for polyphyly

Besides presenting specific evidence for polyphyly, and countering some of the specific cases that have been made for monophyly, a good case must be made for the general plausibility of there being three or more quite separate groups of extant arthropodised animals, for this seems at first glance to be a non-parsimonious hypothesis. Why should all these animals actually look so similar if they are unrelated?

Firstly then we should return to the general similarities shared by all arthropods, set out in table 11.2. The most obvious point, frequently reiterated by polyphyleticists (Schram 1979; Manton 1977), is that these features are not separate and independent. A great many of them reduce to one essential factor, the presence of a continuous stiffened cuticle. The manner in which the possession of an exoskeleton leads to the whole syndrome of arthropodisation is summarised in fig 11.7. The argument runs as follows: if a tough cuticle is present, then the cuticle has to be moulted to allow growth, the moult has to be controlled, and the animal must remain functional while moulting as far as possible, so the moult cycle, the hormones and the tonofibrillae penetrating the cuticle are all inter-dependent features. The cuticularised animal necessarily has articulated legs and a segmented articulated body, otherwise it would be an immovable solid rod; and tagmosis with functional specialisation of particular sections is a predictable consequence of segmentation. All these factors in turn entail changes in nervous system, tendons, and muscle arrangements; and striation of muscles is a normal accompaniment of fixed contraction lengths (chapter 4) which are of course necessitated by a rigid skeleton. Internally the gut is inevitably cuticle-lined at each end, as both fore- and hindgut are invaginated ectodermal tissues (a similar situation occurs in other groups such as nematodes). The respiratory and other exchange systems have necessarily to be modified to get supplies through a relatively impermeable surface. The spacious segmented coelom, traditionally presumed to characterise the ancestral forms, is no longer required for hydrostatic functions and is perhaps disadvantageous as a distribution system with separate blood vessels; so it can readily regress and allow reversion to the open plan haemocoelic circulation, where turgor will be evenly distributed. Loss of coelom entails changes in

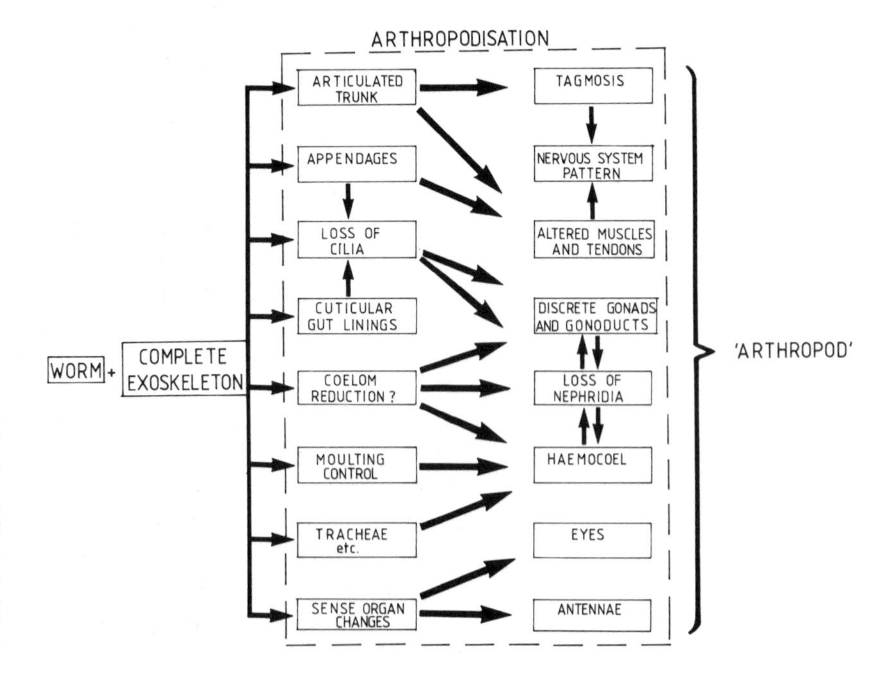

Fig 11.7. A schematic view of the process of arthropodisation; the drastic effects of adding an exoskeleton to a worm-like animal entrain, of necessity, all the other changes that result in an animal we call an 'arthropod'; (further explanation in the text).

reproductive and excretory systems, both traditionally associated with the coelom and surrounding tissues. (Or, if a non-segmented acoelomate ancestor is invoked for some groups, as Anderson's theories require, then the haemocoel could be acquired 'directly' as in molluscs (chapter 2), and reproductive and excretory systems modified from a flatworm stage instead.) Finally, sense organs have to be altered to use the properties of the available hard materials; they can no longer use motile cilia, as these are clearly incompatible with cuticle, and any pores they bear must be reduced or protected. Almost everything that defines an arthropod can therefore be regarded as a necessary and inevitable consequence of the adoption of a cuticle - there is simply no other way to design any creature with a complete exoskeleton.

Thus it can be argued that the acquisition of stiffened epidermal secretions by any worm, segmented or not, will inexorably create an animal of the type we call an arthropod. It is equally apparent that during the later Pre-Cambrian or very early Cambrian periods the selective pressure to acquire hardened exteriors must have been very great (chapter 3), as so many other groups were developing hard mouthparts facilitating predation, or hard coverings giving protection against such attack. This is precisely the period when exoskeletal arthropods appeared in profusion, and if the fossil record has any clear message for arthropod phylogeny it is that multiple experiments in this kind of design were being attempted. Even without this evidence, the very familiar benefits accruing to owners of exoskeletons (mechanical, physiological, and as anti-predator or pathogen defences) should be enough to suggest that this feature would evolve more than once. To propose that it did so cannot reasonably be dismissed merely by recourse to arid parsimony arguments.

A further element in the case for polyphyly reverts again to the issue of convergence. Whilst many early authors regarded this as a problem for any polyphyletic scheme, there is wider recognition now that the extensive convergence inherent in Manton's proposed three extant phyla is hardly a major stumbling block. For one thing, the arguments advanced above suggest that convergence is not only possible but functionally inevitable; and for another, literally any theory of arthropod relationships has to accept a startling component of convergent evolution. Two simple examples should underline this point. Tracheae are indisputably separate inventions in uniramians and the higher arachnids (since they are absent in lower arachnids); yet in the Solifugae in particular they bear a quite remarkable similarity to the insect version. Similar though less elaborate structures are present in some isopod crustaceans, and many classifications (even monophyletic ones) require them to have arisen four or five times independently (see Tiegs & Manton 1958). Quite simply, they represent an obvious solution to the problems of terrestrial respiration in an animal covered by cuticle. A second example is the Malpighian tubules, excretory organs unlike those in any other animal group (being extensions from the gut) yet clearly independently evolved in most insects and in certain arachnids. Again they represent a sensible solution to terrestrial excretory and water-balance problems, though here the convergence is perhaps more surprising as other excretory systems would seem to be equally possible. Many more examples of clearcut convergent evolution amongst the terrestrial groups could be quoted, in terms of physiology, mechanics and ecology; indeed, whatever we believe about the relationsips amongst arthropod groups, these animals provide some of the best possible examples to convince us of the reality and force of convergence as an evolutionary process. It cannot therefore be held to be a serious problem in accepting polyphyly.

Finally, the case for polyphyly is supported by the existence of a number of small and curious groups of animals also possessing cuticles but uncertainly related to the major types of arthropod dealt with so far (fig 11.8). The status of tardigrades (the 'water bears') is a case in point; they are possible uniramian relatives, but have very different cuticle structures and strange coelomic organisation, with some of their features being more reminiscent of the pseudocoelomate assemblage (chapter 9). Similar controversy surrounds the

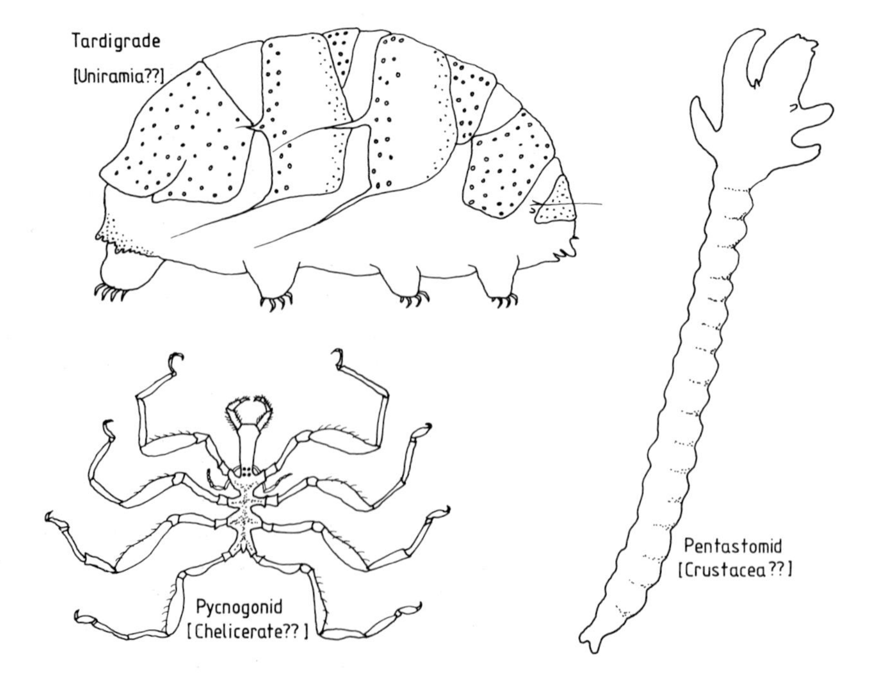

Fig 11.8. Three small groups of partly 'arthropodised' animals, often included in the old phylum Arthropoda. Actual sizes: tardigrades 0.05-1mm; pentastomids 20-100mm; pycnogonids 2-20mm.

pycnogonids or pantopods (sea spiders), frequently said to be related to the chelicerates (Manton 1978), but with evident dissimilarities; and the pentastomids (or linguatulids - tongue worms), whose proper relations are obscured even further by their parasitic habits and specialisations, though some affinities with crustaceans may be indicated (Riley *et al.* 1978; but see also Anderson 1982). Each of these groups tends to be included within the fringes of a phylum 'Arthropoda' by the monophyleticists, but surely the very best that can be said of them is that they make monophyletic schemes more difficult. If they are in any sense helpful, it is by bearing witness yet again to the preponderance of 'arthropodisation' as a process, and of the profound effects of a cuticle in causing convergence of design and function, thus deserving individual phyletic status.

Even the onychophorans could be said to come in the same category. While these 'velvet worms' or 'walking worms' are firmly designated as part of the Uniramia in Manton's scheme, this is not an essential or perhaps even a helpful part of a polyphyletic view. It might be reasonably argued that they too are different enough, in terms of cuticle, jaws, legs, muscle structure, a lack of inhibitory neurones, strange tracheal structure, and the presence of functional cilia, to merit the higher rank of phylum in their own right.

11.4 Conclusions

Before summarising the evidence in this chapter, this is a useful point at which to recall the issues briefly discussed in chapter 1, concerning the effects of methodology on phylogenetic conclusions. Nowhere are such effects more apparent than in the arthropod literature. Manton's works are classics of comparative functional morphology, and while there have been few specific or wide-ranging critiques of her work (see Lauterbach 1973; Weygoldt 1979) there have been many complaints by cladists that her methodology is incorrect and the results therefore invalid. Indeed the only comprehensive work since Manton is that of Boudreaux (1979), which provides a complete contrast in being a thoroughly Hennigian approach using the rigorous analysis of phylogenetic systematics. Boudreaux rejects the Mantonian position largely because of its 'unsound methodology'; he rarely attempts to confound her views, but frequently ignores them. Manton seeks for differences, and not surprisingly believes she finds them and that they necessitate a polyphyletic view; Boudreaux seeks for homologies (synapomorphies), and equally predictably believes he finds them. Boudreaux's methods require that he should preclude convergence whenever he possibly can, and his conclusions are therefore almost inevitably monophyletic; though his particular and detailed conclusions are often rather peculiar (see Kristensen 1979; Schram 1979). A comparison of the works of Manton and of Boudreaux provides an instructive insight into the differing strengths and weaknesses of phylogenetic methods, and of the prejudgements each makes. Such a comparison may also leave us feeling that perhaps the lesser evil is to reject the requirement of minimum convergence, given ample evidence that it *has* repeatedly occurred; and to accept the carefully argued case presented by Manton and her associates, rather than the sometimes tortuous reasoning imposed by a too rigid methodology in the works of Boudreaux and other cladistic authors. This feeling clearly colours the conclusions reached here.

Three complementary approaches have been given that together help establish the polyphyletic origins of the arthropods. While no single approach or isolated feature can ever be conclusive, there is abundant evidence of differences between the main extant arthropod groups, and a sufficiently cogent set of reasons as to why polyphyly is likely and monophyly is less feasible. Polyphyletic origins are supported not only by detailed morphology and the impossibility of functional intermediates, but also by embryological differences, and (increasingly) by the evidence of multiple arthropodisations in the fossil record. Certain features mentioned in the second part of this book might also be recalled: the very different haemoglobins of insects and crustaceans (chapter 4), and the differing intercellular junctions in each group (chapter 6), for example.

The term 'arthropod' is therefore to be taken only as representative of a grade of organisation, arising whenever soft-bodied worms develop toughened cuticles, rather than as a cohesive monophyletic taxon. Whilst many similarities may exist between the groups, these are either necessary consequences of the presence of exoskeleton; or are present in only some of the main groups and absent in others; or are quite clearly convergent anyway, being adaptations to similar environmental problems.

Overall, then, most available evidence suggests that the three major phyla of modern arthropod-type animals - Crustacea, Chelicerata, and Uniramia - cannot be successfully united as a natural group. Each has evolved independently from amongst the spiralian worms, and the crustaceans are perhaps most distant from the other groups on embryological grounds. The uniramians may be derived from a proto-annelid group, whilst crustaceans probably diverged from the stem spiralians earlier, from a flatworm-like stage; chelicerate origins are still enigmatic. The various lesser forms of cuticularised animal (tardigrades, sea spiders, tongue worms, and perhaps velvet worms) cannot reasonably be granted less than phyletic status. A set of relations as shown in fig 11.9 therefore seems to be indicated. Significant new evidence may yet be forthcoming, from fossil beds or from molecular phylogeneticists, to overturn these views; but currently they represent the most logical position we can take about arthropod phylogeny.

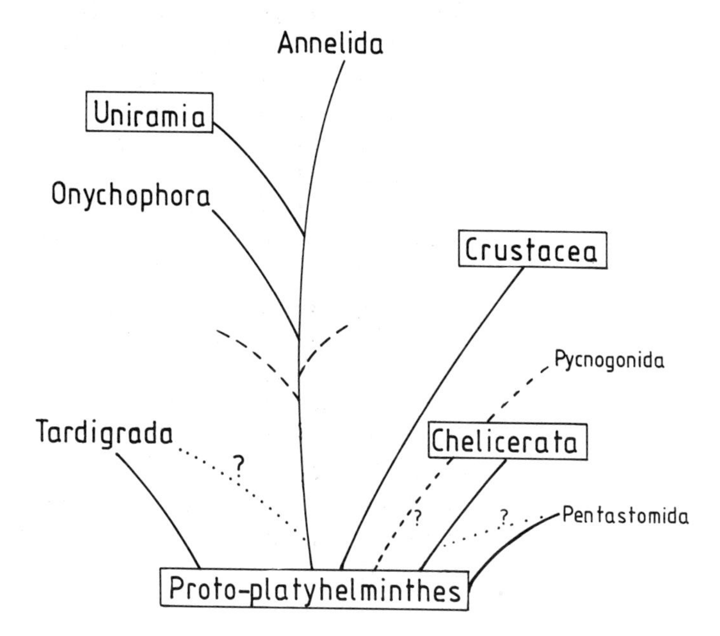

Fig 11.9. Possible status of the various 'arthropod' groups in relation to other 'protostome' invertebrates.

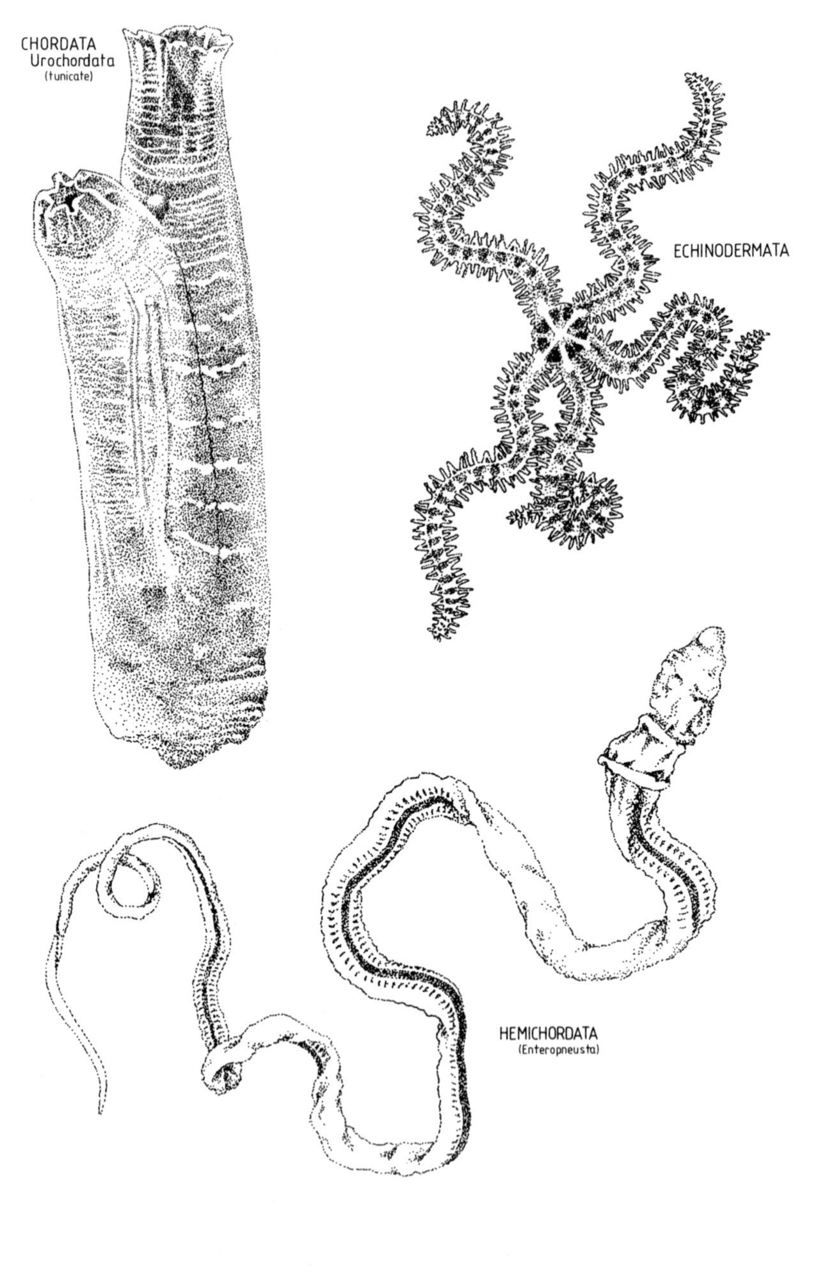
CHORDATA
Urochordata
(tunicate)
ECHINODERMATA
HEMICHORDATA
(Enteropneusta)

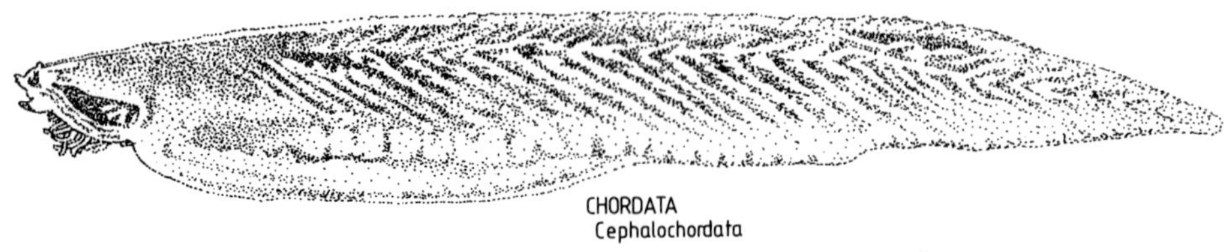
CHORDATA
Cephalochordata

12 The deuterostomes

12.1 Introduction

In the preceding chapters, all the groups discussed were members of the assemblage conventionally labelled (with varying conviction) as protostomes. Now attention turns to the other 'half' of the animal kingdom, the smaller numbers of phyla that are generally held to constitute the deuterostomes, and from amongst which the vertebrates are thought to have arisen. These phyla - the echinoderms, and slightly less familiar groups such as acorn worms, sea squirts and lancelets - are expected to share a suite of developmental features originally discussed in chapter 5, and also certain adult features; an overall diagnosis of the deuterostomes is given in table 12.1. In this chapter, the validity of these concepts is examined. Is there really a distinction between these few invertebrates and all the rest? And are we looking in the right place for the origin of the chordates?

Table 12.1 *Diagnosis of the deuterostomes*

Embryological Features:	
	Radial, indeterminate cleavage Blastopore does not form mouth, which is secondary Mesoderm forms from infolding of gut wall Enterocoelic coelom Dipleurula-type larva, prototroch around the mouth
Adult Features:	
	Tripartite body Intra-epidermal nervous system Mesodermal skeleton Monociliate cells? (Absence of cephalisation) (Absence of chitin)

12.2 Echinoderms

The phylum Echinodermata is one of the few about which we have good evidence for evolutionary patterns and the changes of body form through time, as the calcitic internal skeleton of separate or fused ossicles is readily preserved as fossils and good material is available from the early Cambrian (see chapter 3). We know that the current representatives of the phylum, generally assigned to five different classes, exhibit only a small fraction of the past diversity of these animals, with around twenty classes recognised in total (see Sprinkle 1976; Paul 1977, 1979). Early echinoderms were primarily bilateral, though often with marked asymmetries, and many were stalked and sessile (see fig 12.1). Successive waves of diversification and extinction occurred (fig 12.2); diversity was particularly high at the class level in the Ordovician, and at genus level in the Carboniferous, and it has increased again mainly at the genus level into the Cenozoic (Paul 1979). Just a few of the potential echinoderm designs survived through each extinction phase. The present groups (traditionally five, but recently increased to six classes - Nichols 1986) may therefore be far removed from ancestors that could tell us more clearly about echinoderm phylogenetic relations.

That said, however, it is also true that echinoderms are in many senses about the most uniform and 'recognisable' of all animal phyla, their basic design having changed remarkably little (Hyman 1955; Lawrence 1987). They

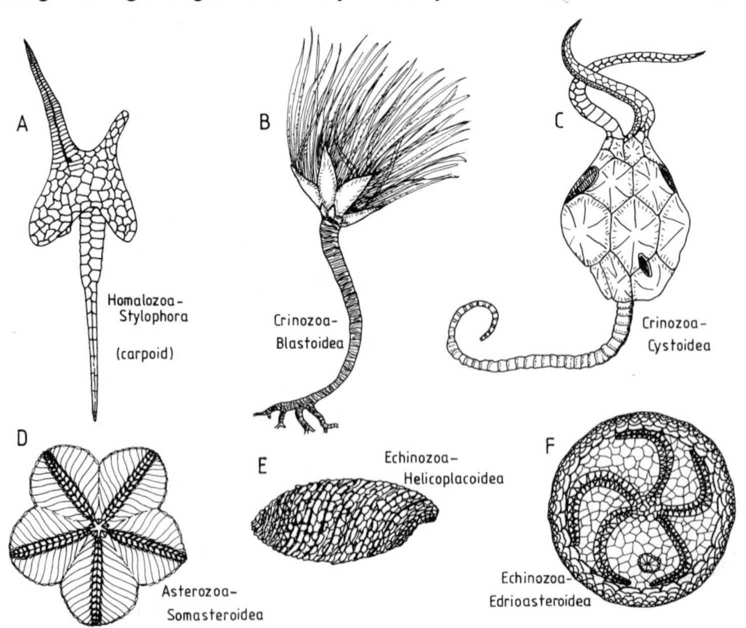

Fig 12.1. Fossil echinoderms from the Cambrian and Ordovician (redrawn from several sources, not to scale).

Fig 12.2. Diversity of classes and of genera of echinoderms, through the Phanerozoic (based on data in Paul 1979).

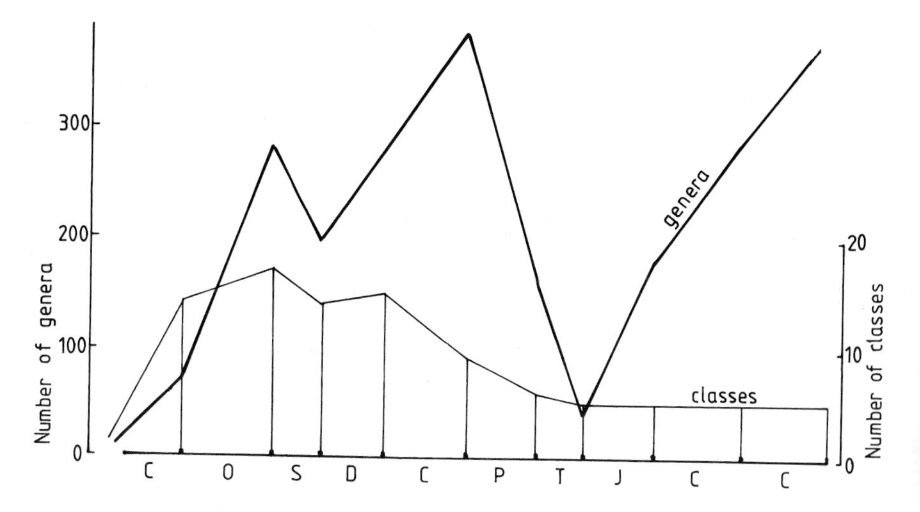

all possess the water-vascular system, with tube-feet (podia) arranged along ambulacral grooves; if an animal has these, it is an echinoderm, and if not, not (even the 'apodous' type of holothurians retain buccal podia, modified as tentacles). This kind of single diagnostic feature is almost unheard of elsewhere in the zoological world (perhaps cnidarians and nematocysts are the only other case). In addition, echinoderm skeletons are completely characteristic of the group, formed within the mesoderm from magnesium-enriched calcium carbonate and with each ossicle as a single crystalline structure hollowed out into a reticulate form by resorptive cells that in life remain within the ossicle as an organic phase, the stroma. These features make echinoderms uniquely easy to define, whether living or fossil, and also make them uniquely set apart from other phyla. This difference is accentuated in modern forms by the ubiquitous, if secondarily acquired, pentamerous symmetry, rendering the phylum so peculiarly unlike any other. Where then do the echinoderms belong in the scheme of things?

The alternatives that have been seriously discussed are relatively few in number. About the only resemblances that can be seen are on the one hand to the hemichordates (a classic deuterostome linkage) and on the other hand to sipunculans (chapter 8), a less traditional link cutting across protostome/deuterostome bifurcations but nevertheless receiving some support from Nichols (1967). This latter view depends on similarities of design and operation of the echinoderm water-vascular system and the ciliated tentacles and coelomic compensation sac system of sipunculan worms (fig 12.3). The echinoderm tube-feet are seen as derivatives of this tentacular system, a view compatible with their primitively respiratory function, with an additional role

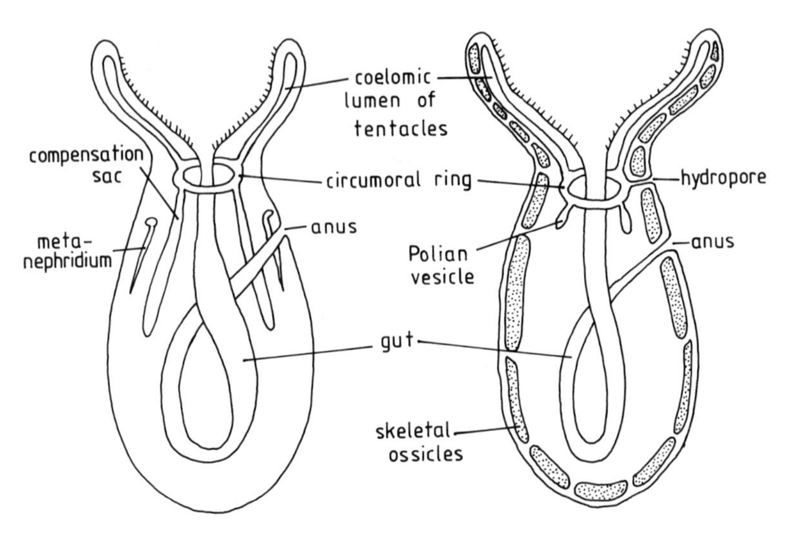

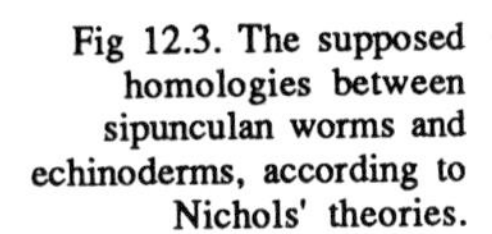
Fig 12.3. The supposed homologies between sipunculan worms and echinoderms, according to Nichols' theories.

in ciliary feeding. Sipunculans and early echinoderms also share the feature of a recurved gut. Sipunculans are used by Nichols as a link between annelids and the lophophorate, echinoderm and hemichordate assemblage, setting aside the embryological differences involved.

More direct links to hemichordates are indicated by a series of possible homologies, noted ever since the work of Grobben (1924) and summarised by Hyman (1959), Ubaghs (1969) and Jefferies (1979, 1986); they are shown in table 12.2. These are largely derivatives of deuterostomy and the tripartite nature of the coelom (so that many German authors regard all these phyla as close to their hypothetical tripartite 'archecoelomate' ancestor, as outlined in chapter 1). The nervous systems of the two groups are also similar (see Cobb & Laverack 1967; Dilly 1975), being of a primitive nature and diffuse within the ectoderm, with the exception of a small zone in the enteropneusts where the nerves are more concentrated as a hollow dorsal cord (see below). Some authorities have regarded the links between the two phyla as sufficient to merit their inclusion in one unit, the 'Ambulacraria' or 'Coelomopora'.

Jefferies (1979, 1981, 1986) stresses the links between these two groups in a rather different fashion from most authors, believing that certain of the early echinoderms (cornutes and mitrates - part of the group commonly called carpoid echinoderms) are closely allied to hemichordates and represent the stock of early chordates. In particular he points to similarities between *Cephalodiscus* and the early cornute *Ceratocystis*, the latter in his interpretation (fig 12.4) showing evidence of gill slits and other essential characteristics of the true Chordata. This controversial view is mainly relevant to the issue of vertebrate origins, and therefore somewhat beyond the theme of this book;

Table 12.2. *Possible homologies of echinoderms and hemichordates*

	Hemichordates	Echinoderms
Embryology:		
	Anus forms from blastopore.	Anus forms from, or at site of, blastopore.
	Enterocoelic coelom, 1, 2, or 3 parts forming initially, always 3 finally by splitting.	Enterocoelic coelom, usually initially 1 part, splitting into 3.
	Dipleurula-type larva (tornaria). Bilateral.	Dipleurula-type larva (various). Bilateral.
Tripartite coelomic compartments:		
	Paired protocoels, with pores.	Axocoel, with pore on left side.
	Pulsatile vesicle associated with protocoel.	Pulsatile vesicle associated with axocoel.
	Paired mesocoels.	Hydrocoel (on left only) forms water vascular system.
	Paired metacoels.	Paired somatocoels.
Gill slits:		
	Numerous in enteropneusts; 0/1 pair in pterobranchs.	Absent (except in some fossils??).
Tentacles:		
	Hollow, monociliated, with extensions of mesocoel.	Usually absent. If present, hollow, monociliated, with extensions of hydrocoel.
Nervous system:		
	Diffuse, ectodermal; no CNS.	Diffuse, ectodermal; no CNS.

aspects of it are considered in section 12.5, but it is otherwise amply reviewed in Jefferies' papers and book, and in critiques by Philip (1979), Ubaghs (1979) and Jollie (1982). Whether correct or not, it indicates the ubiquity of the idea that echinoderms, hemichordates and chordates are closely allied groups.

The fossil record of echinoderms does not greatly assist in deciding on their origins and affinities, for reasons outlined above. The schemes devised by Fell & Pawson (1966), by Ubaghs (1969) and by Paul (1977, 1979) are amalgamated in fig 12.5, giving the general consensus that bilateral forms such as Eocrinoidea represent the ancestral body pattern, perhaps having a tentaculate feeding apparatus; and that the carpoids (dating from the mid-Cambrian) are an early offshoot. If this is true, then it would be relatively easy to derive the first echinoderms from pterobranch-like forms, but the

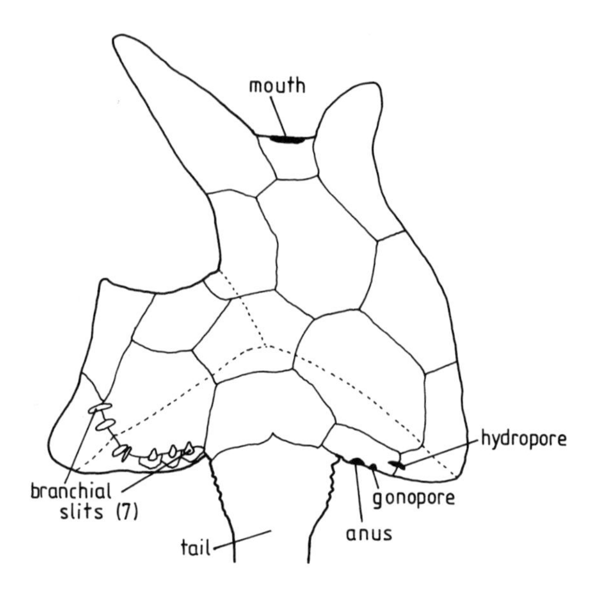

Fig 12.4. The fossil cornute ('calcichordate') *Ceratocystis*, with openings interpreted as evidence for chordate ancestry by Jefferies.

resemblance between *Cephalodiscus* and the cornutes becomes more superficial and secondary than Jefferies would have us believe. On the other hand a more recent evaluation by Paul & Smith (1984) has resulted in a phylogenetic scheme (fig 12.6) for the echinoderms that reverts to the old sub-

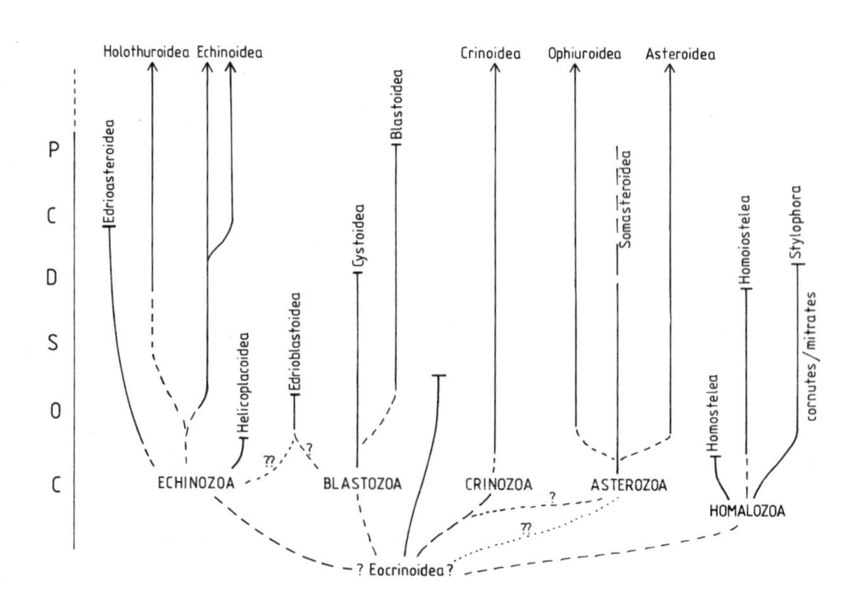

Fig 12.5. An overview of echinoderm radiation, based on works by Fell & Pawson, Ubaghs, and Paul.

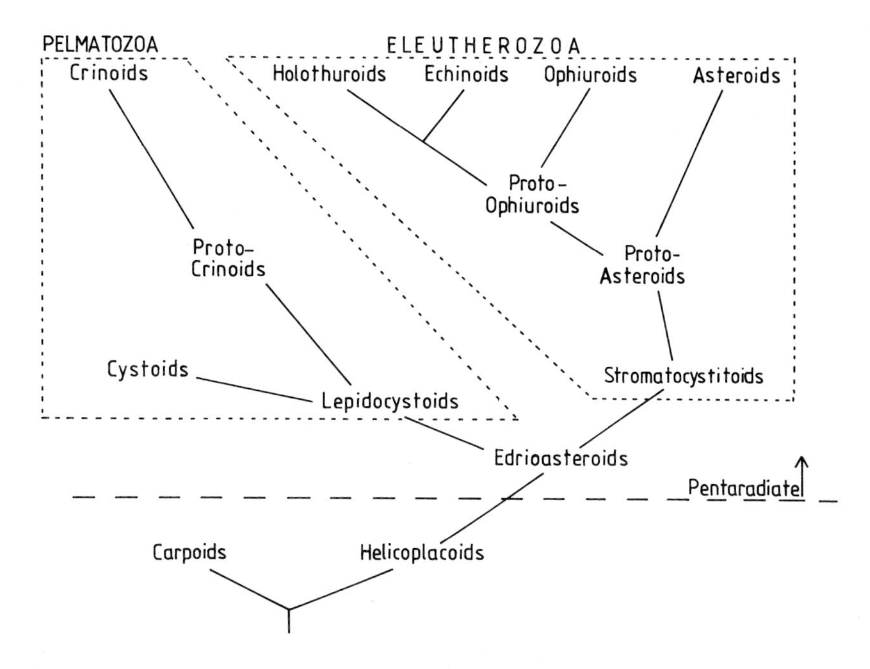

Fig 12.6. A recent reinterpretation of echinoderm evolution by Paul & Smith (1984), with stalked pelmatozoans once again separated from the free-living eleutherozoans.

phyla of Pelmatozoa (attached crinozoans and extinct cystoids) and Eleutherozoa (the other four or five extant groups). These are all seen as descendants of edrioasteroids (fig 12.1*F*), which are in turn derived from helicoplacoids (fig 12.1*E*). The carpoids are said to be the very earliest offshoot from the echinoderm stock, which would make them eminently suitable for Jefferies' schemes.

Embryology, the other traditional source of clues as to phyletic origins and relations, is in one sense not very much more help for echinoderms, for their embryology is almost by definition uniquely and classically deuterostome because the sea urchin was a central tool in defining the syndrome of deuterostomy. Thus they have radial cleavage, a monociliate coeloblastula with invaginating gastrulation, and a posterior blastopore; and the coelom develops from the gut wall, usually as one anterior compartment which then splits to give the classic three pairs with only the anterior coelom being in contact via a pore with the outside world. Their embryology may imply links with hemichordates and chordates, but the necessary features are by no means so clearcut in these latter groups and may be completely absent (vertebrates and perhaps cephalochordates are predominantly schizocoelic, for example). To complicate matters, surviving members of the 'oldest' group, the crinoids, all develop from yolky eggs and possess the vitellaria larval form (see chapter 5). Since this larva occurs in two other groups, it may even represent the ancestral form (Fell 1948), indicating that planktotrophy and the organised cleavage and gastrulation patterns of other modern echinoderms are secondary phenomena. This would accord with the conclusions from a long-standing problem of larval forms in the echinoderms (Fell 1948; Fell & Pawson 1966), whose occurrence seems to contradict ideas of relations between the classes obtained from adult morphology and palaeontology. Thus the pluteus larva occurs in ophiuroids and echinoids, whilst the auricularia type occurs in asteroids and holothuroids (see fig 5.11); yet from the fossil evidence (see figs 12.5 and 12.6) it is reasonably certain that asteroids and ophiuroids are actually two rather closely related extant groups, both descended from Somasteroidea. It is also usually accepted (though not in the Paul & Smith scheme of fig 12.6) that echinoids have been widely separate from other classes since their inception, so that the echinopluteus and the ophiopluteus must be convergent

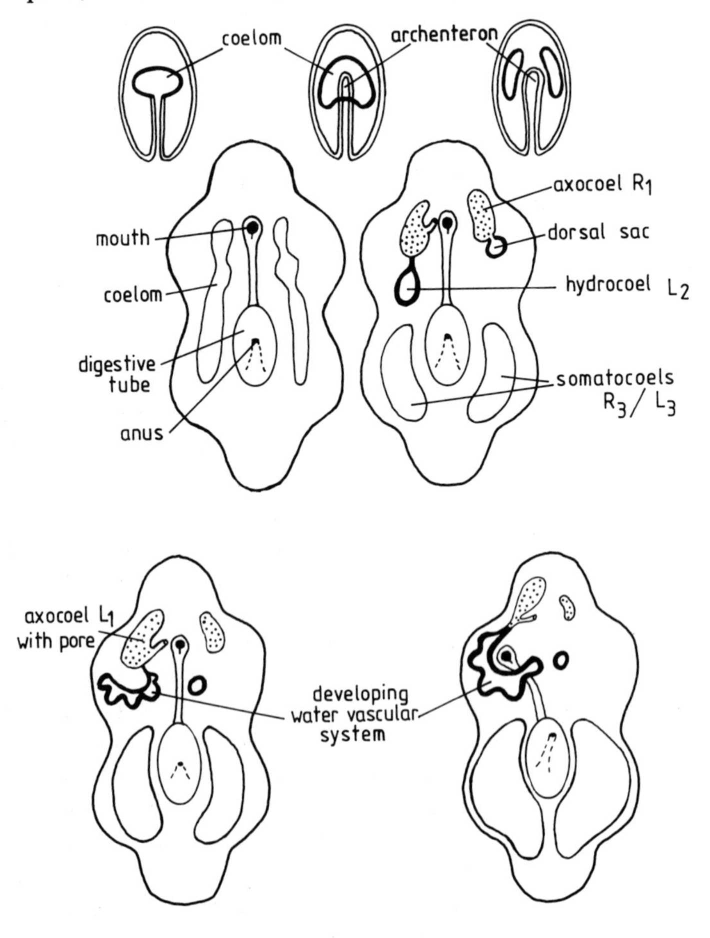

Fig 12.7. The development of the coelomic pouches in an echinoderm, with left-side hydrocoel forming the water vascular system.

phenomena. These factors, together with very marked divergence of larval form amongst some groups (particularly ophiuroids), testify in no uncertain terms to the unreliability of larval evidence in phylogenetic debates.

One other point worth noting about echinoderm embryology is that most of the larval forms are distinctly bilateral in organisation (so the group is clearly fundamentally bilateral, and earlier postulated links with the 'radiate' cnidarians can be discarded); and they have imposed on this a degree of asymmetry, apparent from the time of coelom formation (see Ubaghs 1969), in which the left-side organs are dominant with right-side structures atrophied (fig 12.7). There is no way of telling at what point during the evolution of the phylum this asymmetry was established, though its effects on the subsequent morphology of the group have been profound. Jefferies (1979, 1981, 1986) at least attempts to account for it with his theory of a 'dexiothetic' stage when a preceding hemichordate-like form fell onto its right side on a muddy substrate and subsequently the left-side organs took over the major functions. Though this sounds very odd, it does 'explain' the left-side dominance that is evident in echinoderms, and that is also present (at least in terms of timing of ontogenetic events) in other deuterostome groups (see below).

With little assistance in deciphering echinoderm origins coming from either fossils or embryological evidence, other than an implied link with hemichordates, the best solution is to resort once again to a worm-like form as an ancestor. Echinoderms can be seen as one of many experiments in adding hard (protective?) components to a soft worm-like body, being unusual in that the calcite was deposited within the tissues. (Paul 1979 likens this design to a mediaeval castle, where the orifices within the skeleton represent 'slit windows' through which the animal breathes and feeds without admitting invaders; contemporaneously, many other animals devised incomplete external shells where the castle was not perforated by windows but had a draw-bridge to allow intermittent access to a mantle cavity and other soft surfaces.) Having the perforated endoskeleton limited the subsequent design and diversification of echinoderms, and may explain much of their peculiar biology. It restricted them to a largely benthic marine existence while promoting an unusual degree of multifunctionality in all those structures that protrude from the theca (tube-feet that are locomotory, respiratory, sensory and filter-feeding; spines for defence, locomotion and feeding, and so on).

Echinoderms appear in the early Cambrian already possessed of their distinctive features (though often in a rather inefficient form - Paul 1979), and can probably be derived, as above, from some preceding soft-bodied stock with a tentaculate filter-feeding mechanism. Sipunculans are one possibility if Nichols' ideas are accepted; the tentaculate groups such as phoronids immediately offer further possibilities as they seem to have a primitively tripartite body and their embryology is not too diametrically opposite or 'spiralian' (chapter 13). The latter group is favoured for these reasons. This link between the lophophorate phyla and the roots of the echinoderms and hemichordates is certainly gaining ground with zoologists, even those who start from very different phylogenetic premises (see for example the works of Remane, Siewing, Salvini-Plawen, Clark and Nichols). The lophophorate assemblage has often been accorded a position somewhere between traditional 'protostomes' and the echinoderm-dominated invertebrate 'deuterostomes', and such a lophophorate/hemichordate/echinoderm connection would allow them to be not just intermediate but potentially transitional (chapter 13).

12.3 Hemichordates

The two parts of this phylum (free-living acorn worms of the class Enteropneusta, and the even less familiar sessile and tentaculate class Pterobranchia) have had a chequered history. Initially acorn worms were identified as a form of sea cucumber, and their chordate affinities were only recognised in the middle of the nineteenth century, when they were allotted the status of full members of the Chordata alongside tunicates, *Branchiostoma* (amphioxus) and the vertebrates. However they are now generally given phyletic status on their own as Hemichordata, because they show most, but not quite all, of the agreed chordate features as summarised in table 12.3; that is, they always lack the one central feature of a notochord. Therefore the hemichordates are not chordates, but are presumably close relatives and may give indications of a stage in chordate ancestry. They are often included in an irregular group called the 'protochordates', with the tunicates and

Table 12.3 *Characteristics of the chordates*

Deuterostomy	(numerous features - Table 12.1)
Notochord	(dorsal rod-like stiffening)
Dorsal hollow nerve cord	(by epidermal inrolling)
Pharyngeal clefts	(primitively filtering, ± respiratory)
Post-anal tail	(due to movements of blastopore lip)
Endostyle	(iodine-secreting tissue)
Blood flow pattern	(dorsal backwards, ventral forwards)

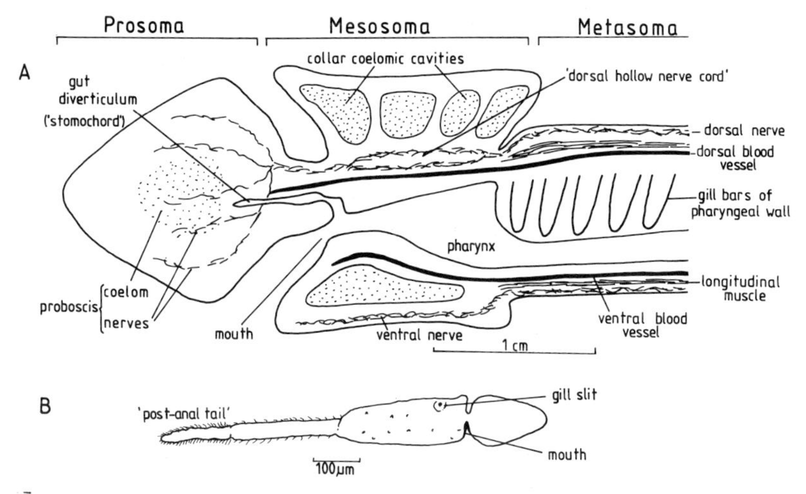

Fig 12.8. *A*, transverse section of the head end of an enteropneust hemichordate, showing the deuterostome and chordate features present; *B*, the larval stage of an enteropneust, with indication of a post-anal 'tail' or stalk.

cephalochordates (see Barrington 1965; Barrington & Jefferies 1975).

The hemichordates have been given a central role in cementing together the deuterostome assemblage because, while possessing some chordate characters, at the same time their embryology and larval structure seem to provide a clear link with the echinoderms (Ubaghs 1969; and see table 12.2). They have radial cleavage, gastrulation by invagination and an anus formed from the clearly posterior blastopore (see Godeaux 1974). Their tripartite body contains enterocoelic coeloms (single in the prosoma, paired in the mesosoma and metasoma) developing in just the same way as those of an echinoderm (Hyman 1955, 1959). As was discussed in chapter 5, the tornaria larva of the enteropneusts is almost identical with the echinoderm auricularia (fig 5.12). Thus the young stages of the hemichordates clearly ally them with the invertebrate echinoderms and, as discussed in section 12.2, there are also a number of features of adult anatomy, at least in the pterobranchs, that point to the same link.

It is their adult form, this time particularly in the enteropneusts, that also allies them with chordates. The adult acorn worm (fig 12.8*A*) has numerous U-shaped gill slits on the trunk, and a nervous system that for at least part of its course in the collar region could be designated as dorsal and hollow. It also has a repetition of structure suggestive of segmentation, for example of the gill slits, and of gonads in the trunk, and the trunk has only longitudinal muscle (as do most chordates, in effect). A post-anal tail as a juvenile (fig 12.8*B*) may also indicate links with chordate anatomy (though some believe this structure is merely the homologue of the pterobranch stalk). However, substantial differences occur; the pattern of blood flow in hemichordates is forwards dorsally, like other invertebrates, whereas in chordates it flows forwards ventrally and backwards dorsally. And earlier attempts to homologise the dorsal gut diverticulum ('stomochord' or 'hemichord') with the notochord structure of chordates cannot be upheld. Indeed some zoologists have maintained that all the adult 'chordate' characters are dubious and given certain mechanical constraints could well be convergent (e.g. Godeaux 1974).

These reservations notwithstanding, the whole construct of 'deuterostomes' seems to be held together rather satisfactorily, so long as it is accepted that the two classes of hemichordates do indeed belong together. It could therefore plausibly be supposed that both echinoderms and early hemichordates evolved from common sessile ancestors that fed externally with ciliated grooves, as all extant hemichordates still do and as early fossil echinoderms probably did. In addition, the presumably more primitive pterobranch class may provide the link between these early deuterostomes and the lophophorate groups covered in the next chapter, having a rather similar organisation and tentaculate feeding apparatus. Fig 12.9 shows two of the only three genera of pterobranchs: *Cephalodiscus* and *Atubaria* each have 1 pair of gill slits, and *Rhabdopleura*, very much smaller, has no gill slits. The now extinct graptolites (fig 12.9*C*) probably also belong with this assemblage, being very like extant colonial pterobranchs such as *Rhabdopleura* in form and growth (Rickards 1975, 1979). Many authors (see above) now seem to converge on a view that all these animals arose from a lophophorate phoronid-like ancestor in the late Pre-Cambrian; this is also a view much repeated in modern vertebrate texts (e.g. McFarland *et al.* 1979).

This all makes a neat story. However, one or two reservations should be

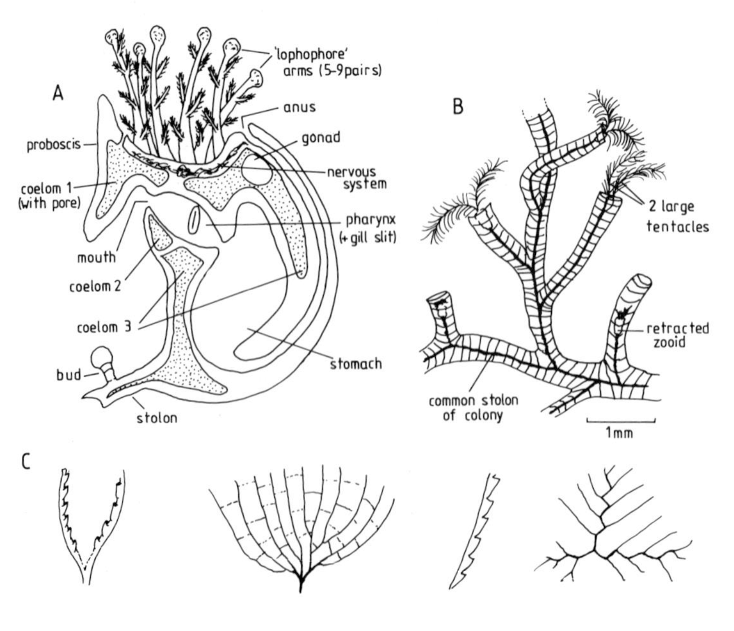

Fig 12.9. Structure of the pterobranch hemichordates and graptolites: *A, Cephalodiscus*; *B, Rhabdopleura*; *C*, various graptolite growth forms.

borne in mind at this stage. Firstly, the whole issue of the validity of deuterostome embryological features has been raised before (chapter 5), and further criticisms of this dogma will be considered below; it leads writers such as Løvtrup to regard the hemichordates as irrelevant to the whole issue. Secondly, section 12.2 pointed out the very considerable problems raised by the diversity of echinoderm larval forms. If it is true that these provide clear evidence that larval similarities are frequently convergent and misleading, and perhaps even meaningless, then the resemblance between the tornaria and the auricularia could be entirely fortuitous. One cannot have it both ways, after all; if larval similarity amongst the echinoderms is unhelpful in their internal classification, then the similarity between just one of those echinoderm larvae and the larva of another phylum can hardly be taken as proof of phyletic relations.

On first sight, then, the hemichordates are a vital and unifying link in the deuterostome story. On further examination, the issues may once again be a little more cloudy than elementary texts allow, and an element of cautious reservation should be retained, at least until the rest of the protochordate assemblage has been considered.

12.4 Chordates

Traditionally the phylum Chordata exhibits a very clear set of features as in table 12.3, including all the characteristics of the hemichordates but in addition having the notochord as a dorsal stiffening rod (precursor of the vertebral column) and an endostyle (precursor of the thyroid gland). The phylum consists of three sub-phyla, by far the most familiar being the vertebrates (Vertebrata or Craniata). The origin of these animals from amongst the invertebrates remains hotly contentious, but there are many good recent reviews so that an extensive summarising and argument here is unnecessary. This section concentrates instead on the (admittedly related) issue of the affinities of the two less familiar groups that can properly be termed invertebrate.

Urochordates, or tunicates

Tunicates are most commonly met with in the form of the ascidians, familiarly called sea squirts. These are rather abundant creatures readily to be found in the lower littoral and sub-littoral zone, where they form part of the filter-feeding community. They are remarkable partly for looking exceedingly unlike animals at all, and partly for having a greater efficiency as filterers than almost any other group, taking the concept of gill slits ('pharyngotremy') to extremes in pursuit of this distinction. It is hard to see much connection between the squat, sessile, filtering tunicates, with simple nervous sytems, musculature and circulations, and the vertebrates to which they are supposed to be related.

Indeed, as adults the link is exceedingly tenuous, and theories of tunicates as vertebrate cousins have never had much to do with the adult sea squirt. Rather, and principally since the analyses of Garstang (1928*b*), this part of the story of animal phylogeny has relied heavily on embryology and larval biology and invocations of neoteny/paedogenesis. The fate map of the ascidian tadpole is very like that of amphioxus and of craniates, though the neurula-type embryo becomes determined very early on (unusual in deuterostomes - chapter 5) and the larva being non-feeding possesses no blastopore-derived anus. The larval sea squirt is nevertheless most convincingly a chordate (Katz 1983), of an appropriate tadpole-like form to foreshadow the actively swimming early vertebrates. This tadpole larva (fig 12.10) has a relatively short life during which it actively swims to seek out a new resting site, before settling, resorbing parts of its own body, and taking up life as a plankton-filtering machine. To this end, it possesses a rudder-like post-anal tail, stiffened by a notochord formed from a row of enlarged vacuolated cells, with strong muscles, and a battery of suitable sense organs associated with a simple and hollow dorsal nerve cord. This transforms by a process of torsion and tail loss into the sessile adult, losing the locomotory and nervous specialisations and developing (with a degree of asymmetry) an ever greater array of pharyngeal slits.

Most of the deuterostome and chordate characters are therefore clearcut only in the larval form, and it has become conventional to regard chordates as derivatives of a tadpole larva, argument merely continuing (for many authors, anyway) as to whether this tadpole was that of an ascidian, or could have occurred earlier in a more generalised and primitive urochordate (see Berrill 1955; Bone 1979). An alternative view maintaining that the tunicate tadpole is a recapitulation of an ancestral adult condition is supported by Jefferies (1986) as part of his calcichordate theory, but this still keeps the tunicates firmly in the chordate origins story.

A few authors have questioned the idea of urochordates as chordate

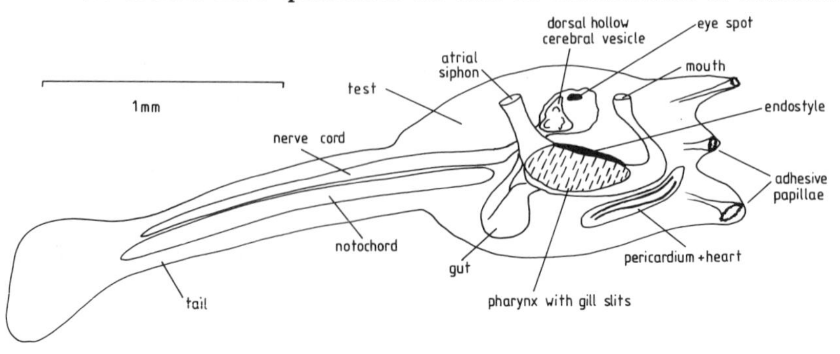

Fig 12.10. The 'tadpole larva' of an ascidian tunicate (sea squirt), with all the true chordate characters present.

ancestors though, and there are certainly a few problems raised by their embryology, by the absence of any spacious coelomic cavities and hence of enterocoely, and by a lack of segmentation. Carter (1957) disputes the whole concept of neotenic evolution (without which tunicates perhaps look unlikely to be relevant), whilst Bone (1960, 1979, 1981) and others believe the neotenic transformation did occur but must have been intitiated earlier still, on grounds of functional plausibility, since the tunicate tadpole has only a very short non-feeding pelagic life and is highly specialised for site-selection. Such views lead many authors to regard the hemichordates as more direct chordate ancestors, a long-living tadpole larva having arisen from the tornaria larva and gradually specialised to exploit surface plankton for longer periods, and tunicates are thus set aside from the main line of chordate evolution.

As there are no tunicate fossils earlier than the Permian, and none at all of convincing larval forms, there is no way of directly judging between these possibilities. Nearly all authors, however, continue to regard the tunicates as descendants of a hemichordate-like animal, and in this sense their status is hardly in dispute. Whether or not they are part of the story of vertebrate origins, they are offshoots from the tripartite worm-like stock of 'proto-hemichordates'; and it seems more plausible on functional grounds, following Bone (1960, 1981), to keep them separate from the true vertebrate lineage (fig 12.11).

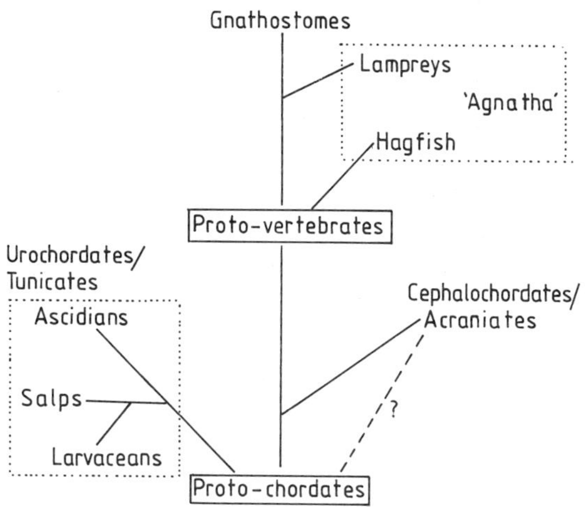

Fig 12.11. Possible position of the tunicates amongst the chordate assemblage, on a separate line from the cephalochordates (acraniates) and vertebrates.

Cephalochordates, or acraniates

These are the creatures commonly called lancelets or referred to by the now obsolete generic name 'amphioxus'. They are marine, benthic and microphagous, despite a rather tadpole-like appearance suggestive of a more active lifestyle. For much of the last hundred years they have been seen as bridging the gap between invertebrates and vertebrates; now their main interest lies in whether they are seen as a direct stage in the ancestry of vertebrates, or as a separate and somewhat degenerate offshoot.

Branchiostoma is strikingly similar in structure to a vertebrate (Godeaux 1974), with notochord, nerve cord, and perforated pharynx occupying similar positions and with segmented zig-zagging striated trunk musculature (myotomes). The notochord is a rod-like stack of vacuolated cells surrounded by a collagen sheath, reminiscent of the notochord structure in tunicate tadpole larvae and in higher chordates (Flood 1975). It is formed, though, from modified muscle cells, and can change its properties (shape and/or stiffness) by muscular activation, so that direct homology of notochords in the different protochordate groups is unlikely though similar ontogeny occurs. The nerve cord is hollow and of similar structure to that of craniates (Guthrie 1975). The gill slits also have similar ciliation patterns, and develop in similar fashion, to those of tunicates and lower craniates, with the characteristic peculiarity that left-side slits appear earlier (see Jefferies 1986 for a modern review of this issue). The embryology of the group also indicates a rather close relationship with craniates, as after typically deuterostome beginnings the amphioxus embryo is of the classic 'neurula' type (fig 12.12) unknown in other invertebrates but absolutely characteristic of vertebrates. Coelomic pouches 2 and 3 form early on from the endodermal roof of the archenteron, but many more form thereafter by schizocoely, with the first pouch appearing rather late; so the animal can never be described as truly tripartite. Later development is again very like that of vertebrates, though with an extra degree of asymmetry when many of the left-side organs grow earlier and are dominant for much of the period of metamorphosis.

However a number of odd and/or primitive features of an adult amphioxus should also be noted that are unlike the craniates. The epidermis is of simple mucoid cells, there is a persistent notochord and no other true skeleton, and the nervous system has no brain, spinal ganglia or major sense organs. There are segmentally arranged solenocyte-type protonephridia, and numerous segmental gonads open into the atrium. All these features set the cephalochordates well apart from the true vertebrates.

Thus the similarities between amphioxus and the vertebrates are of a rather general type, and are most noticeable early on. However there are some clearly 'primitive' features present in the lancelet, and there are few traces of 'degenerated' systems that would indicate secondary loss due to the sessile lifestyle. It therefore seems likely that the cephalochordates are a separate and specialised branch from the stock that gave rise to vertebrates, and not a degenerative offshoot from the main vertebrate stock itself (fig 12.11). Apart from certain specialisations associated with their mud-dwelling lifestyle, they are probably little modified from this stock, as compared with tunicates. Their

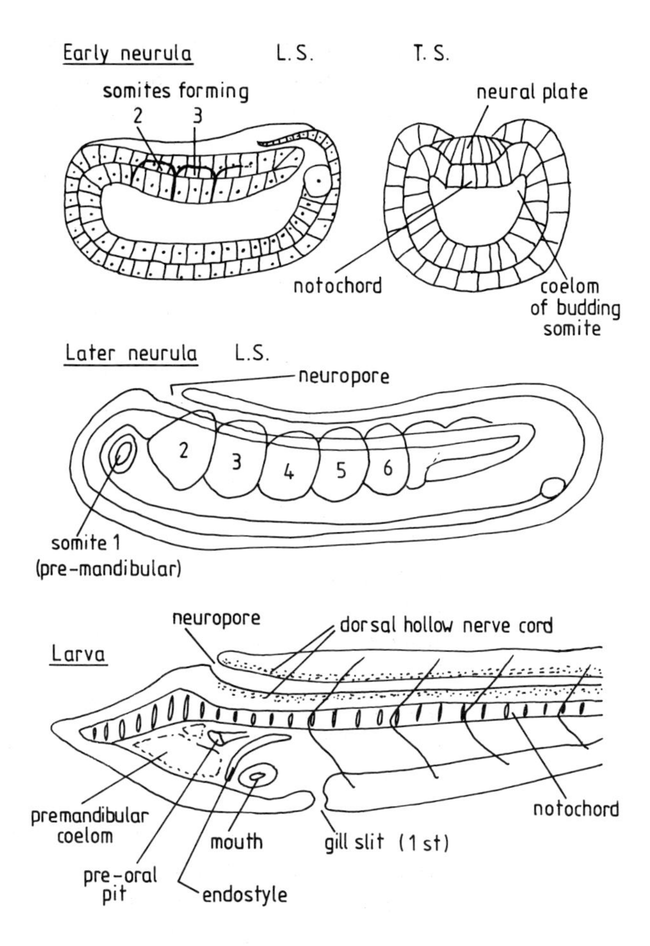

Fig 12.12. Embryology in cephalochordates, closely resembling that of vertebrates from the early neurula stage.

striking resemblance to the earliest known chordate fossil (*Pikaia* - see chapter 3, fig 3.7) would certainly suggest that this is so, this animal preceding by tens of millions of years the first bony vertebrate traces. And since the 'conodont animal' may also have been of a rather cephalochordate-like design (chapter 3), and perhaps akin to the jawless hagfish, it seems quite likely that the ability to deposit apatite evolved in a common cephalochordate-vertebrate stock, firstly as teeth-like denticles and later as dermal ossicles and true bone.

12.5 Other deuterostome groups?

Two further phyla need some consideration here, because many text-books and invertebrate family trees will be found that include the pogonophorans and the chaetognaths within the deuterostome assemblage. Do either of them merit this status?

The pogonophorans were first encountered in chapter 8, where a little of their history was given. With the discovery of their tail-piece (opisthosoma), having a septate condition and chitinous chaetae, their status as tripartite deuterostomes somewhat similar to hemichordates seemed to be dealt a death-blow and many authors incorporated them into the protostome assemblage instead (e.g. Southward 1975; van der Land & Nørrevang 1975), sometimes even within the annelids. However it is worth noting here that some writers still prefer to accord them a position retaining deuterostome affinities (e.g. Ivanov 1975), or set them in the centre of the animal kingdom having affinities with each of the traditional super-phyla (e.g. Cutler 1975; Siewing 1975).

As deuterostome features the following should be considered: a body dominated by three coelomic compartments, a simple intra-epidermal nervous system, and a blastopore at the rear end of the body near where an anus would be if the animals had one. Coelom formation may also have elements of enterocoely (Ivanov 1975), despite the lack of gut in the adults; and the cavities so formed may not be homologous with those of annelids (Jones 1985). But to set against this, the pogonophorans share with the protostomes (and/or with the annelids specifically) the possession of chitin and a particular kind of chaeta structure, the structure of the cuticle, and a metameric opisthosoma. Then again, they have some features that are intermediate in character (cleavage patterns, larval type, both mono- and multiciliated cells (table 6.1)), and some features that are simply aberrant (absence of a gut, 'coelom' lacking a peritoneum and of uncertain ontogeny, and tentacles borne on the first (prosomal) segment). On balance there is certainly no good evidence here that they belong amongst the main deuterostome phyla; their position may not be as close to annelids as some would like to believe, but is likely to be quite separate from that of the hemichordates.

Chaetognaths are a very different matter, as nobody has ever felt confident of their status; de Beauchamp described them as 'possibly the most isolated group in the animal kingdom', and they have at various times been allied with molluscs, nematodes, spiders, or brachiopods (see Ghirardelli 1968). They have radial cleavage, a blastopore at the rear end of the body, and a post-anal tail; and the larva possesses a classic enterocoelic coelomic cavity, though in two rather than three parts. In the adult, however, the cavity system is in three parts but lacking a peritoneum and should perhaps more accurately be described as pseudocoelomic. The chaetognaths were in fact considered briefly in chapter 9, as they share certain other features with some pseudocoelomate groups (longitudinal muscles only, and a thick cuticle perhaps akin to that of rotifers and acanthocephalans); but the muscles are very different in structure to those of nematodes, and the cuticle is not moulted. Their nervous system is described as more like that of protostomes than deuterostomes (Rehkämper &

Welsch 1985), and they possess chitin which is probably a protostome trait (chapter 4); but their bristle receptors are said to show some resemblance to the lateral-line receptors in fish, and their fin-rays may be structurally and chemically similar to the actinotrichs of fish (Reisinger 1970).

Clearly the association with deuterostomes is extremely tenuous, largely involving embryological similarities. Given that in some respects the chaetognaths are quite unique (for example, their photoreceptors and their cell-cell junctions are unlike those of any other group - see chapter 6), independent evolution of the peculiar chaetognath design from a very early stage, probably no further advanced than the acoeloid or proto-platyhelminth form that may be at the roots of the Metazoa, seems the most plausible explanation for this phylum's origins.

Thus neither of these groups is likely to be relevant to the central question of deuterostome affinities and evolution; they will be given no further consideration here.

12.6 Relationships between the groups

A summary of all the features that have been invoked as characteristic of deuterostomes, and their occurrence in the various 'core' groups, is given in table 12.4. What weight should be put upon them, and what do they reveal about the relationships and phylogenetic sequences in this part of the animal kingdom? Most importantly, are these features 'good' ones, and are the groups really related at all?

Firstly, all the problems with the actual characters that define a deuterostome should be recalled. In chapter 5 the dubiety of some of the embryological features was analysed, and Løvtrup (1975, 1977) has used these concerns to attack the whole idea of a deuterostome super-phylum, claiming that the echinoderms, protochordates and vertebrates share very few good taxonomic characters and that the vertebrates are in fact closer to molluscs and arthropods. However, Løvtrup's whole approach is to abandon morphological characters and base his classification on his choice of 'non-morphological' or 'functional' characters, which he maintains are usually disregarded; and he believes that the probability of convergence in such characters is 'so low that it should be of negligible consequence'. It is surely evident from common sense, and from much of the argument presented throughout this book, that functional characters are precisely those *most likely* to be convergent and confusing, arising from common adaptations to solve common problems in pursuit of common lifestyles in common habitats. Thus Løvtrup's list of features uniting vertebrates and molluscs (including details of connective tissues, gills, circulations, digestive and excretory systems, nervous and

Table 12.4 *Possible homologies in the deuterostome groups*

	Hemichordates		Echinoderms	Chordates		
	Ptero	Ent		Tun	Ceph	Vert
Blastopore at or near anus	+?	+	+	+	+	+
Cleavage						
Radial	+	+	+	+	+	+
Indeterminate	+	+	+	no	+	+
Coelom						
Enterocoelic	+	+	+	(1)	+(2)	no
Tripartite	+	+	+	(1)	no	no
Dipleurula larval type	+	+	+	no	no	no
Left-side dominance	no	no	+	+	+	+
Gill slits	+/no	+	no/+*	+	+	+
Notochord	no	no	no	+#	+	+
Post-anal tail	no	+?#	no	+#	+	+
Endostyle	no	no	no	+	+	+
Hollow dorsal nerve cord	no	+?	no	+#	+	+
Blood forwards ventrally	no	no		+/no	+	+
Tentacles on mesocoel	+	no	+/no	no	no	no
Monociliate	+	+/no	+	+/no	+	no

+ - present
# - only in larvae or juveniles
* - only (possibly) in fossils

(1) - no coelom present
(2) - enterocoely and schizocoely occur

hormonal systems, and reproductive traits) ends up looking profoundly unconvincing. Further criticisms of his methodology and choice of characters is given by Jefferies (1986).

Evidence that the groups conventionally called deuterostomes really do belong together is not too hard to come by. To begin with embryology, perhaps most convincing of all is the fact that every member of the three central phyla (echinoderms, hemichordates, chordates) exhibits one of the core features of the deuterostome dogma: they do have radial (or rather, never have spiral) cleavage. As discussed in chapter 5, cleavage patterns are not infallible, and many problems of interpretation arise; yet it seems probable that spiral cleavage is a simple (thermodynamically most stable) and perhaps primitive mode, so that its complete absence may be indicative of an overriding genetic development in some common stock, and hence does seem to indicate a degree of relatedness. Other features of the deuterostome syndrome also suggest links between these phyla; however none is as convincing, given that the blastopore fate is rather variable, most of the chordates have anomalous schizocoely, larvae even within groups are demonstrably unreliable, and at least the urochordates are highly determinate. But there is one further developmental feature that looks convincingly homologous between most of the groups. This is the element of asymmetry and left-side dominance in the ontogeny of all but hemichordates, which has led Jefferies (1986) to unite echinoderms and chordates in a new monophyletic group called Dexiothetica. The reasons for this phenomenon may remain uncertain, but it seems unlikely that it could have arisen repeatedly and convergently as it has little functional effect or advantage. Presumably it arose once in the embryological development of a pterobranch-like organism, and has been inherited by all the derived groups.

It is necessary to move away from embryology to find other uniting features. Adult homologies fortunately seem to strengthen the case. The notochord is of rather similar construction in all groups where it occurs (though perhaps derived from different cell precursors in acraniates as discussed above). The gill slit ontogeny (developing in a horseshoe-shaped fashion), and an asymmetry in the ciliation patterns around the gills (Bone 1979), are similar throughout the group. Barrington (1979) points out that the ciliary filter-feeding mechanisms of all deuterostomes form a logical evolutionary sequence, with external ciliated grooves visible in early fossil echinoderms and in hemichordates gradually giving way to ciliated gill slits and internal food collection. A strong link is also supported by the apparent homology of the enteropneust pre-oral ciliary groove, the cephalochordate pre-oral pit, and the vertebrate adenohypophysis (Welsch & Welsch 1978). One further feature not yet mentioned is the endostyle, a mucus-secreting and iodine-binding tissue most noticeable in the branchial basket of tunicates and cephalochordates; many early authors suggested this was homologous with the thyroid gland of vertebrates, where iodinated proteins are also produced. This has been confirmed (Thorpe & Thorndyke 1975); in fact the endostyle of the larval lamprey actually transforms into the adult thyroid. Detailed features of this kind tend to suggest that other similarities between the three traditional sub-phyla of the Chordata are not merely convergence, as some authors have maintained (see above). Patterns of intercellular junctions (chapter 6), of musculature, nervous systems, and neuromuscular innervation (protochordate muscles usually sending processes to nerves), and certain biochemical features such as the absence of chitin and occurrence of sialic acids and of creatine phosphate (Watts 1975; and see chapter 4), also testify to a probable link between all the deuterostome phyla. So, despite reservations when we had only considered echinoderms and hemichordates, it is hard to avoid the conclusion that these groups do all belong together.

There are then four main possibilities as to the sequence of relations between all these taxa, shown diagrammatically in fig 12.13. Traditionally, echinoderms were set well apart near the base of the deuterostome tree, while the hemichordates were close to, but did not quite fully qualify as, chordates (fig 12.13*A*). This pattern will be found in many elementary texts, and was reproduced from varying sources in several earlier figures in this book (e.g. figs 5.13, 6.2, 6.11).

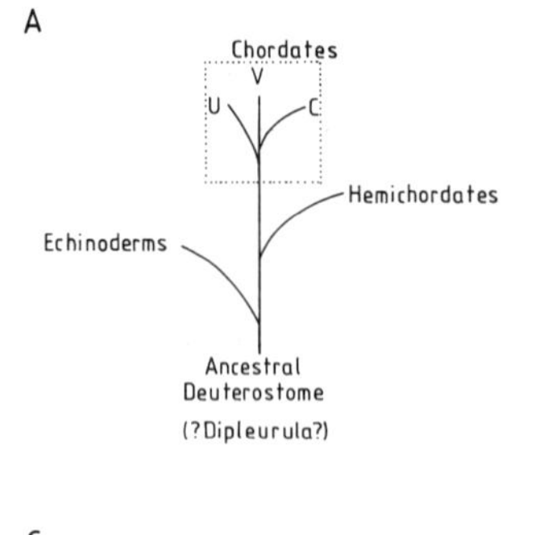

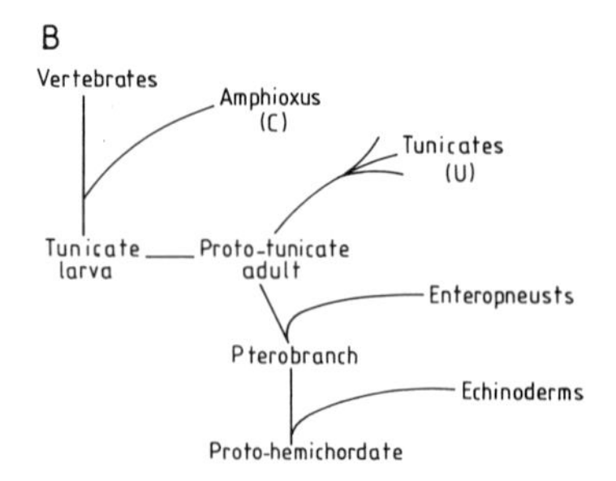

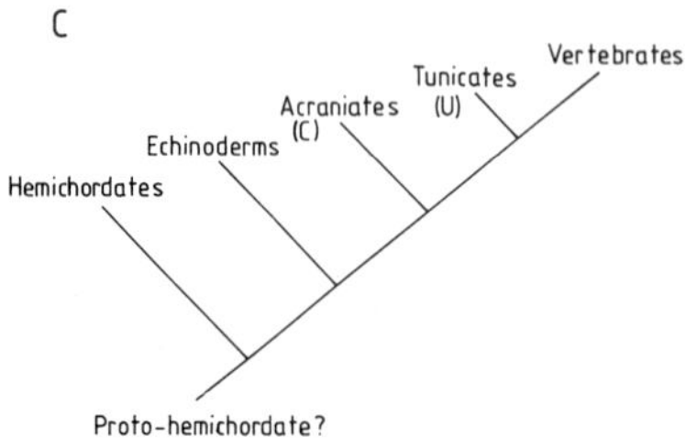

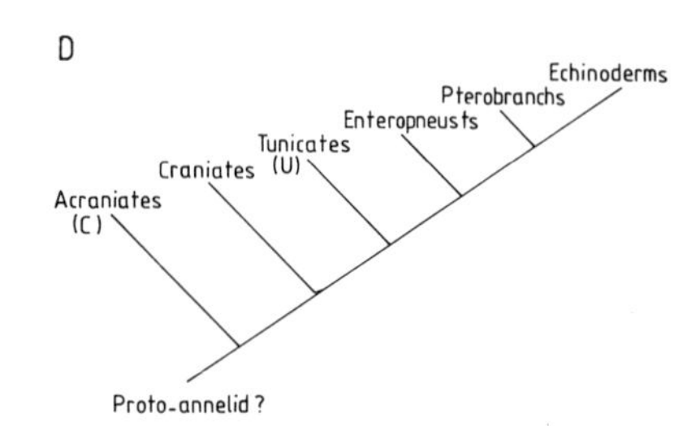

Fig 12.13. Different possible schemes for the phylogenetic relationships of the deuterostomes: *A*, traditional view with hemichordates only slightly removed from the true chordates; *B*, currently popular view in which all deuterostomes stem from a hemichordate-like animal and vertebrates arise from a neotenous tadpole larva; *C*, Jefferies' view in which echinoderms give rise directly to chordates, via the cornutes and mitrates; *D*, the view of Gutmann and others, with vertebrates as direct descendants of segmented worms, and other deuterostomes as later developments.

Certainly hemichordates look more 'like' chordates, but logically as we have seen it is more likely to have gone the other way, with hemichordates (pterobranchs in particular) as the primitive and basal stock from which both echinoderms and chordates evolved. Echinoderms appear more different from their ancestors largely because of the profound torsional metamorphosis they undergo in association with sessility, radiality and pentamery. Further argument then centres on whether the echinoderms are a quite separate offshoot from the base of the tree, with hemichordates or tunicates giving rise to higher chordates by a process of neoteny, as most authors would now hold (fig 12.13*B*); or whether they are an integral part of the route to the chordates, via the cornutes and mitrates, as Jefferies believes (fig 12.13*C*). This argument is destined to continue for some time. A good defence of the traditional schemes is given by Bone (1979, 1981), incorporating Garstang's idea that notochord and gill slits are adult adaptations, pushed back gradually into the larva and thus retained in vertebrates, whilst the dorsal hollow nerve cord could be a primary larval feature, arising by the conjunction and inrolling of larval ciliated bands (the 'auricularia theory'). Such schemes have considerable attraction for many modern authors, though they require an act of faith in accepting neotenic evolution *per se*, and complex larval-adult transitions for many characters. The fact that neoteny is demonstrably quite common in protochordates (it occurs in larvacean tunicates, and trends towards neoteny can be seen in some cephalochordate larvae) certainly adds to the plausibility of such schemes though.

On the other side of the argument, Jefferies has by no means convinced most current vertebrate and echinoderm experts. His scheme, though it explains the apparent presence of gill slits in some early echinoderm fossils and neatly accounts for left-side dominance in so many groups, is criticised by others on several fronts. It is not quite convincing in mechanical terms that the tail (often rather narrow, and with little room for substantial musculature once a notochord has been housed between the covering plates) is supposed to achieve propulsion by pushing through surface sediment; nor does the feeding system look very well designed if this is a backwards movement as Jefferies maintains, with a large mouth at the trailing end of the animal in churned-up mud. The scheme requires rather remarkable repetitious gain and loss of features (notably the calcitic skeleton), and also the resorption and remaking of tails in mitrates to achieve the correct relation between putative nerve cords and notochords; it gives the cornutes and mitrates surprisingly elaborate brains at a very early stage when they were supposed to be rather inactive filter-feeders; and it upsets the usual view that vertebrates have a segmented head. Many of these criticisms have been considered by Jefferies, and counter-arguments of varying conviction are set out in his book (1986). And it should certainly be noted to Jefferies' credit that his scheme does have the merit of being based on actual and accessible fossil material, so that argument need not proceed in quite such a vacuum as is customary in phylogenetic controversy!

All these views take an oligomeric (tripartite) condition as a starting point, and the gill slit/branchial basket as the first stage in chordate evolution. There is another and quite different view promulgated in the literature, that of Gutmann, developed in concert with Bonik, Grasshoff, Peters and others of this German school, and summarised in Gutmann (1981). He accepts that all the groups discussed here do belong together, but believes on mechanical/functional grounds that the evolutionary sequence goes exactly the other way around from that conventionally accepted. The cephalochordates are seen as closest to the stem group (fig 12.13*D*), having evolved by development of a notochord in a metameric worm-like organism (thus potentially re-uniting all the main metamerically segmented animals). From this acraniate organisation, whose primitive excretory systems and gonads are neatly accounted for by such a scenario, the true vertebrates developed as a major branch, with greater cephalisation and motility. Tunicates were a separate and advanced offshoot in which the branchial basket became a dominant feature and other chordate features only persisted in the larva (hence no invocation of neoteny is required, and sudden larval transformations such as acquiring a tail are not needed). Enteropneust hemichordates specialised for burrowing instead, thus losing the notochord (the stomochord is held to be a vestige of this structure) and developing the proboscis and collar region (with a vestige of the hollow neural tube) to aid penetration of mud and direct feeding currents. Pterobranchs arose from enteropneusts, losing the branchial basket and developing tentacles as adaptations to life in tubes. Echinoderms were the final step, derived from pterobranchs, specialising for sessile life and tentaculate feeding. Links between echinoderms or hemichordates and the lophophorates are therefore discounted.

While Gutmann's ideas do attempt a sound functional analysis of the problem, and give reasonable explanations for certain anomalies (protonephridia in amphioxus, and the presence of only longitudinal muscles in the trunk of hemichordates), they also give rise to serious anomalies, particularly for the hemichordates (for example why should they have 'lost' a good central nervous system, and turned their blood vascular system back to the annelid pattern?). Most zoologists will no doubt continue to find the central premise that chordates evolved directly from metameric worms, and the correlate that pterobranchs and echinoderms are actually the most 'advanced' deuterostomes, unacceptable.

Finally, whatever the relationship between echinoderms, hemichordates, tunicates and cephalochordates, there does seem to be rather a large gap between all these groups and the actual vertebrates. Vertebrate cells when mature are peculiar in being multiciliate (chapter 6; see figs 6.2 and 6.3), though ontogenetically their cells may be monociliate at first; whereas the deuterostomes as a whole retain monociliate cells throughout life. There are just a few cases of adult multiciliation in special cells in hemichordates and tunicates, and none as yet known in cephalochordates. This does not give much support to ideas of a neotenic transformation in the evolution of vertebrates (rather the reverse!); indeed, all larvae so far tested within the group are monociliated. Vertebrates also stand well apart from all other animals tested, including cephalochordates and tunicates, in their DNA composition (Russell & Subak-Sharpe 1977; fig 4.3), though hemichordates were not included in that analysis. If anything, these points may be indicating that vertebrates were evolved rather directly from the hemichordates, with tunicates as a sideline from them and with cephalochordates also somewhat separate. It certainly seems likely that, in whichever group it occurred, the evolution of the full vertebrate body plan was a rather precise, 'sudden' and unitary event.

12.7 Conclusions

Most of the evidence considered here, notwithstanding the Løvtrup heresies, confirms that the three phyla usually designated as deuterostomes do indeed belong together as a monophyletic unit. They share too many features, both embryologically and as adults, for other conclusions to be admitted.

Beyond this, there is no overall consensus as to the directions of evolutionary change within the group. Various authors even in recent years have taken echinoderms, hemichordates, tunicates or acraniates to be the ancestral stock. Of all the possibilities, most now converge on seeing the pterobranch hemichordates as closest to the origins of the deuterostomes, and much of the available evidence would seem to support this view, giving a phylogenetic picture something like that in fig 12.13*B* or *C*. Links to the lophophorate phyla are also now widely supported, for reasons that will be discussed further in chapter 13. However, the alternative view that the deuterostomes are a separate and rather direct offshoot of a cnidarian-like ancestor is still promoted by adherents of the archecoelomate concept (for example Remane, Marcus, Jägersten, Siewing) and by certain polyphyletic authors (e.g. Inglis 1985, for whom the enteropneusts are a suitably worm-like trimeric starting point). Such views have the merit that the deuterostomes as a whole are monociliate, as are cnidarians (cf. fig 6.2*B*), but are rather difficult to reconcile with the common occurrence of homeobox genetic sequences both in the coelomate deuterostomes and in the coelomate spiralians descended from flatworms (see chapter 4).

The additional question of vertebrate origins from amongst the deuterostomes remains particularly controversial. Whether or not neoteny/paedogenesis was involved, and if so when and where the requisite tadpole-like larva first occurred, is highly debatable, but does not appear to be an absolutely necessary part of a plausible scheme. Given the rather heavily armoured, sluggish appearance, and bottom-dwelling habit of the earliest-known (early Ordovician) vertebrates, the often-repeated need for neoteny to give a 'suitably motile' ancestor rings a little hollow. The conodonts (probably associated with a very chordate-like creature, as discussed in chapter 3, and in existence from the earliest Cambrian) seem likely to be incorporated in vertebrate origin schemes in the near future, together with soft Cambrian fossils such as *Pikaia*. One way or another (via Jefferies' route, or via soft cephalochordate-like ancestral types?) it seems probable that palaeontology will prove to be the key to relationships within the deuterostomes.

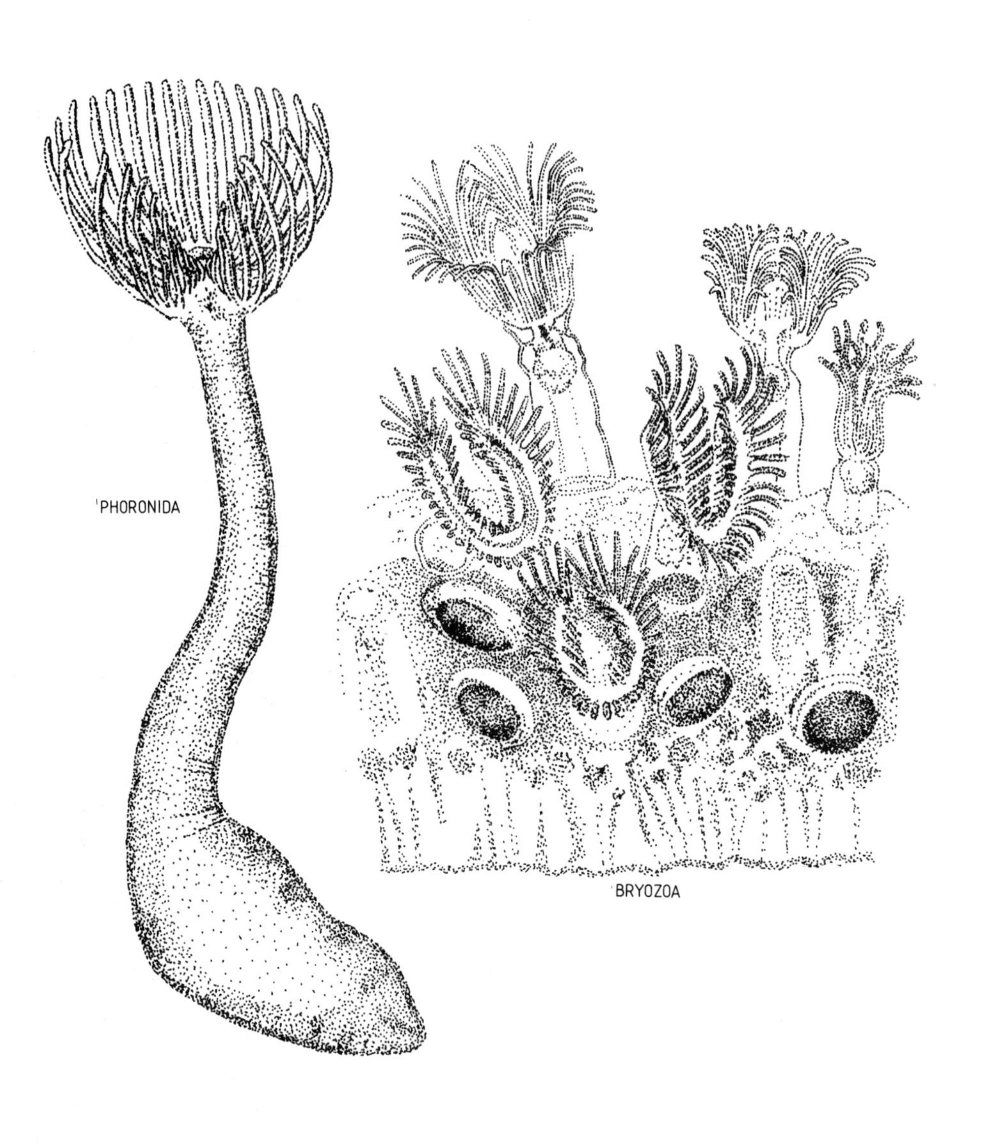
PHORONIDA
BRYOZOA

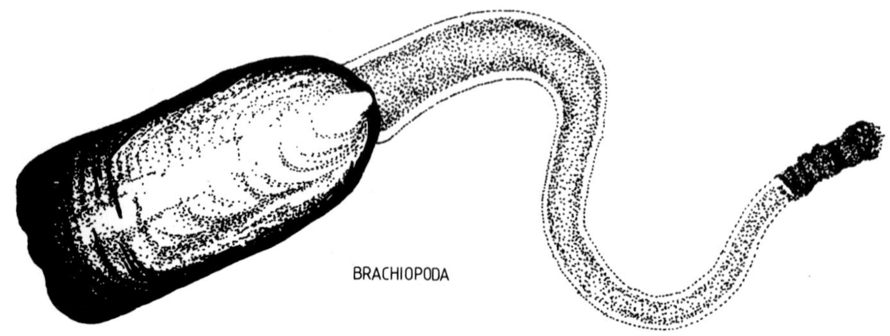
BRACHIOPODA

13 The lophophorates

13.1 Introduction

The lophophorates are a group of three phyla, sometimes given taxonomic status collectively as the Lophophorata or Tentaculata, and having in common the presence of a special kind of feeding organ, made up of a ring of ciliated tentacles surrounding the mouth, termed a lophophore. They are not particularly familiar to the layman, and rarely receive attention in introductory zoology courses, but are nevertheless of considerable importance in interpreting the pattern of the animal kingdom, so are justifiably given a chapter to themselves in this book.

The most important of the three phyla nowadays is the moss animals (**bryozoans** or ectoprocts), individually tiny but with the zooids assembled as often substantial colonies, diverse in form but commonly sheet-like or erect and branching. The bryozoans (reviewed by Ryland 1970; Woollacott & Zimmer 1977) are actually amongst the larger phyla in terms of species number. They are very frequently met with in the lower littoral zone, where they may dominate the filter-feeding community, and they may also be washed up as dried remains on the strandline; most people will therefore have met specimens of these creatures, though they may have misinterpreted them as plant remains. A second group is the lamp shells or **brachiopods**, reviewed by Rudwick (1970). They resemble molluscan bivalves but have two dissimilar calcareous shell valves that are dorsal and ventral (rather than identical valves that are bilateral), and within which a spiral array of tentacles may be visible. The brachiopods are not abundant now, but have been dominant in previous eras and their fossilised remains are invaluable to geologists for dating purposes. The third phylum included here is the **phoronids**, one of the smallest and most abstruse groups of animals, so rarely met with by non-experts that they have no common name. There are perhaps only ten species in total (Emig 1982), usually mud-dwelling and supported by a chitinous tube, with a fan of tentacles protruding.

At least two of the lophophorate groups, then, are of considerable practical importance; but what of the theoretical phylogenetic implications of the assemblage as a whole? This chapter considers what more can be learnt of the relationship between various groups of the animal kingdom from a knowledge of these unfamiliar but fascinating animals.

13.2 Lophophorate features

The three groups briefly described above certainly do not appear conspicuously similar, and the assertion that they are related needs some careful justification. Certain key features are said to unite them in the lophophorate assemblage, and these should be analysed rather carefully as a first step to sorting out the group's relationships. An 'archetype' for the lophophorates, incorporating the following points, is shown in fig 13.1.

The lophophore. The lophophore is a feeding organ designed to filter small particles from water; it is a horseshoe-shaped or circular array of tentacles. The classic features required of a true lophophore (as distinct from other tentaculate feeding organs - Hyman 1959) are that each tentacle is hollow and coelomate, and the tentacles are arranged to surround the mouth but not the anus, this being set apart outside the ring of tentacles. In addition, the pattern of ciliation

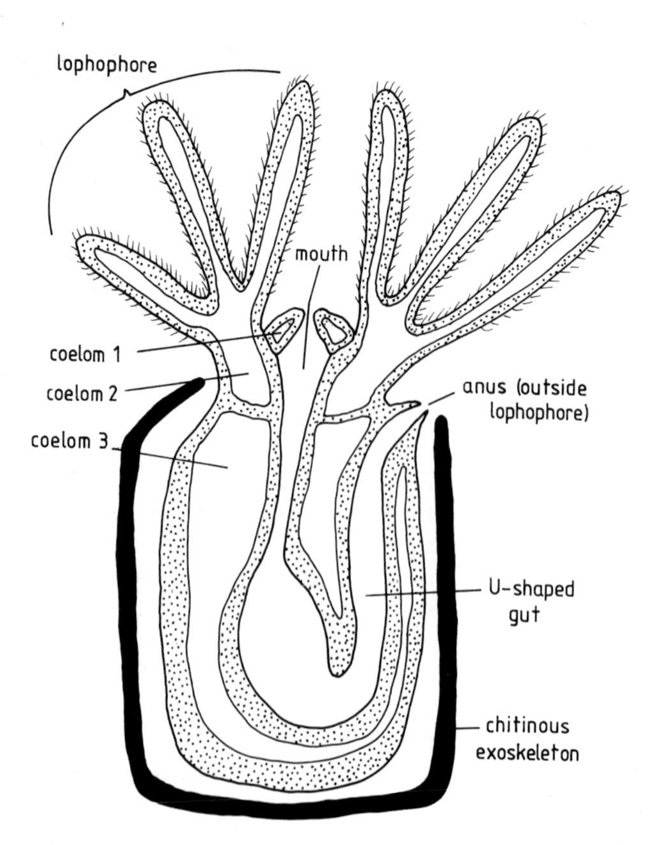

Fig 13.1. An entirely hypothetical archetype for the lophophorate groups.

is characteristic (fig 13.2), lateral cilia creating the water movement and frontal cilia trapping food and carrying it to the mouth. Currents thus pass downwards into the centre of the lophophore and out ventro-laterally over the anus. Hence the actual feeding mechanism of all lophophores is similar (Bullivant 1968; Winston 1977). Possibly it should also be part of the definition that a true lophophore is borne on the second segment of the body, as discussed below.

In the phoronids, usually regarded as the most primitive of the three lophophorate groups, the tentacles are relatively few in number and, whilst in just one species they form a complete oval, they are normally arranged as a double horseshoe, with each arm of the horseshoe sometimes spiralled on itself (Emig 1982). The brachiopods carry this to extremes, the lophophore becoming a complex two-armed and three-dimensional spiral, taking up much

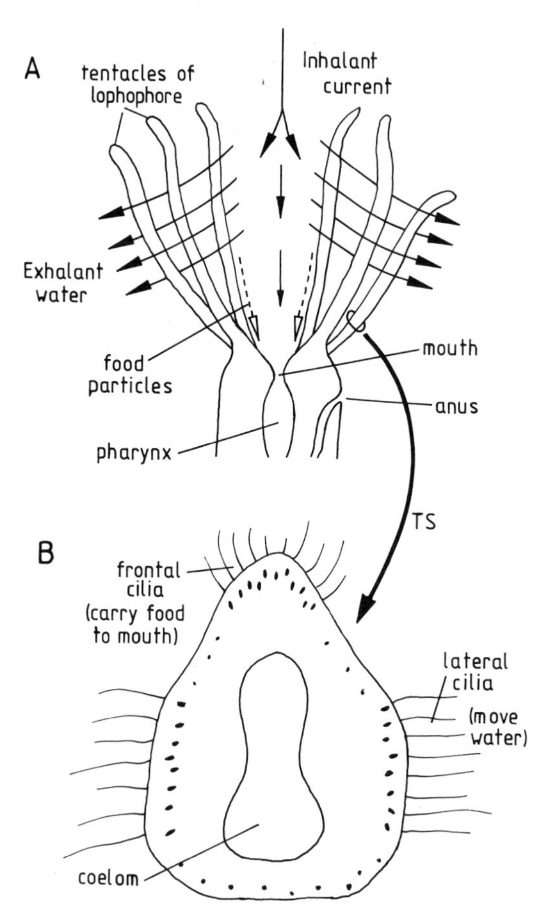

Fig 13.2. *A*, Structure of the lophophore, showing the water currents and feeding tracts; *B*, the relevant ciliation on an individual tentacle.

of the space within the valves of the shell and efficiently separating this space into inhalant and exhalant chambers. In these two groups the lophophore as a whole is not retractable. By contrast, the bryozoans can pull the whole structure back within the casing in which each zooid sits, when they are not actively feeding. Of the three classes that make up the phylum, only one has a horseshoe lophophore (the fresh-water Phylactolaemata), the other two classes having circular arrays of tentacles.

Many other groups of animals have ciliated tentacles for feeding. Pogonophorans, entoprocts, sipunculans and holothurian echinoderms, together with some annelids (fanworms) and cnidarians could be mentioned; but in these groups the feeding apparatus is not normally termed a lophophore. In some cases the tentacles are solid rather than coelomate, or are clearly a de novo acquisition or modification within a phylum; in other cases the patterns of water flow are quite different; and in others again the ring of tentacles surrounds both the mouth and anus, thus involving very different embryological and ontogenetic patterns. There is a reasonable consensus that the structure as exhibited in the three phyla Phoronida, Bryozoa and Brachiopoda is a different phenomenon from all these, and is homologous, evolving perhaps from the simpler condition in phoronid worms to more elaborate patterns in the other two phyla. The mode of development, and patterns of ciliation, are so similar in each of the three phyla as to make this very likely. The only other group that might reasonably be said to have a lophophore, as narrowly defined, is the hemichordates, met with in chapter 12; the tentacles are on the second segment, and operate in much the same fashion as in the conventionally 'lophophorate' groups.

Chitinous exoskeleton, and other chemical features. All the lophophorate groups have some form of exoskeletal material, whether a simple tube, heavy bivalved shells, or box-like or bottle-like cases around tiny zooids. It has long been recognised that in all three groups chitin is involved in these structures, and this molecule is principally a protostome feature (chapter 4; fig 4.1). Minerals are incorporated in many bryozoans and all brachiopods, and the resulting fine structure of the exoskeletal material in these groups is very similar, and may be a good character indicating their phylogenetic affinity (Tavener-Smith 1973; Sandberg 1977; Williams 1984).

In the Brachiopoda, the mineral chemistry of the exoskeleton has changed through evolutionary time, for shells were phosphatic in early fossil brachiopods and are still partly so in the few remaining genera of the class Inarticulata, but became predominantly calcareous in the geologically later and now more abundant Articulata. Indeed, the presence of apatite in the phosphatic shells is of possible phylogenetic interest, as this compound is only otherwise found in abundance in vertebrate bone and in the fossil conodonts, now interpreted as chordate (or in at least one case lophophorate-like) animals (chapter 3).

Other aspects of lophophorate chemistry and architecture also seem to imply mixed protostome and deuterostome affinities. For example they share with protostomes a lack of sialic acids, and their ribosomal RNA has the characteristic protostome central 'break' in its structure (Ishikawa 1977). However their DNA is very similar in overall composition to that of a number of deuterostomes (Russell & Subak-Sharpe 1977; fig 4.3). They have intercellular junctions rather like those of deuterostomes. And the haemoglobin present in phoronids is very much intermediate in character between the two super-phyla (fig 4.2).

Tripartite coelomate body. It has often been asserted that all three of the lophophorate phyla show a primitive trimeric state, with the same division of prosoma, mesosoma and metasoma as was met with in the deuterostomes (chapter 12), each part having its own coelom. The lophophore is borne on the mesosoma, and the metasoma forms the trunk or main part of the body. If this is true, it is certainly not always clear what has happened to the prosoma or first segment, and its existence is somewhat controversial. It can be identified with reasonable conviction in the phoronids, as a small wedge of tissue above the neck of the lophophore confusingly given the name 'epistome'; this prosomal segment is rather larger in the early developmental stages when it clearly has its own coelom. The phoronids, then, probably are tripartite. But in bryozoans the identification of a prosoma is more difficult, and it can only be seen with any ease in the class Phylactolaemata. The Stenolaemata have perhaps a trace of it, but in the most common class Gymnolaemata it is totally absent. This, together with the simpler horseshoe form of the lophophore and of its retraction mechanisms, and the non-calcified nature of the skeleton, leads most authors to regard the Phylactolaemata as the most primitive group of bryozoans, despite their unusual fresh-water habit. Thus a primitive trimery can be identified with some confidence, coupled with a trend towards losing the prosoma in more advanced groups, perhaps due to a progressive miniaturisation involved in the evolution of the phylum. The adult brachiopods have no trace of a first segment, but this could be due to the complete loss of a 'head end' once enclosed in a shell, as may be seen in bivalve molluscs. There are indications of trimery in the larvae, at least of inarticulate forms.

U-shaped gut. Each of the lophophorate phyla has a recurved gut, the anus opening somewhere near the mouth, just peripheral to the lophophore. This is hardly surprising, however, being clearly associated with the sessile, filter-feeding habit. It is most obvious in the phoronids, with a reasonably sized and elongate body; in bryozoans the whole body is so small that the gut curvature is much less apparent, though the anus is distinctly 'dorsal', whereas in many brachiopods (the class Articulata) there is no anus and the gut is blind-ending and scarcely recurved.

Embryological features. The embryology of lophophorates remains one of the most controversial aspects of their biology, and also one of the most instructive. Through much of this century, they were regarded as close to the protostome assemblage by virtue of their embryological patterns. Hyman (1959) reiterated and strengthened this view, pointing to the primary mouth formed from the blastopore in some groups, and to traces of spiral cleavage, schizocoely and a trochophore larva; she allotted the three phyla a place close to the main protostomes, but acknowledged that the link was not close, as the lophophorates also showed deuterostome features and should perhaps be given a separate intermediate status of their own as 'the lophophorate coelomates'. More recently, though, new analyses of embryology in an increasing number of lophophorate species have led to a substantial reassessment of their affinities. Most authors therefore now include them with, or very close to, the deuterostomes; the group have moved from one half of the animal kingdom to the other.

A review of the embryological features involved is shown in table 13.1, based on a summary by Zimmer (1973). Some of these points need further explanation. Only one group, the phoronids, has a mouth formed from the blastopore in the protostome fashion; in bryozoans the mouth is clearly secondary, and this seems also to be true of brachiopods. As regards cleavage, the traces of spirality detected by Hyman have been largely discounted by other observers (e.g. Zimmer 1973); there may be imperfect radial patterns in phoronids (Emig 1977, 1982), but the bryozoans and brachiopods are now conventionally referred to as having standard radial cleavage. Matters are more complicated when it comes to mesoderm and coelom formation though. We have mentioned phoronids previously (chapters 2, 5) as having a special mode of body cavity formation all of their own, and this is certainly true of some of the species studied (Emig 1982). In effect, the gut wall thickens up locally, and a cavity appears as a split within the thickening. This is most peculiar, for by virtue of the tissue involved (endoderm, rather than the 4d mesentoblast derivatives) the process should be called enterocoely, but since it involves splitting rather than pouching it more nearly resembles schizocoely. Some authors have consented to call it 'modified enterocoely'; but if anything it seems to be a novel mode of arriving at a secondary body cavity, tending to confirm earlier suggestions (chapter 2) that coeloms are thoroughly

Table 13.1. *Protostome and deuterostome features of lophophorates*

	Phoronids	Bryozoans			Brachiopods	
		Phylo	Steno	Gymno	Inartic	Artic
Embryological features						
Blastopore	P	D	D	D	?P	?P
Cleavage	?D	D	D	D	D	D
Mesoderm	P/D	?	?	?	?P	D
Coelom	P/D	?	?	?	?P	D
Larva	?P	-	-	??P	???P	???P
Adult features						
Cephalisation	D	D	D	D	D	D
Nervous system	D	P	P	P	D	D
Body regionation	D(3)	D(3?)	D(3?)	D(2?)	D(3?)	D(3?)
Skeletal system	P	P	P	P	P	P
Blood flow	P	-	-	-	?	?
Biochemistry	P/D	P/D	P/D	P/D	P/D	P/D

polyphyletic. Passing on from the phoronids, the details of mesoderm and coelom origin in bryozoans remain uncertain, as the animals are so very small and cell movements and derivations hard to see (Zimmer 1973); this even leads to some dispute as to whether the bryozoans really have a coelom at all (see below). In the brachiopods another odd situation arises in that the inarticulate forms have a process most nearly resembling schizocoely (mesoderm proliferates from the gut wall, then later splits to form a cavity), whilst the articulates (descended from them according to fossil testimony) are clearly and classically enterocoelic. This could again (with more difficulty) be taken as evidence for polyphyly of body cavities, but it looks more like good evidence that coelom formation can be modified from one mode to another, giving rather a body-blow to the protostome/deuterostome debate.

Moving on to later stages of development, the problem of larvae and interpretation of trochophores is again raised. Many of the lophophorates have direct development, and free planktotrophic larvae are rather rare; but where they do occur, the traditional view, again reinforced by the authority of Hyman (1959), was to regard them as 'modified trochophores'. In the case of the actinotroch larva of phoronids (fig 13.3*A*) this at first seems not too unreasonable, as there is a certain resemblance. But by the more rigorous definitions of Salvini-Plawen (1973, 1980*b*; see chapter 5) the elongate actinotroch is clearly not a trochophore, having the aberrant features of a large pre-oral hood, a ring of tentacles, and a highly developed locomotory telotroch; it may even be incorrect to regard it as possessing a real prototroch (Nielsen 1977*a*). If this is to be allowed the status of a trochophore, then virtually *any* larva could be so called. However, having allowed the actinotroch to receive this designation, the other lophophorates (being clearly related) also had to be given similar status. Thus the situation arose where the cyphonautes (fig 13.3*B*), a flattened triangular larva with chitinous shell-valves, found in some gymnolaemate bryozoans, and the even more peculiar larva of brachiopods (fig 13.3*C*,*D*), were also termed 'modified trochophores', albeit with decreasing conviction. Again, it is only rather recently that the whole situation has been reviewed and put on a safer footing; these lophophorate larvae are *not* trochophores, and they really provide no evidence for affinities to either protostomes or deuterostomes. It is not particularly helpful that Emig (1982) now requires that the actinotroch must be a 'modified dipleurula'!

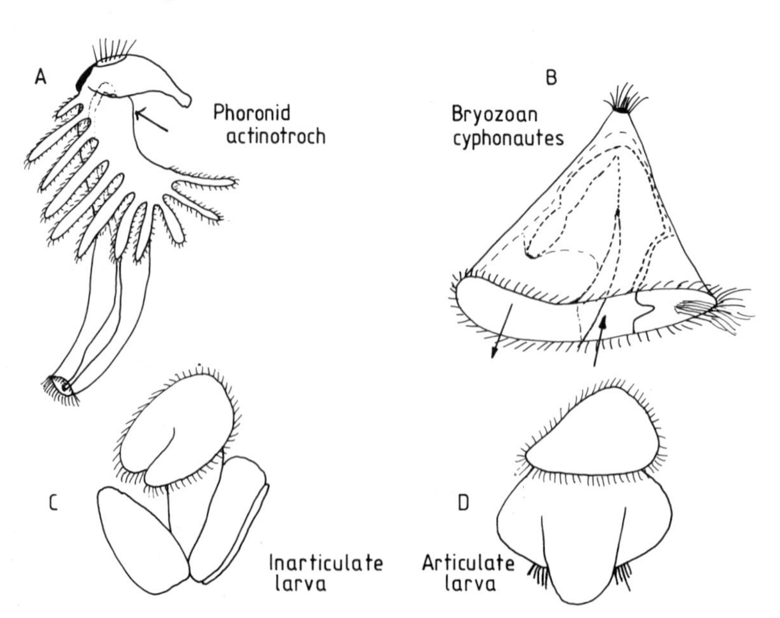

Fig 13.3. The various larvae of lophophorate groups.

On balance then the embryology of these three groups does not give a clear indication of where they belong; they have just one clear protostome feature between them (the primary mouth of phoronids), and a good number of deuterostome features, but there are also many intermediate and/or indeterminate characteristics. Considering the adult features appended to table 13.1, the deuterostome links become even clearer, with a primitively tripartite condition (the strongest deuterostome link in the view of many), a lack of cephalisation, and in two of the groups an intra-epidermal nervous system; only the presence of an external chitinous skeleton, and the direction of blood flow in phoronids, are clearly protostomian. It looks very much as if the lophophorates do belong somewhere 'between' the two main super-phyletic groups, but may be considerably closer to the deuterostomes. They have indeed changed their position rather radically over the last thirty years.

13.3 The lophophorate assemblage as a phyletic entity

Do the three phyla belong together?

The features outlined above, and in particular the detailed similarities of the lophophore itself, are now widely taken as evidence of common ancestry for these three groups of animals, even by those holding quite different views on general metazoan phylogeny (e.g. Marcus 1958; Hyman 1959; Jägersten 1972). At the same time each group is normally given phyletic status in recognition of obvious dissimilarities; though Emig (1977, 1982) now treats each as a class within the phylum Lophophorata.

It could of course be argued, rather as has been done elsewhere in this book, that the lophophore is merely a convergent similarity of no phylogenetic significance, resulting from the constraints on designing an efficient filter-feeding aparatus. However in this particular case an argument of that type looks rather thin. Firstly this feature involves several precise and potentially independent similarities in each of the groups; there is no reason why all filtering tentacles that have a coelom should also have identical ciliation patterns, for instance, or why the anus should then always lie outside the ring of tentacles. Secondly there are reasonably clearcut changes in the form of the lophophore through the three groups (horseshoe-shaped to circular), following the same evolutionary trend from phoronid to both bryozoan and brachiopod as is usually thought to be indicated by other features, yet without essential characteristics of the lophophore being changed. Thirdly, and most obviously, there *are* many other examples of differently designed tentacular feeding systems in the animal kingdom; it is in fact not a particularly tightly constrained feature, so that it may be all the more significant that in three of the groups bearing such a feeding apparatus the design is so similar. Given other similarities between these phyla, a close relationship seems very likely.

Do any other groups belong with them?

The view given in the previous section, though now very common, is not quite universal. Nielsen (1971; 1977*a*,*b*) has repeatedly argued that another group, the entoprocts, also merits consideration as part of a loose lophophorate assemblage, and links this group very closely with the bryozoans. The two types of animal certainly do look rather similar, being small, sessile and colonial, with a degree of radiality and a crown of filtering tentacles. (A word of explanation should be added here; these two groups were for a long time united just as Nielsen now requires, and were together given the name Bryozoa or Polyzoa, but when their dissimilarities were noted they were split apart into two phyla, the Endoprocta or Entoprocta having the anus within the ring of tentacles and the Ectoprocta having it outside. Subsequently the name Bryozoa has been reinstated as a term for the latter group only, and is used in that sense here since it is less easily confused with the entoprocts; but in some texts its older usage may still be met with.) Entoprocts have already been covered in chapter 9, as not the least of their differences from the bryozoans is that in so far as they have a body cavity at all (see fig 9.8) it clearly lacks a peritoneum and is of a pseudocoelomic character. Furthermore, to most authors they differ considerably in the details of the 'lophophore'. The ciliation patterns are exactly reversed, so that the water currents pass in at the sides of the tentacles and leave upwards through the middle of the ring; the tentacles themselves are solid (fig 9.8); and the whole structure is non-retractable but merely curls up like a clenched fist when not feeding.

Nevertheless, Nielsen dismisses these differences. He points out that the body cavity of an ectoproct bryozoan is hardly spacious, and barely rates the designation 'coelom' at all; in many respects these animals do resemble the entoprocts in having a rather solid body, being so small that a cavity is scarcely needed. For him, the entoprocts and bryozoans are the closely related groups, and belong somewhere near the annelids and molluscs in a spiralian assemblage; and the lophophores of the phoronids and brachiopods are to be explained as convergent, since he leaves these two groups alongside the deuterostomes. He supports this view with a proposed evolutionary sequence that derives ectoprocts directly from colonial entoprocts, citing similarities in budding processes, larval eye structure, larval form (though even from his own figures this resemblance is hard to see), and spermatozoa.

However, entoprocts are indisputably spiral in their cleavage patterns, and have 4d mesoderm (Nielsen 1977*b*), neither feature applying to bryozoans; and they have multiciliation on the tentacles, whereas bryozoan cells are nearly always monociliate. Given all the other differences in the lophophore mentioned earlier, it seems that the differences considerably outweigh the similarities Nielsen claims to have found; and a conclusion that entoprocts do not belong with the lophophorates, but should remain in their conventional place somewhere amongst the pseudocoelomates, as in chapter 9, seems fully justified.

One other group that might plausibly be regarded as part of a lophophorate assemblage is the hemichordates, particularly the pterobranchs (chapter 12); their relationship to the three main phyla under review is considered further below.

What is the evolutionary sequence between groups?

The most popular view of relations between the lophophorates is that phoronids are close to the basal stock, and gave rise to bryozoans (Hyman 1959; Clark 1964). However, one of Nielsen's major points in defence of his ectoproct/entoproct grouping is that ectoprocts cannot readily be derived from phoronids, because their life-cycles are incompatible and intermediates cannot be imagined; he is, as discussed in earlier chapters, a supporter of the Jägersten-type overview of entire life-history phylogeny, with pelago-benthic life-cycles as a central feature. However Nielsen's point is amply overcome by the scenario suggested by Farmer (1977), who derives the entire bryozoan life-cycle quite satisfactorily from that of phoronids. The cyphonautes larva is derived from the actinotroch by modifications giving it a prolonged planktonic life and better motility; this is fully consonant with the miniaturisation of the adult as it develops epifaunal and ultimately colonial habits (fig 13.4). This scheme is particularly plausible given that the first bryozoan fossils appear in the Ordovician (one of the few groups to apparently have no Cambrian history; see Larwood & Taylor 1979), at a time when stabilised marine conditions may have led to greater planktonic abundance favouring the development of a larger and much more diverse filter-feeding community.

Bryozoans, then, can be very plausibly derived from phoronids, which accords well with the various more 'primitive' features to be found in the phoronids as discussed in section 13.2. Very much the same can be said of the brachiopods, seen by most recent authors as the outcome of phoronid-like animals acquiring a pair of mineralised shell valves (Wright 1979). The only flaw in this story is the puzzling occurrence of 'primitive' uniflagellate sperm

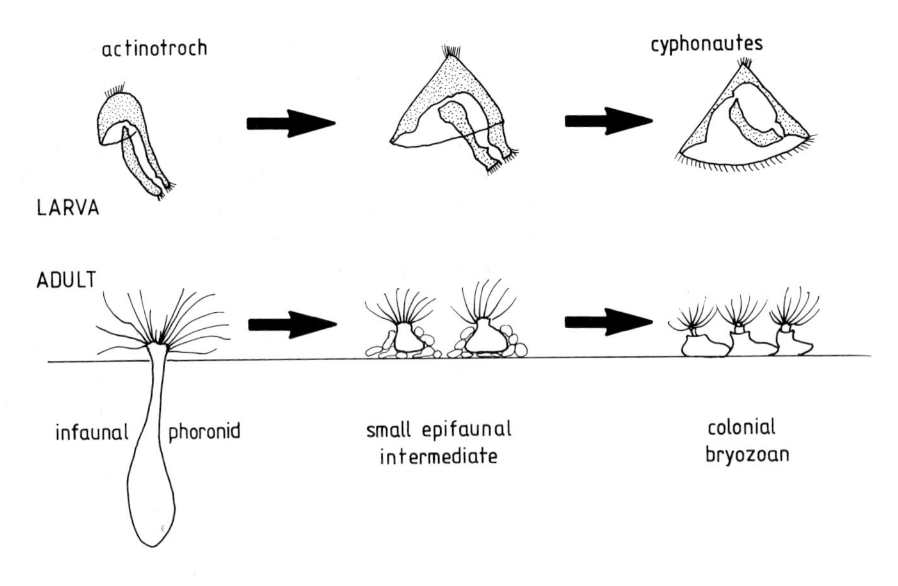

Fig 13.4. A model of the evolution of bryozoan form, and life-cycle, from those of phoronids (based on Farmer 1977).

in the brachiopods, when these are absent in the supposedly ancestral phoronids (see chapter 6).

Although some authors see the brachiopods as monophyletic (e.g. Williams & Hurst 1977), there is some evidence that this shell acquisition may have occurred many times over during the Cambrian era. Wright (1979) therefore regards the early inarticulate brachiopods (those having no real hinge between the two valves, and a muscular stalk or pedicle) as a polyphyletic assemblage (fig 13.5), representing many separate experiments in putting a shell on a phoronid-like precursor. From within this group, one type gave rise to the much more successful and probably monophyletic Articulata, where the two valves are held together by a complex tooth-and-socket hinge, the pedicle is solid, there is no anus, and the lophophore is supported by a mineralised skeleton (the brachidium). In an analysis of this monophyletic/polyphyletic problem, which is very much a re-run of the cladists versus evolutionary taxonomists debate, Forey (1982) complains (as a cladist, of course) that the polyphyletic brachiopod school are using inadequate methodology. They erect too many groups because 'acceptable ancestors have not yet been found' from which all could be derived; and they attempt to provide causal explanations for the multiplicity of convergent groups, which 'does not solve the problem'. It may not do so; but the argument is surely that acceptable ancestors could not have existed because they would be functionally (ecologically, or mechanically) non-viable, not merely that such ancestors' fossils have not been found. There is the usual impasse, met with when considering arthropod phylogeny in chapter 11; the evolutionary taxonomist can allow and attempt to explain convergent evolution, but the cladist cannot admit it and sees the associated explanations as mere speculative scenario-building.

Whether the brachiopod design was achieved once or several times, it is apparent that two of the lophophorate groups can be successfully derived from the traditionally more primitive phoronids; and the whole assemblage can be seen as a primitively tripartite stock, having certain features intermediate between protostomes and deuterostomes. Bryozoans differ largely as a consequence of their small size and colonial habit, lacking the spacious coelomic cavities and without the organised circulatory, muscular, and regulatory systems of their phoronid ancestors. Brachiopods are more highly protected by the shell valves, and can therefore develop more elaborate filtering lophophores, again with reduced circulations, guts, and body wall muscles.

There are a few dissenting voices to this view though. For example Emig (1982), a major phoronid expert, regards this group as the most advanced of the lophophorates, and runs the evolutionary sequence in the other direction. As evidence he cites the 'larger proportion of evolved features' in the

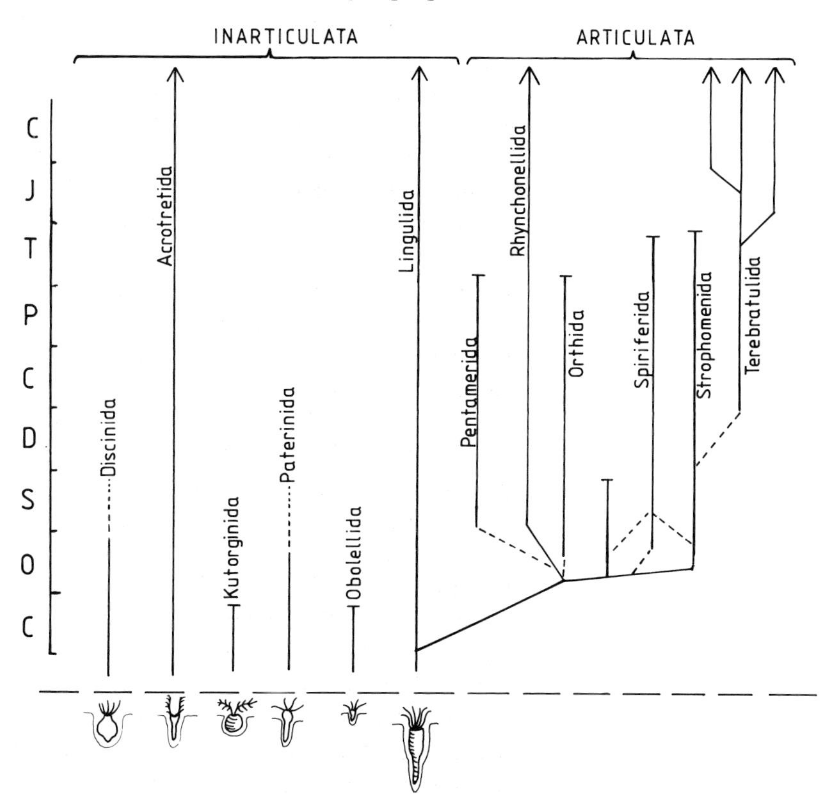

Fig 13.5. The evolution and radiation of brachiopods, showing the probable polyphyletic status of the inarticulate group (based on Wright 1979).

phoronids, and points out that the one phoronid (*Phoronis ovalis*) with an oval lophophore most like that of bryozoans is also the most primitive in many respects and is like the earliest known (Devonian) indisputably phoronid fossil. Thus he implies that the circular lophophore comes first, and believes brachiopod-bryozoan-phoronid to be the correct evolutionary sequence, the first two groups being blind side-branches of a monophyletic line of infaunal worm-like animals. This scheme does have the merit of plausibly allowing brachiopods their primitive sperm, and it could still commence with an ancestor not unlike modern phoronids in general design.

Another dissenter is Jägersten (1972), who also takes the brachiopods as the primitive lophophorates, and endows them with a cyphonautes-type larva, since he thinks this to be the least specialised for planktotrophy of all the lophophorate larvae. His view seems to be incompatible with functional considerations, and is in any case contradicted by recent demonstrations that the cyphonautes is actually highly efficient as a planktonic dispersal stage (see Farmer 1977).

Finally the views of Gutmann *et al.* (1978) should be mentioned. They are not exactly dissenting from the commonly preferred sequence within the lophophorates, again believing the phoronids to be the most primitive of their group 'Tentaculata'; but they derive them most unusually from metameric polychaete-like forms, and present an ingenious mechanical argument as to how the brachiopods could be derived from such a body plan. This view provides a neat explanation for the otherwise disconcerting presence of annelid-like setae in the brachiopods (chapter 6). It may be recalled that a similar annelidan derivation of chordates was proposed by this group of workers, and discussed in chapter 12.

13.4 The status of the lophophorates

The three core lophophorate groups seem legitimately to belong together, and to stand somewhat apart from other phyla. Most importantly, they show a very complex mixture of the set of features that are normally used to define protostomes and deuterostomes, thus giving them an apparently anomalous position somewhere in the 'middle' of the animal kingdom. There are three possible meanings of this (fig 13.6), each of which has important consequences for an interpretation of the relationship of the Metazoa as a whole. Possibly the lophophorates are a genuinely transitional group, indicative of an evolutionary sequence that could run either from protostomes to deuterostomes (fig 13.6*A*; and cf. fig 1.4), or in exactly the reverse direction (fig 13.6*B*; and cf. fig 1.3), these three phyla having not fully effected the transition. Or they may be regarded as proof that the two

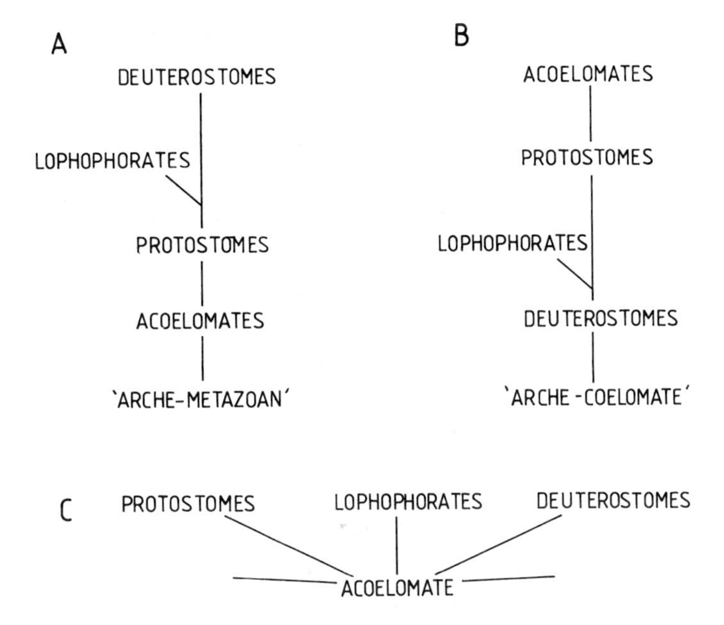

Fig 13.6. Three possible alternatives for the status of lophophorates: *A*, as an intermediate stage from protostomes to deuterostomes; *B*, as an intermediate of a transformation in the other direction; *C*, as a quite separate offshoot from acoelomate ancestors.

traditional super-phyla have little or no meaning anyway, given that each phylum of lophophorates exhibits such a mixed set of features and show the possibility of more than just two basic permutations of these key characters; they would then merit quite separate status with no significant links to other phyla (fig 13.6*C*; and cf. Clark's scheme shown in fig 1.7). Each of these hypotheses will be examined in turn.

Lophophorates as a link from protostomes to deuterostomes

If it is thought that the spiralian or protostome assemblage includes the more primitive invertebrates, then the lophophorates could be the transitional stage in the 'invention' of the deuterostomes. This is the assumption that underlies some recent views of the animal kingdom, notably the carefully worked ideas of Salvini-Plawen (1982*a*; fig 1.4*B*). A more detailed phylogeny based on his views is shown in fig 13.7. The phoronids, with their primary mouth and larval protonephridia, and an element of schizocoely, link to the protostomes in one direction, and to the ancestral deuterostomes (sessile, trimeric, and tentaculate) in the other. This scheme actually suggests a paedomorphic step from the spiralians to the early oligomeric lophophorates, the larval tentacles being retained and the larval dermal nerve net (intra-epithelial) used instead of the adult sub-epidermal orthogonal nervous system (see chapter 2).

Since it appears probable that the worm-like phoronids (or at least infaunal creatures very like a modern phoronid) are the most primitive of the lophophorate phyla, they should give the key to any such scheme. They could perhaps be derived from a worm rather like a priapulid or sipunculan, as Salvini-Plawen requires, of spiralian character and (unlike actual sipunculans) endowed with an ability to manufacture and secrete chitin. And they could link rather directly not just to the two modified lophophorate groups but also to the hemichordates at the base of the deuterostome lineage. As discussed in the last chapter (and above), the pterobranch hemichordates are possessed of a tentaculate feeding system, on the second segment of the tripartite body, and they could only with difficulty be excluded from a formal group to be termed the Lophophorata.

Given that the phoronids make this morphological link by their general design, it is satisfying that they are the only one of the three lophophorate phyla to have any clearcut protostome embryological feature, deriving their mouth from the blastopore (a possible sequence for the transition was shown in fig 5.7). However both they and the brachiopods are fully monociliated, and bryozoans are the only one of the three phyla so far known to possess both mono- and multiciliated cells (the latter in larvae), possibly indicative of a transitional state (fig 6.2 *A,B*). Similarly bryozoans are the only one of the three phyla to have the sub-epidermal (protostome) nervous system. So not all the characters fit in quite the orderly pattern needed to substantiate the evolutionary sequence that has been proposed by Salvini-Plawen. Indeed if

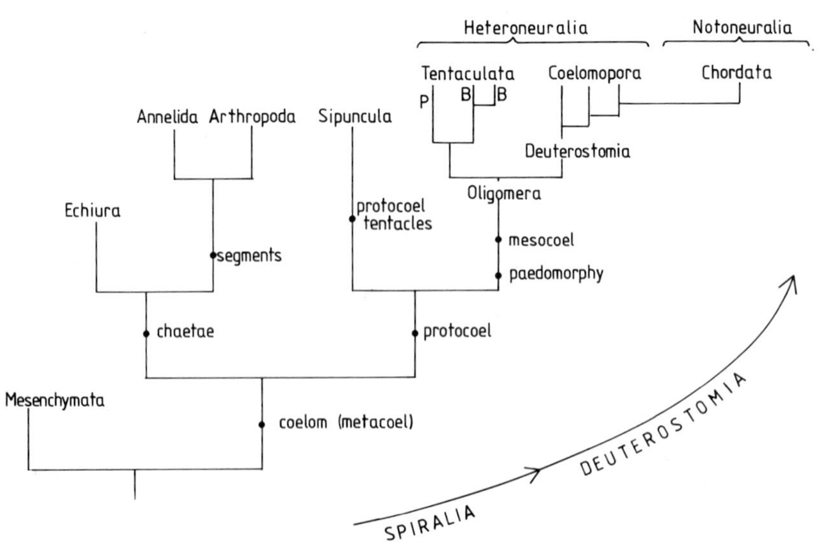

Fig 13.7. Salvini-Plawen's scheme (1982*a*) in which lophophorates bridge the gap between spiralian protostomes (where sipunculans are their nearest relatives) and deuterostomes, (equivalent to fig 13.6*A*).

Emig is right and phoronids are actually the most advanced of the lophophorates, then since they are the 'most protostome-like' of the three phyla the whole construct of the lophophorate as a link group from spiralians to deuterostomes looks a little suspect. (This was no doubt intended, since Emig in essence supports the opposing archecoelomate idea, though preferring the term Archimerata for his ancestral tripartite animals.)

Lophophorates as a link from deuterostomes to protostomes

The traditional German view of an archecoelomate ancestor for all invertebrates, met on numerous previous occasions in this book, almost inevitably contains the assumption that the tripartite coelomate ancestral form gave rise first to the similarly tripartite and enterocoelic deuterostomes, with schizocoelomate spiralians and acoelomates as secondary developments (see the works of Remane, Nielsen, Siewing, etc.). Thus the lophophorates could become a transitional group, but in the reverse direction. An example of this kind of phylogeny is shown in fig 13.8, based on the works of Siewing (1976, 1980). The archecoelomate gives rise to tripartite deuterostomes directly, whilst lophophorates gradually lose the three-part body, switch to schizocoely and show the first traces of spiral cleavage, and thus are close to the rootstock of the whole spiralian line. A number of arguments against such

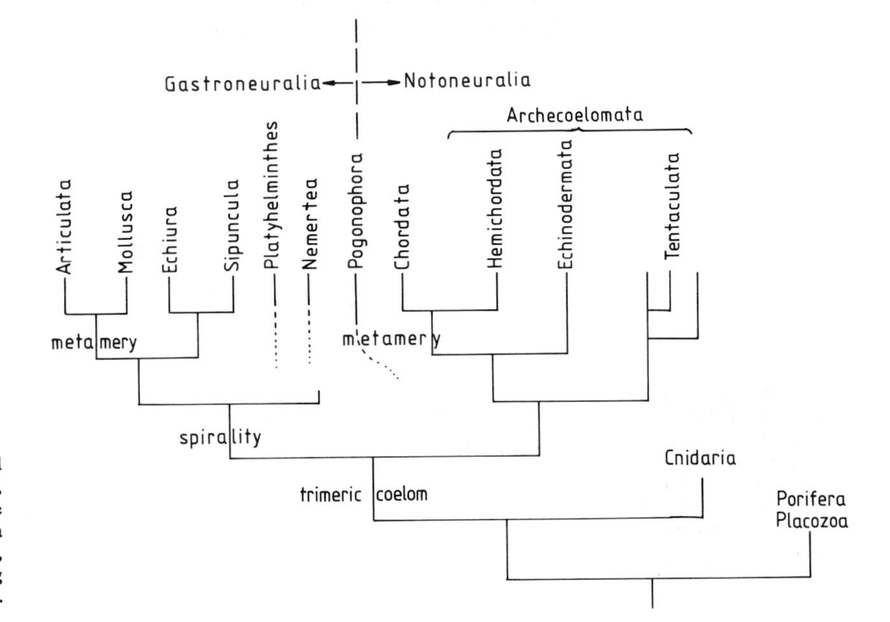

Fig 13.8. A scheme based on Siewing (1976, 1980), in which lophophorates are the nearest Archecoelomata relatives of the spiralians, (roughly equivalent to fig 13.6*B*).

views have been advanced elsewhere in this book, and will be summarised in the concluding chapter.

Separate independent status for the lophophorates

Perhaps the mixture of embryological and adult features displayed by the lophophorate groups is actually indicating something much more fundamental about relationships in the Metazoa than either of the above schemes would admit. They do show a number of 'protostome/spiralian' characters, and a slightly larger number of 'deuterostome' characters; but instead of making them a transitional assemblage, this may in fact mean that the usual distinction of these two super-phyla is unacceptable. Indeed the lophophorates can be used as a major argument against the whole concept of Protostomia and Deuterostomia.

Firstly, they very clearly indicate that the features used to define the dichotomy do not fall into just two neat sets; other combinations are possible. On the one hand this slightly undermines the critical views of Løvtrup (1977; see chapter 5) that all these embryological characters are invalid because they are non-independent and inevitably occur together, determined merely by the one underlying feature of the presence or absence of embryonic membranes. But at the same time it suggests that we should be looking, if anything, for multiple super-phyletic groupings, rather than trying to squash slightly aberrant phyla inappropriately into an artificial dichotomy.

Secondly, there are a number of details shown in table 13.1 that are instructive. The novel 'intermediate' method of making a coelom in phoronids is of interest, suggesting that transitions between the two 'normal' modes can and do occur, and/or that coeloms are polyphyletic and their mode of formation (culled from limited possible modes) potentially misleading as a phylogenetic character. These points are further supported by the brachiopods, where there is the intriguing situation that even within one traditional phylum two classes of animals can form their coeloms in almost exactly 'opposite' fashion. This must surely indicate that, unless the brachiopods really are two or more distinct and independent phyla, transitions between schizocoely and enterocoely can be fairly readily made. Such a conclusion leaves one wondering how many other times a similar switch may have occurred and remained undetected.

The history of lophophorate larvae, once all dignified as 'modified trochophores' but now seeming to have little affinity with other larval forms, may also have a lesson to teach us.

Given that the pattern of occurrence of the various features critical to the protostome/deuterostome debate within the lophophorates does not entirely accord with any evolutionary sequence giving them transitional status in either direction, it is probably safer to conclude, as did Hyman (1959) and Clark (1964, 1979) that they deserve separate status as a super-phylum in their own right, developed independently from a very early stage in the metazoan story. They could plausibly be set quite apart from both the traditional assemblages, with the earliest phoronid worms derived as one of many experiments in making secondary body cavities, starting from an initially acoelomate worm (fig 13.9). However a plausible alternative might still set them entirely apart from the spiralian assemblage, covered in chapter 8, but give them a common proto-coelomate ancestor with the deuterostomes via the pterobranchs (as in fig 13.10), with whom they share some rather specific and functionally unconstrained characters and a broad suite of similarities. The possible relationships of the fossil conodonts, discussed in chapter 3, may be of relevance here, as apatite structures reported in association with both lophophorate-like and chordate-like fossils.

13.5 Conclusions

It seems that the main conclusions to be drawn from this chapter are rather general ones about the whole patterns of the animal kingdom and the

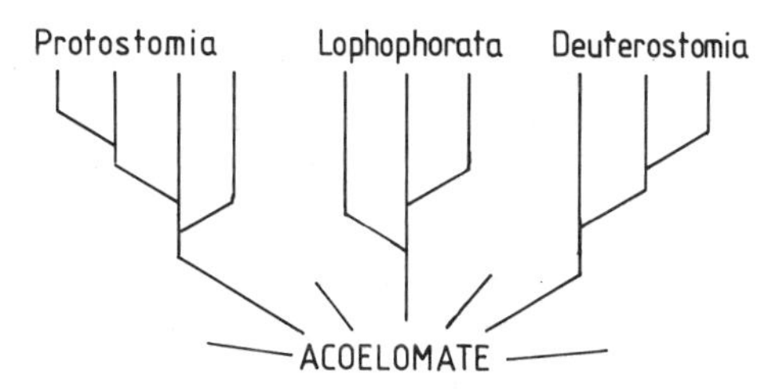

Fig 13.9. Lophophorates as quite separate independent offshoots from an acoelomate stage, coincidentally showing characters intermediate between protostomes and deuterostomes.

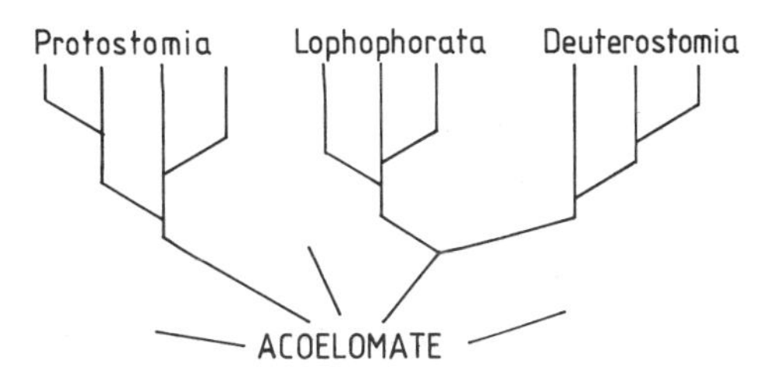

Fig 13.10. An alternative scheme recognising the affinities of lophophorates with deuterostomes above the acoelomate level, while avoiding implications that they are a genuine 'intermediate' between the other two major super-phyla.

characters used to sub-divide it. The lophophorates more than any other animals may leave us with some doubts about the conventional categories Protostomia and Deuterostomia, not least because they have so markedly, and with such apparent ease, changed their own position in relation to this dichotomy. They make it abundantly clear that some of the features used to define these super-phyla are not in fact dichotomous but may have intermediate states; that particular suites of those features are not the only possible combinations; and that transitions between one mode of doing things and another (a character and its direct opposite) are perfectly possible, within closely related phyla, and even perhaps within one phylum. Thus we may have to give the lophophorates a transitional status, believing that protostomes gave rise to deuterostomes via the lophophorates or vice versa. Alternatively we must accept that the concept of a pair of widely applicable (or indeed all-embracing) super-phyla is obsolete, there being instead several or perhaps many other super-phyletic assemblages of equal status, having no real relation to one another. This latter proposal is favoured, given that no evolutionary sequence linking protostomes and deuterostomes by way of lophophorates fits well with a majority of the trends in particular features that have been identified.

The lophophorates also have lessons for us when it comes to the issue of body cavities. They are conventionally all coelomate, though even this is in dispute for bryozoans, indicating how difficult it can be to decide in practice. Even if they are allowed coelomic status, they provide good evidence for the hazards in treating the coelom as a monophyletic phenomenon; some are classically enterocoelic, but others show varying degrees of schizocoely, with the split occurring in cells proliferated off from the gut or still attached to it. There are clearly many ways of achieving a secondary body cavity. Perhaps these methods can be seen as a continuum, as some have maintained, though in that case using enterocoely and schizocoely as good characters in defining the super-phyla debate is rendered suspect. Or they may in fact be fairly irrelevant to phylogeny, each group of animals using whatever ontogenetic method comes to hand, according to its size and relative developmental rates in different tissues, when devising a coelom.

Then there is the question of embryological and larval evidence in general, and its value in phylogenetic reconstruction. Lophophorates show such a mixture of the conventionally valuable embryological characters that they may unsettle the whole fabric of belief in such matters; and they only too clearly show the possibility of achieving any permutation of these features, whether only two possible character states or a continuum between them exist. The lophophorates certainly do not seem to give much support to the notion that embryology must be weighted more heavily than adult similarity in deciding matters of relatedness. Indeed, considering the controversies that have surrounded the status of lophophorate larvae (which - of course! - must be 'modified trochophores' or 'modified dipleurulas') one may justifiably wonder how far, and for how long, the attempts of some zoologists to squash all extant phyla into two neat super-phyletic pigeon-holes must hamper an overall appreciation of metazoan relationships.

Finally, it would be as well to recall the issue of convergence and its prevalence in invertebrates. From a study of the lophophorates, this message becomes unavoidable. Probably these three phyla belong together, and most other groups with tentaculate feeding systems are merely convergently similar; in particular the entoprocts, for many decades united with bryozoans (but now commonly allied with pseudocoelomates), are startlingly like them in general form and habits. Alternatively, the entoprocts and bryozoans do belong together as Nielsen maintains, but then the profound similarities of the lophophore in brachiopods and phoronids to that of bryozoans are indicative of convergence to an extraordinary degree. Thus convergent evolution is indisputable in this matter, one way or the other; and this factor will have to form part of any conclusion to be reached about invertebrate relationships in general.

The lophophorates, then, have some important lessons for any student of these topics. The three groups discussed here probably belong together in any phylogenetic scheme, and perhaps as a rather close-knit assemblage. They cannot reasonably be included in either of the conventional Protostome/Spiralia or Deuterostome/Archecoelomate super-phyla of animals, and give good evidence that those concepts have been over-worked in the past and in some recent invertebrate family trees. It seems unlikely that they are a transitional group between the handful of phyla that do properly belong to the 'protostomes' and the three that may constitute the 'deuterostomes'.

The lophophorates perhaps do have some link to the deuterostomes, via a common tripartite 'proto-coelomate', because it is rather difficult to exclude the pterobranch hemichordates from any formally defined lophophorate grouping. A simple trimeric tentaculate 'worm', only slightly advanced over a flatworm stage, makes a plausible common ancestor. But thereafter, and hence at a very early stage in metazoan evolution, the two assemblages diverged quite markedly; so the three phyla dealt with in this chapter should be set apart as a major super-phylum in their own right, of roughly equivalent status to any other that has been proposed.

14 A phylogenetic overview of invertebrates

14.1 Introduction

Every one of the existing phyla of invertebrates has now been considered, and its relationships to other groups analysed. Some types of animal seem to be very isolated, with little resemblance to and no convincing homologies with any other known forms. Other types of animal appear to be part of fairly clearcut super-phyletic groups, though membership of these is apparently always restricted to a rather small number of phyla. We therefore need only to review the major issues raised by the evidence presented in earlier chapters, and assess critically the existing phylogenetic schemes. It should then be possible to achieve an overview of relationships in the animal kingdom, and to present, somewhat reluctantly, an actual phylogenetic tree of the invertebrates for the sceptical reader to disagree with.

14.2 Major issues

Status of the 'lower' groups

In respect of complexity of organisation, the 'lowest' groups of the Metazoa are traditionally the sponges, cnidarians, and ctenophores, together with the placozoans and mesozoans. However, the evidence discussed in chapters 2 and 7 makes it clear that these phyla are not necessarily at all closely related to each other, and in particular that the old super-phyletic groupings of Radiata (as opposed to Bilateria) and of Diploblastica (as opposed to Triploblastica) are unhelpful, inaccurate and eminently worth discarding. Furthermore, most of these 'lower phyla' may not be in any sense close to the starting point of animal radiation; no obvious homologies unite them with the rest of the metazoans.

Sponges are certainly rather primitive in their design and are organised with limited coordination between cells. They are not as uniquely different from other Metazoa as was formerly thought, since their embryology has been reinterpreted to accord with conventional patterns and their choanocytes have been associated with the microvillar monociliate cells found in many other phyla. However, their links with all other groups are slight, and independent origin from flagellate protists (Choanoflagellata?) seems beyond doubt. In fact, if Bergquist's view that the glass sponges (Hexactinellida) represent a separate phylum is correct, then the sponge design has been achieved independently at least twice, and possibly three times if the extinct archaeocyathans are also deemed to have a roughly poriferan body plan.

Cnidarians *may* have a fossil record right back in the Pre-Cambrian that would indicate a position near the base of the metazoan kingdom; but the Ediacaran traces described as sea-pens and jellyfish are becoming increasingly dubious (chapter 3), and a very early origin for this phylum seems particularly unlikely in view of the cnidarians' largely carnivorous habit. Again a separate and slightly later ancestry seems more plausible; and schemes that derive cnidarians from a form like the present planula larva offer the best solution, whereas those that regard the design as a retrogressive development from 'triploblastic' flatworms are functionally awkward. Any scheme that requires close links with the ctenophores is most unlikely, the only shared feature being the approximately two-layered body wall and even that being very dubious (chapters 2 and 7); differences in musculature, epidermal structures, tentacle anatomy, and above all in ciliation patterns must set these two phyla well apart. Ctenophores can only be derived with any degree of conviction from a planula form; possibly entirely independently, or perhaps as an offshoot of an early, non-spiral, multiciliate proto-platyhelminth stock.

Placozoans and mesozoans probably both represent further independent attempts at a multicellular condition. Placozoans can readily be accommodated in a general polyphyletic planula theory, as a close descendant of one of several planuloid ancestors; mesozoans require a multiciliate ancestral stock, and (though they might yet prove to be 'degenerate flatworms') they could most plausibly be derived independently from a unicellular ciliate (not from modern Ciliophora, but perhaps from opalinid or other multiciliated protistan ancestors).

Every one of the 'lower' phyla is therefore best derived separately and uniquely from protozoans, via planula or gastrula stages; and few links with the acoelomates can be discerned.

Acoelomates, coelomates and archecoelomates

Consideration of the origins of Metazoa, in chapter 7, gave a strong indication that whatever else may be true, the idea promulgated in German literature of an archecoelomate as the starting point of metazoan radiation cannot be upheld. Many lines of evidence throughout this book have indicated that the resulting concept of acoelomates as a secondary and regressive development from larger ancestral coelomates is untenable. For example the patterns of evolution of the mesoderm, of the nervous system, and of the blastopore/mouth/anus relationships may all be interpreted as consistent with an acoelomate to coelomate progression, as discussed in chapter 2 and by Salvini-Plawen (1980*a*, 1982*a*). Flame cell morphology shows a similar progression (chapter 6). The embryological switch from spirality to radiality, perhaps in part due to the evolution of embryonic membranes (chapter 5), is more readily acceptable than the reverse switch, though the latter would be required by a traditional archecoelomate scheme that makes tripartite deuterostomes close to the base of the animal kingdom. Fossil evidence (chapter 3) also accords (somewhat tentatively) with an earlier origin for protostomes than for lophophorates or deuterostomes. We have also met mechanical and functional arguments, particularly those of Clark. Firstly, the sealing off of gut pouches in a cnidarian to make the coelom seems most unlikely, as the digestive surfaces would be so drastically reduced; and secondly the coelom has such profound mechanical advantages that its loss in groups such as flatworms and ctenophores, lacking even the 'excuse' of being interstitial, is inconceivable. A review of existing coelomates makes the polyphyletic origins of the coelom itself almost certain, and this is clearly not consistent with an archecoelomate view. Arguments that a large body size must be primitive to allow for pelago-benthic lifestyles (see chapter 5) lose their force unless the unproven primitive status of larvae is accepted uncritically. Relationships within the molluscs (chapter 10) suggest small acoelomate ancestors, not regression from a fully coelomate stage. Nor do relationships within the platyhelminths accord with an ancestral large coelomate condition (chapter 8). Finally there is the fact of the existence of monomeric coelomates (notably the sipunculans), despite the fact that archecoelomate theories begin with a three-segmented animal and link the coelom inextricably with segmentation.

In the face of all these kinds of evidence, and given very flimsy evidence on the reverse side of the issue (the recent structural analyses of mesoderm in flatworms, discussed in chapter 2, and of groups such as lobatocerebrids, and nemerteans, in chapter 8), archecoelomate theories are decisively rejected. The alternative view that acoelomate flatworms came first remains completely convincing.

Radiation from acoelomates

For the further evolution of animal diversity, then, all available evidence suggests acoelomate flatworms, derived from a planula, as a starting point. However the planula, a simple ovoid ciliate organism is of such simple construction that it could readily have been 'invented' several times, as discussed above and in chapter 7, so that the planula evolving into a flatworm may have had little to do with the planulas giving rise to the other lower groups as discussed above. But according to recent analyses, there are actually three quite distinct existing lineages within the phylum Platyhelminthes, and few if any links can be traced between them (Smith *et al.* 1985). In addition there are the gnathostomulid worms, also acoelomate and of very similar basic design, with the distinctive features of monociliation and chitinous teeth, but perhaps no more different from platyhelminths than the three sub-groups of platyhelminths are from each other. So possibly the flatworm assemblage is itself something of a polyphyletic hotch-potch; after all there are so few possible features to vary in a small, flat, bottle-gutted acoelomate creature, if it is to remain functional, that they all look similar and are all classified together. Where variation can occur - in development, or in sperm type, for example - it does. Thus we may merely be looking at a grade of organisation, arrived at not once but repeatedly from the planula form of early metazoan; some being multiciliate, others monociliate.

From various points within this rather muddled assemblage of acoelomate worms, and probably mainly from those few groups possessing 'normal' monoflagellate spermatozoa, the remaining metazoan phyla arose in an explosive radiation of larger animals. These were possessed of a variety of body cavities, with a continuum from acoelomate through pseudocoels and haemocoels, and with varying acquisition of secondary intra-mesodermal coeloms (chapter 2). Many of them also acquired hardened mineralised jaws or protective covers, which further contributed to the great increase in diversity.

Super-phyla and dichotomies

We have seen that larger animals, particularly coelomates, are a secondary development, of later origin than the acoelomate design, and derived from it. What can be said of the various attempts to classify the 'higher' phyla into larger super-phyletic groupings; and indeed in some cases to extend the super-phyla to include the acoelomate and pseudocoelomate groups? Throughout the earlier chapters of this book the traditional super-phyla Protostomia (Spiralia) and Deuterostomia have been critically discussed (originally summarised in table 5.1); here only a brief revision is needed.

Protostomes/spiralians; and pseudocoelomates. Within the acoelomate flatworms and early spiralians, varying embryological patterns are met. These could be readily envisaged as a hierarchy of increasingly organised development (Anderson 1982; Inglis 1985), and may thus give clues to the status and evolution of the spiralian assemblage as a whole.

Firstly, one set of phyla show only limited indications of spiral cleavage (sometimes the pattern is better described as duet cleavage); the 4d mesoderm character is not expressed, and no trochophore larva occurs. These include some platyhelminths, plus the crustaceans and chelicerates discussed in chapter 11. From chapter 9 several of the pseudocoelomate phyla also belong here, notably the related trio of gastrotrichs, nematodes and nematomorphs, but also perhaps rotifers and the acanthocephalans. The stem group for all of these was presumably made up of multiciliate proto-platyhelminths lacking tightly organised spiral cleavage and with mesodermal tissues derived from several sources (discussed in chapter 2). In some offshoots from this stem, development of cuticles occurred; where these were chemically cross-linked and hardened, segmented crustaceans and chelicerates arose, whereas less stiff cuticles characterised the gastrotrich/nematode/nematomorph line and the rotifers. These lines are all, almost certainly, quite independent offshoots from the flatworms, sharing no synapomorphies not also present in the acoelomate ancestors.

Secondly, certain other groups show full spiral cleavage, though not always of a precisely patterned type, and have 4d-derived mesoderm. Notable here are some platyhelminths, all entoprocts, nemerteans, and sipunculans, the molluscs (chapter 10), and perhaps also the tardigrades (chapter 11). Where a secondary body cavity occurs in such animals, it is generally schizocoelic. But these groups, according to Salvini-Plawen (1980*b*), lack the true trochophore larva. A sub-group of the platyhelminths with full spirality and specific 4d mesoderm probably gave rise to these groups, again perhaps as quite separate derivatives; though sipunculans and molluscs may be loosely allied by the presence of a 'molluscan cross' in the embryo.

Finally, a few phyla probably share a more definite common ancestry - the annelids, echiurans, and pogonophorans, plus the uniramians discussed in chapter 11, and possibly the onychophorans. These groups show (or their embryology can be derived from) a common pattern of spiral cleavage and of precise cell determination, particularly the derivation of mesodermal tissues from the 4d cell; and where the embryo is not too yolky a trochophore larva occurs. All possess a schizocoel, and all may show elements of segmentation (though this is less than clear in echiurans).

Only these final groups, that are clearly coelomate, could reasonably be designated as Protostomia in the strict sense. The non-coelomate groups might be included in a looser assemblage for which the term Spiralia is more appropriate. But the fact that modern platyhelminths have members at at least two of the levels of embryological sophistication described here makes rigorous application of any terminology rather problematical, and again points to the possibility that the platyhelminths themselves might be a polyphyletic assemblage.

With respect to pseudocoelomates, there can be no doubt that multiple origins from amongst the flatworm assemblage are indicated, as in fig 9.10; some of the phyla probably do have links to the spiralians as above, though Nielsen (1985) refutes most of the reports of 'spiral' cleavage in these animals. In any case other pseudocoelomates seem quite unrelated to normal spiralians, and at least one of the phyla (the priapulids) is apparently closer to the deuterostomes at least in the possession of standard radial cleavage.

Deuterostomes; and lophophorates. The deuterostome assemblage could be said to be united by the reverse of all the features considered above for spiralians. In many ways they do differ from the protostome groups, and form a rather tightly knit cohort of phyla; their embryology is generally distinct, and the three phyla share many similar chemical features, ultrastructural details, and adult morphological characters.

Within the deuterostomes, the most obvious discontinuity seems to lie between vertebrates and all the other groups, in respect of a number of features outlined in chapter 12. This is disconcerting in terms of defining 'chordates' as a phyletic unit, and makes theories of vertebrate origins particularly controversial, but links between vertebrates and the hemichordates and echinoderms are nevertheless fairly well established.

Whether the three core deuterostome phyla are linked to another major super-phylum, the Lophophorata, remains uncertain. Separate origins for the three tentaculate groups, via a phoronid-like ancestor derived from flatworms, is an attractive hypothesis, but similarities between the phoronid and the hemichordate cannot be denied, and it is quite possible that a tripartite coelomate worm lies at the root of both these major super-phyletic groupings. Similarities between lophophorates and deuterostomes include not only major design features such as the trimeric condition and the ciliated tentacle array on the second segment, but also more precise features such as DNA properties, eye structures, intercellular junctions, and predominantly monociliated cells. However, nearly as many characters link the groups to spiralians; for example the primary mouth of phoronids (chapter 13), the presence of chitin and absence of sialic acids, and the structures of RNA and of actins (chapter 4). On balance it might therefore be sensible to keep the lophophorates as a quite separate super-phylum of tripartite coelomates, linked to deuterostomes just above the acoelomate flatworm (this time essentially monociliate) from which they can most readily be derived.

Convergence and polyphyly

Most of the analyses of relationships given briefly above, and in more detail in earlier chapters, have tended to favour keeping groups apart rather than joining them in neat linear sequences. This is not a random favouritism of 'splitting' rather than 'lumping', but a conscious recognition of one overriding message to be learnt from invertebrate phylogeny - the fact that convergent origins of characters are incredibly prevalent. Detailed structures, and gross

Table 14.1. *Convergent features of invertebrates - a review*

Biochemical features		
	Respiratory pigments	Tubulins??
	Collagens??	
Cell features and cell types		
	Collar cells?	Setae
	Photoreceptors	Cardiac muscle
	Multiciliate cells	Striated muscle
	Adhesive gland cells	Cuticles
Organ systems		
	Blood vascular systems	Calcitic shells
	Tracheae	Gills
	Malpighian tubules	Digestive glands
Major design features and mechanisms		
	Segmentation	Medusae
	Coeloms	Sex
	Pseudocoeloms	Multicellularity
Body plans		
	Arthropods	Brachiopods ?
	Bryozoans/entoprocts	Platyhelminths?
	Sponges ?	

morphologies, occur repeatedly and in many cases are of incontrovertibly convergent nature; therefore, a polyphyletic view of the invertebrate world is quite logically to be favoured.

In case the predominance of convergence in invertebrate design still needs further underlining, a quick resumé of examples from earlier chapters, generally drawn directly from the opinions of the particular expert authors who have discussed their favourite features, is given in table 14.1. Convergent features range from cellular structures, to whole cell types, to multicellular organs and features of body plans. Even the whole animal 'Bauplan' can be convergent; perhaps this applies to sponges, bryozoans (in the broader sense), brachiopods and platyhelminths, but most strikingly of all it seems to apply to arthropods.

There can be no reasonable doubt; convergence is rife, and polyphyletic origins of major and minor features must surely reflect the polyphyletic origins of whole groups of invertebrates. With this in mind, the main schemes of invertebrate relationships first outlined in chapter 1 can now be critically reviewed.

14.3 Major schemes

Traditional dichotomous trees

There *is* a distinction between the two traditional halves of the animal kingdom; too many features, from biochemistry to cellular structures and embryology, accord with this split for its existence to be doubted, as the brief analysis above confirms. However, there is also no doubt that the dichotomy has been taken much too far in the traditional schemes displayed in most text-books (fig 1.2), giving the impression that there are effectively only two primary branches to the invertebrate tree. In fact, the evidence reviewed here suggests that only four phyla are strictly protostomian. Rather more belong loosely to a spiralian assemblage, but these are almost certainly separately descended from parts of an ancestral flatworm stock. Similarly there are only three real deuterostome phyla, probably also descended from the flatworms, though the lophophorate groups *may* be close to this deuterostome alliance, linked at an early coelomate stage.

The remaining phyla (twenty five to thirty of them) have no real place in a simple dichotomous scheme, and such schemes are therefore unhelpfully oversimplified. They conceal trends rather than revealing them.

Deuterostome and archecoelomate ancestry

For all the reasons given in section 14.2, archecoelomate schemes are implausible, if not impossible. Cnidarians are not part of the ancestry of most invertebrate groups. The archecoelomate is a myth, the coelom is not monophyletic, and tripartite coelomates did not precede all other invertebrates. Deuterostomes are not the forerunners of protostomes, and acoelomates are not a secondary regressive development. Schemes such as those of Remane, Siewing (fig 1.3) and other German authors have little to offer.

Spiralian ancestry

The idea that spiralians are close to the rootstock of the invertebrates is not unreasonable, given the occurrence of spiral cleavage in many modern platyhelminths, and the scheme devised recently by Salvini-Plawen (figs 1.4*B* and 13.7) is in many respects a very sensible one, giving good functional analyses of possible sequences of change in morphology and embryology. It neatly accounts for the 'mixed' features found in lophophorate groups, by invoking a paedomorphic descent of lophophorates and deuterostomes from spiralians, with sipunculans as an intermediate. But the pronounced linearity of the scheme takes no account of the evidence of convergence in invertebrates, and therefore requires an enormous number of switches and reversals of condition for particular characters, notably such things as ciliation. If a monophyletic scheme for the invertebrates is to be insisted upon, this one makes more sense than most, but perhaps a polyphyletic scheme will better explain the observations amassed in earlier chapters.

Polyphyletic schemes

A number of polyphyletic schemes were considered in chapter 1, varying from the 'polyphyly of coelomates' views of Clark and Valentine to the 'totally polyphyletic from protistan' views of Nursall, with Inglis' and Anderson's recent schemes somewhere between these positions. The first point that can be made is an important issue raised in chapter 4, namely that because many of the higher phyla (including all the major coelomates) share one clear genetic feature in the presence of homeobox sequences, they cannot reasonably be independently derived from protistans. The homeobox sequence does not occur in protozoans, or in 'lower' metazoans, or in many platyhelminths (and it is most unlikely that *all* these 'primitive' groups lost it secondarily). It may therefore have arisen within part of the 'phylum' Platyhelminthes, and been inherited by most of the descendant groups. Polyphyly of coelomates remains perfectly plausible, indeed likely, but this must be from an acoelomate stage and not right back at a protozoan stage. Nursall's (1962) scheme, and also elements of the more recent polyphyletic schemes devised by Anderson (1982) and more specifically by Inglis (1985), are therefore implausible.

Schemes that take coelomates to be polyphyletic from an acoelomate stage are however very much supported by all the evidence accumulated here. Clark (fig 1.7) has four main coelomate super-phyla, (metameric, unsegmented, and two derived from a common trimeric stock). Valentine (fig 1.8) has five, keeping the Sipuncula and the Mollusca separate; the designation of sipunculans as a distinct unit is clearly a good idea, as they are monomeric and show no very clear links to the protostomes. However as seen in chapter 10 the molluscs are better not regarded as coelomate at all, but rather as a separate lineage from the flatworm stage with no other close relatives. A scheme having something in common with these versions of polyphyly looks eminently plausible and attractive.

14.4 A final view

Many kinds of invertebrate do appear to have been 'invented' several times over, with particular designs reappearing repeatedly. Given the lack of any convincing homologies amongst lower groups, and the obvious advantages of a larger size concomitant on multicellularity, one cannnot avoid the inference that this also applies to the most fundamental design feature of all: metazoan status itself was achieved more than once, so that 'animals' as a whole are polyphyletic.

Perhaps some of the early forms were of a gastraea type of organisation, but for many of the simpler metazoans derivation from a planula form presents fewer functional problems. Thus placozoans, cnidarians and sponges probably arose separately, each from monociliate ancestors like modern flagellate protozoans; the flatworm design may have been reached more than once, via both monociliate and multiciliate planulas derived from flagellate or ciliate protists; and the ctenophores may have been an offshoot of a flatworm or have had their own separate ciliate origins.

From the acoelomate grade, most other metazoan phyla can readily be derived, by a whole series of separate acquisitions of many different forms of body cavity. The larger animals arose as part of an explosive radiation; an 'evolutionary wave' in Inglis' terminology (1985). The modern acoelomates show a series of grades of complexity, particularly relating to embryology, and to acquisition of the developmentally important homeobox genetic sequence (to which other genetic homologies will probably be added in the near future). All the 'higher' groups can be derived from different points within this apparent hierarchy, some expanding the remnants of their blastocoel to varying extents, others acquiring a secondary cavity by a variety of means. However, any attempt to produce a real hierarchy within these groups, with nested sets of features, is fraught with difficulties. In fig 14.1 the attempt is made using the embryological features outlined earlier in this chapter, and at first sight this seems satisfactory, so that one could suppose nemerteans were a later and more 'advanced' development than gastrotrichs, molluscs were more 'advanced' than nemerteans, and so on. But when the one extra feature of homeobox sequences is added in to the scheme (the dotted line), major problems occur. The homeobox itself (a common genetic sequence of about 180 base pairs) surely cannot be a convergent phenomenon; yet 4d mesoderm occurs in groups with and without the homeobox, and so too does radial cleavage. Thus several of the key embryological traits are revealed as necessarily polyphyletic, and the hierarchical concept begins to look shaky. Clearly it would only get worse as more characters were brought into consideration.

The only solution must be to avoid the temptation to draw hierarchies above the acoelomate level, and accept that so many characters were occurring repeatedly that the evolving 'higher' phyla received very mixed permutations of characters from their already rather polyphyletic acoelomate ancestors. With convergence so very prevalent, any attempt to impose a hierarchy of such features and achieve a higher level cladistic type of classification for all these

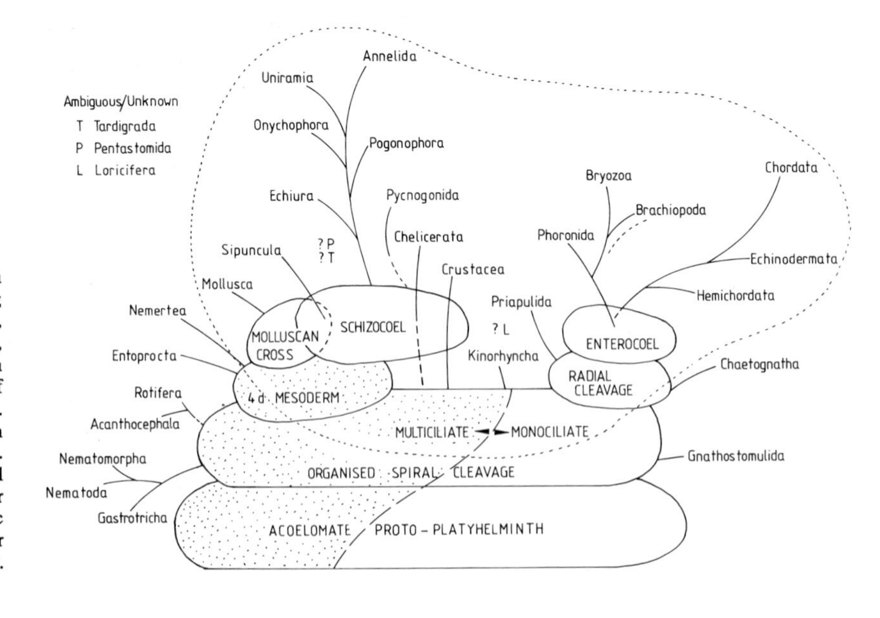

Fig 14.1. An amalgamation of earlier summarising diagrams (see figs 8.10, 9.10, 10.9, 11.9, 12.13*B*,*C*, and 13.10), with a possible hierarchy of embryological features. Dotted area: the modern 'phylum' Platyhelminthes. Dotted line: encloses all groups with similar homeobox genetic sequences. Further discussion in the text.

groups would be utterly artificial, if not impossible. Schemes that do this (e.g. Salvini-Plawen 1982*a*, 1985; Nielsen 1985; Jefferies 1986) involve considerable selectivity with the evidence, and in Nielsen's case some rather improbable hypothetical ancestors and transitions. It is very hard to believe that animal evolution really did proceed in such orderly patterned fashions as these schemes require.

Only small sub-sets of the invertebrate phyla show real affinities beyond an acoelomate stage, then, and these can sensibly be given super-phyletic status: the four phyla of protostomes, three of deuterostomes and three of lophophorates, plus perhaps one group of three, and another pair, of the pseudocoelomate phyla. Virtually all the other phyla are best seen as separate and independent offshoots from the acoelomate grade.

A phylogenetic scheme for the invertebrates, based on all the evidence amassed here, can therefore be tentatively drawn as in fig 14.2. Quite clearly it cannot be a vine, or a neatly dichotomous tree (cf. fig 1.10). It must branch like a field of grass low down (and presumably many other attempts at multicellularity may have existed and gone extinct - perhaps the Ediacaran faunas are evidence of this). But it must also have a degree of coherence further up, with some common branchings where homologies are clearcut. The overall effect is therefore not that of a neat lawn of grass, but rather of an old-fashioned meadow, where a few hardy perennial designs flourish and branch amongst the grasses. These common perennial 'branches' do not necessarily unite groups that 'look similar', of course; the 'arthropods' are well separated, in recognition of their polyphyletic origins and constrained

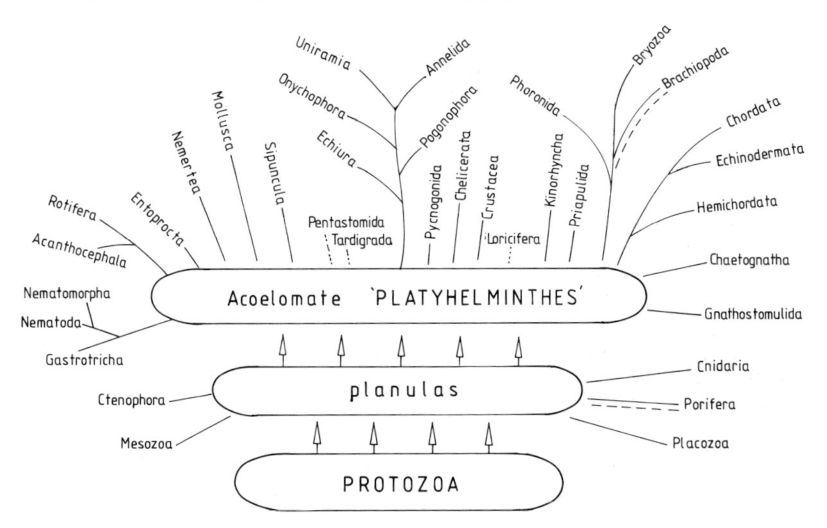

Fig 14.2. A final approximation to an invertebrate phylogeny based on the evidence discussed in this book. A few recognised 'phyla' (dashed lines) may themselves be polyphyletic. As discussed in the text, the phylogeny resembles a field of grass in which just a few designs particularly thrive and branch. (Especially enigmatic groups are shown with dotted lines.)

convergence. And some other traditional groupings are completely split up, notably the pseudocoelomates.

I doubt if this represents the 'right answer' to the question of invertebrate phylogeny, mentioned at the start of this book. In fact, this scheme is undoubtedly wrong, and new evidence from molecular geneticists will hopefully start to show up its errors fairly soon. Such schemes only exist to be attacked, after all. It is to be hoped, though, that as a phylogeny it is less inaccurate, and less purely speculative, than some of the other schemes, and that it firmly establishes the prevalence of convergent evolution in the minds of some interested evolutionary biologists. And of course, above all, I hope that it will inject a dose of healthy scepticism into students of invertebrates who have been brought up on more traditional schemes (in which either superficial similarities of design, or simplified interpretations of embryological and larval stages, have been unduly stressed), thus stimulating them to start seeking new invertebrate relationships for themselves.

References

Adams E. (1978). Invertebrate collagens. **Science 202** 591-8

Adouette A., Claissé M., Maunoury R. & Beisson J. (1985). Tubulin evolution: ciliate-specific epitopes are conserved in the ciliary tubulin of Metazoa. **J. Mol. Evol. 22** 220-9

Afzelius B. A. (1969). Ultrastructure of cilia and flagella. In: **Handbook of Molecular Cytology,** (ed. A. Lima de Faria), pp. 1220-42. North Holland, Amsterdam.

Aldridge R. J., Briggs D. E. G., Clarkson E. N. K. & Smith M. P. (1986). The affinities of conodonts - new evidence from the Carboniferous of Edinburgh, Scotland. **Lethaia 19** 279-91

Altner H. & Prillinger L. (1980). Ultrastructure of invertebrate chemo-, thermo- and hygroreceptors, and its functional significance. **Int. Rev. Cytol. 67** 69-139

Anderson D. T. (1973). **Embryology and Phylogeny in Annelids and Arthropods.** Pergamon Press, Oxford.

Anderson D. T. (1979). Embryos, fate maps, and the phylogeny of arthropods. In: **Arthropod Phylogeny,** (ed. A. P. Gupta), pp. 59-105. Van Nostrand Reinhold, New York.

Anderson D. T. (1982). Origins and relationships among the animal phyla. **Proc. Linn. Soc. NSW. 106** 151-66

Ax P. (1963). Relationships and phylogeny of the Turbellaria. In: **The Lower Metazoa,** (ed. E. C. Dougherty), pp. 191-224. University of California Press, Berkeley.

Ax P. (1985). The position of the Gnathostomulida and Platyhelminthes in the phylogenetic system of the Bilateria. In: **The Origins and Relationships of Lower Invertebrates,** (eds. S. Conway Morris, J. D. George, R. Gibson & H. M. Platt), pp. 168-80. Clarendon Press, Oxford.

Baccetti B. (1970) (ed.). **Comparative Spermatology.** Academic Press, New York.

Baccetti B. (1979). Ultrastructure of sperm and its bearing on arthropod phylogeny. In: **Arthropod Phylogeny,** (ed. A. P. Gupta), pp. 609-44. Van Nostrand Reinhold, New York.

Baccetti B. & Afzelius B. A. (1976). The biology of the sperm cell. **Monogr. Dev. Biol. 10** 1-254

Bakke T. (1980). Embryonic and post-embryonic development in the Pogonophora. **Zool. Jb. Anat. 103** 276-84

Bambach R. K. (1975). What is the pattern of change in species diversity with time? **Geol. Soc. Am. Abstr. 7** 987-8

Bambach R. K. (1977). Species richness in marine benthic habitats through the Phanerozoic. **Palaeobiology 3** 152-67

Barber V. C. (1974). Cilia in sense organs. In: **Cilia and Flagella,** (ed. M. A. Sleigh), pp. 403-33. Academic Press, London.

Bardele C. F. (1983). Comparative freeze-fracture study of the ciliary membrane of protists and invertebrates in relation to phylogeny. **J. Submicrosc. Cytol. 15** 263-7

Barnes R. D. (1987). **Invertebrate Zoology.** (5th ed.) Saunders College, Philadelphia.

Barnstaple C. (1983). How molecular is neurobiology? **Nature 306** 14-6

Barrington E. J. W. (1965). **The Biology of Hemichordata and Protochordata.** Oliver & Boyd, Edinburgh & London.

Barrington E. J. W. (1979). **Invertebrate Structure and Function.** (2nd. ed.). Nelson.

Barrington E. J. W. & Jefferies R. P. S. (1975) (eds.). Protochordates. **Symp. Zool. Soc. Lond. 36** Academic Press, London.

Bartolomaeus T. (1985). Ultrastructure and development of the protonephridia of *Lineus viridis* (Nemertini). **Microfauna Marina 2** 61-83

Batson B.S. (1979). Body wall of juvenile and adult *Gastrodermis boophthorae* (Nematoda, Mermithidae): ultrastructure and nutritional role. **Int. J. Parasitol. 9** 495-503

Beklemishev V. N. (1963). On the relationship of the Turbellaria to other groups of the animal kingdom. In: **The Lower Metazoa**, (ed. E.C. Dougherty), pp. 234-44. University of California Press, Berkeley.

Beklemishev V. N. (1969 trans.). **Principles of Comparative Anatomy of Invertebrates,** (2 volumes). University of Chicago Press, Illinois.

Benton M. J. (1987). Conodonts classified at last. **Nature 325** 482-83

Bereiter-Hahn J., Matoltsy A. G. & Richards K. S. (1984) (eds.). **Biology of the Integument, Vol 1, Invertebrates.** Springer-Verlag, Berlin.

Berg G. (1985). *Annulonemertes* gen. nov., a new segmented hoplonemertean. In: **The Origins and Relationships of Lower Invertebrates**, (eds. S. Conway Morris, J. D. George, R. Gibson & H. M. Platt), pp. 200-9. Clarendon Press, Oxford.

Bergquist P. R. (1985). Poriferan relationships. In: **The Origins and Relationships of Lower Invertebrates,** (eds. S. Conway Morris, J. D. George, R. Gibson & H. M. Platt), pp. 14-27. Clarendon Press, Oxford.

Bergström J. (1975). Functional morphology and evolution of xiphosurids. **Fossils Strata 4** 291-305

Bergström J. (1979). Morphology of fossil arthropods as a guide to phylogenetic relationships. In: **Arthropod Phylogeny**, (ed. A. P. Gupta), pp. 33-56. Van Nostrand Reinhold, New York.

Berrill N. J. (1955). **The Origin of Vertebrates.** Clarendon Press, Oxford.

Boaden P. J. S. (1975). Anaerobiosis, meiofauna and early metazoan evolution. **Zool. Scripta 4** 21-4

Boaden P. J. S. (1977). Thiobiotic facts and fancies; aspects of the distribution and evolution of anaerobic meiofauna. In: **The Meiofauna Species in Time and Space,** (eds. W. Sterrer & P. Ax), pp. 45-63. Mikrofauna Meeresboden.

Boaden P. J. S. (1985). Why is a gastrotrich? In: **The Origins and Relationships of Lower Invertebrates,** (eds. S. Conway Morris, J. D. George, R. Gibson & H. M. Platt), pp. 248-60. Clarendon Press, Oxford.

Bock W. J. (1981). Functional adaptive analysis in evolutionary classification. **Am. Zool. 21** 5-20

Bone Q. (1960). The origin of chordates. **J. Linn. Soc. Lond. Zool. 44** 252-69

Bone Q. (1979). **The Origin of Chordates.** (2nd. ed.). Carolina Biological Supply Co., Burlington, N.Cal.

Bone Q. (1981). The neotenic origin of chordates. **Atti. Conv. Lincei 49** 465-86

Bonik K., Grasshoff M. & Gutmann W. F. (1976). Die Evolution der Tierkonstruktionen. **Natur. Mus. 106** 132-43, 178-88

Bonner J. T. (1982) (ed.). **Evolution and Development.** Springer-Verlag, Berlin.

Boudreaux H. B. (1979). **Arthropod Phylogeny with Special Reference to Insects**. John Wiley & Sons, New York.

Bouligand Y. (1967). Les soies et les cellules associées chez deux annelides polychets. **Z. Zellforsch. Mikrosk. Anat. 79** 332-63

Bouligand Y. (1972). Twisted fibrous arrangements in biological materials and cholesteric mesophases. **Tissue & Cell 4** 189-217

Brasier M. D. (1979). The Cambrian radiation event. In: **The Origin of Major Invertebrate Groups**, (ed. M. R. House), pp. 103-59. Academic Press, London & New York.

Brasier M. D. (1986). Why do lower plants and animals biomineralize? **Palaeobiology 12** 241-50

Briggs D. E. G., Clarkson E. N. K. & Aldridge R. J. (1983). The Conodont animal. **Lethaia 16** 1-4

Brown W. M. (1983). Evolution of animal mitochondrial DNA. In: **Evolution of Genes and Proteins,** (eds. M. Nei & R. K. Koehn), pp. 62-88. Sinauer, Sunderland, Mass.

Bullivant J. S. (1968). The method of feeding of lophophorates (Bryozoa, Phoronida, Brachiopoda). **N. Z. Jl. Mar. Freshwater Res. 2** 135-46

Burr A. H. (1984). Evolution of eyes and photoreceptor organelles in the lower phyla. In: **Photoreception and Vision in Invertebrates,** (ed. M. A. Ali), pp. 131-78. Plenum Press, New York.

Cain A. J. (1982). On homology and convergence. In: **Problems of Phylogenetic Reconstruction**, (eds. K. A. Joysey & A. E. Friday), pp. 1-19. Academic Press, London & New York.

Candia Carnevali M. D. & Ferraguti M. (1979). Structure and ultrastructure of muscles in the priapulid *Halicryptus spinulosus* : functional and phylogenetic remarks. **J. Mar. Biol. Ass. UK 59** 737-44

Carter G. S. (1957). Chordate phylogeny. **Syst. Zool. 6** 187-92

Charig A. J. (1982). Systematics in biology: a fundamental comparison of some major schools of thought. In: **Problems of Phylogenetic Reconstruction**, (eds. K. A. Joysey & A. E. Friday), pp. 363-440. Academic Press, London & New York.

Cisne J. L. (1974). Trilobites and the origin of arthropods. **Science 186** 13-8

Cisne J. L. (1982). Origin of the Crustacea. In: **The Biology of Crustacea, Vol.1** (ed. D. E. Bliss), pp. 65-92. Academic Press, London.

Clark R. B. (1964). **Dynamics in Metazoan Evolution; the Origin of the Coelom and Segments.** Clarendon Press, Oxford.

Clark R. B. (1969). Systematics and phylogeny; Annelida, Echiura, Sipuncula. In: **Chemical Zoology, Vol. IV,** (eds. M. Florkin & B. T. Scheer), pp. 1-68. Academic Press, London & New York.

Clark R. B. (1979). Radiation of the Metazoa. In: **The Origin of Major Invertebrate Groups,** (ed. M. R. House), pp. 55-102. Academic Press, London & New York.

Clark R. B. (1981). Locomotion and the phylogeny of the Metazoa. **Boll. Zool. 48** 11-28

Clarke K. U. (1979). Visceral anatomy and arthropod phylogeny. In: **Arthropod Phylogeny,** (ed. A. P. Gupta), pp. 467-549. Van Nostrand Reinhold, New York.

Clarkson E. N. K. (1979). **Invertebrate Palaeontology and Evolution.** George Allen & Unwin, London.

Clément P. (1985). The relationships of rotifers. In: **The Origins and Relationships of Lower Invertebrates,** (eds. S. Conway Morris, J. D. George, R. Gibson & H. M. Platt), pp. 224-47. Clarendon Press, Oxford.

Clemmey H. (1976). World's oldest animal traces. **Nature 261** 576-8

Cloud P. (1968). Pre-Metazoan evolution and the origins of the Metazoa. In: **Evolution and Environment,** (ed. T. Drake), pp. 1-72. Yale University Press, New Haven.

Cloud P. (1976*a*). Beginnings of biospheric evolution and their biogeochemical consequences. **Palaeobiology 2** 351-87

Cloud P. (1976*b*). Major features of crustal evolution. **Trans. Geol. Soc. S. Africa. 79 A** 1-33

Cloud P. (1978). **Cosmos, Earth and Man. A Short History of the Universe.** Yale University Press, New Haven.

Cloud P. & Glaessner M. F. (1982). The Ediacarian Period and System: Metazoa inherit the earth. **Science 217** 783-92

Cobb J. L. S. & Laverack M. S. (1967). Neuromuscular systems in echinoderms. **Symp. Zool. Soc. Lond. 20** 25-51

Conway Morris S. (1979). The Burgess Shale (middle Cambrian) fauna. **Ann. Rev. Ecol. Syst. 10** 327-49

Conway Morris S. (1985). Non-skeletalized lower invertebrate fossils : a review. In: **The Origins and Relationships of Lower Invertebrates,** (eds. S. Conway Morris, J. D. George, R. Gibson & H. M. Platt), pp. 343-59. Clarendon Press, Oxford.

Conway Morris S. & Crompton D. W. T. (1982). The origins and evolution of the Acanthocephala. **Biol. Rev. 57** 85-115

Costello D. P. & Henley C. (1976). Spiralian development; a perspective. **Am. Zool. 16** 277-92

Cowan N. J. & Dudley L. (1983). Tubulin isotypes and the multigenic tubulin families. **Int. Rev. Cytol. 85** 147-73

Crisp D. J. (1974). The role of the pelagic larva. In: **Perspectives in Experimental Zoology,** (ed. P. S. Davies), pp. 145-55. Pergamon Press, New York.

Crowe J. H., Crowe L. M. & Chapman D. (1984). Preservation of membranes in anhydrobiotic organisms; the role of trehalose. **Science 223** 701-3

Cutler E. B. (1975). Pogonophora and Protostomia - a procrustean bed? **Z. Zool. Syst. Evol.** Sonderheft **1** 112-22

Davidson E. H., Thomas T. L., Scheller R. H. & Britten R. J. (1982). The sea urchin actin genes, and a speculation on the evolutionary significance of small gene families. In: **Genome Evolution,** (eds. G. A. Dover & R. B. Flavell), pp. 177-91. Cambridge University Press.

Dayhoff M. O. & Park C. M. (1969). Cytochrome *c*: building a phylogenetic tree. **Atlas Prot. Seq. Struct. 4** 7-16

de Beer G. (1958). **Embryos and Ancestors.** (3rd ed.) Oxford University Press.

Degens E. T., Johanneson B. W. & Meyer R. W. (1967). Mineralization processes in molluscs and their palaeontological significance. **Naturwiss. 54** 638-40

Dickerson R. E. (1971). The structure of cytochrome *c* and the rates of molecular evolution. **J. Mol. Evol. 1** 26-45

Dickerson R. E. (1980). Cytochrome *c* and the evolution of energy metabolism. **Sci. Am. 242** 98-110

Dillon L. S. (1981). **Ultrastructure, Macromolecules and Evolution.** Plenum Press, New York.

Dilly P. N. (1975). The pterobranch *Rhabdopleura compacta*; its nervous system and phylogenetic position. **Symp. Zool. Soc. Lond. 36** 1-16

Durham J. W. (1978). The probable metazoan biota of the Precambrian as indicated by the subsequent record. **Ann. Rev. Earth Planet. Sci. 6** 21-42

Eakin R. M. (1963). Lines of evolution of photoreceptors. In: **General Physiology of Cell Specialisation,** (eds. D. Mazia & A. Tyler), pp. 393-425. McGraw Hill, New York.

Eakin R. M. (1979). Evolutionary significance of photoreceptors; in retrospect. **Am. Zool. 19** 647-53

Eakin R. M. & Brandenburger J. L. (1974). Ultrastructural features of a gordian worm (Nematomorpha). **J. Ultrastruct. Res. 46** 351-74

Ehlers U. (1985*a*). Phylogenetic relationships within the Platyhelminthes. In: **The Origins and Relationships of Lower Invertebrates**, (eds. S. Conway Morris, J. D. George, R. Gibson & H. M. Platt), pp. 143-58. Clarendon Press, Oxford.

Ehlers U. (1985*b*). **Das Phylogenetische System des Plathelminthes**. Gustav Fischer, Stuttgart & New York.

Eldredge N. & Cracraft J. (1980). **Phylogenetic Patterns and the Evolutionary Process. Method and Theory in Comparative Biology.** Columbia University Press, New York.

Emig C. (1977). Embryology of Phoronida. **Am. Zool. 17** 21-38

Emig C. (1982). The biology of Phoronida. **Adv. Mar. Biol. 19** 1-89

Farmer J. D. (1977). An adaptive model for the evolution of the ectoproct life cycle. In: **Biology of Bryozoans,** (eds. R. M. Wollacott & R. L. Zimmer), pp. 487-517. Academic Press, London & New York.

Fedonkin M. A. (1982). Precambrian soft-bodied fauna and the earliest radiation of invertebrates. **Third. N. Am. Palaeont. Conv. Proc. 1,** 165-7

Fell H. B. (1948). Echinoderm embryology and the origin of chordates. **Biol. Rev. 23** 81-107

Fell H. B. & Pawson D. L (1966). General biology of echinoderms. In: **Physiology of Echinodermata,** (ed. R. A. Boolootian), pp. 1-48. Wiley Interscience, New York.

Fenchel T. M. & Riedl R. J. (1970). The sulfide system; a new biotic community underneath the oxidized layer of marine sand bottoms. **Marine Biol. 7** 255-68

Firby J. B. & Durham J. W. (1974). Molluscan radula from earliest Cambrian. **J. Palaeont. 48** 1109-19

Flood P. R. (1975). Fine structure of the notochord of amphioxus. **Symp. zool. Soc. Lond. 36** 81-104

Florkin M. (1966). **A Molecular Approach to Phylogeny.** Elsevier, Amsterdam, London.

Ford T. D. (1979). Precambrian fossils and the origin of the phanerozoic phyla. In: **The Origin of Major Invertebrate Groups,** (ed. M. R. House), pp. 7-21. Academic Press, London & New York.

Forey P. L. (1982). Neontological analysis versus palaeontological stories. In: **Problems of Phylogenetic Reconstruction**, (eds. K. A. Joysey & A. E. Friday), pp. 119-57. Academic Press, London & New York.

Fortey R. A. & Jefferies R. P. S. (1982). Fossils and phylogeny - a compromise approach. In: **Problems of Phylogenetic Reconstruction**, (eds. K. A. Joysey & A. E. Friday), pp. 197-234. Academic Press, London & New York.

Fransen M. E. (1980). Ultrastructure of coelomic organisation in annelids. I. Archiannelida and other small polychaetes. **Zoomorphologie 95** 235-49

Franzén Å. (1967). Remarks on spermiogenesis and morphology of the spermatozoon amongst the lower Metazoa. **Ark. Zool. 19** 335-42

Franzén Å. (1970). Phylogenetic aspects of the morphology of spermatozoa and spermiogenesis. In: **Comparative Spermatology**, (ed. B. Baccetti), pp. 29-46. Academic Press, New York.

Franzén Å. (1977). Sperm structure with regard to fertilisation biology and phylogenetics. **Verh. Dtsch. Zool. Ges. 1977** 123-8

Freeman G. L. (1982). What does the comparative study of development tell us about evolution? In: **Evolution and Development**, (ed. J. T. Bonner), pp. 155-67. Springer-Verlag, Berlin.

Garstang W. (1922). The theory of recapitulation: A critical re-statement of the Biogenetic Law. **Zool. J. Linn. Soc. 35** 81-101

Garstang W. (1928*a*). The origin and evolution of larval forms. **Report, 1928, Brit. Ass. Adv. Science.**

Garstang W. (1928*b*). The morphology of the tunicata and its bearing on the phylogeny of the chordata. **Quart. J. Microsc. Sci. 72** 51-187

Garstang W. (1951). **Larval Forms and other Zoological Verses.** Blackwell, Oxford.

George J. D. & Southward E. C. (1973). A comparative study of the setae of Pogonophora and polychaetous Annelida. **J. Mar. Biol. Ass. UK 53** 403-24

Gerhart J. C. *et al.* (group report) (1982). The cellular basis of morphogenetic change. In: **Evolution and Development,** (ed. J. T. Bonner), pp. 87-114. Springer-Verlag, Berlin.

Ghirardelli E. (1968). Some aspects of the biology of the chaetognaths. **Adv. Mar. Biol. 6** 271-375

Gibson R. (1972). **Nemerteans.** Hutchinson & Co., London.

Glaessner M. F. (1983). The emergence of Metazoa in the early history of life. **Precambrian Res. 20** 427-41

Glaessner M. F. (1984). **The Dawn of Animal Life. A Biohistorical Study.** Cambridge University Press.

Godeaux J. E. A. (1974). Introduction to the morphology, phylogenesis and systematics of lower Deuterostomia. In: **Chemical Zoology, vol. VIII,** (eds. M. Florkin & B. T. Scheer), pp. 3-60. Academic Press, New York & London.

Goodman M., Weiss M. L. & Czelusniak J. (1982). Molecular evolution above the species level: branching pattern, rates, and mechanisms. **Syst. Zool. 31** 376-99

Goodrich E. S. (1945). The study of nephridia and genital ducts since 1895. **Quart. J. Microsc. Sci. 86** 113-392

Goodwin B. C., Holder N. & Wylie C. C. (1983) (eds.). **Development and Evolution.** Cambridge University Press.

Goto T., Takasu N. & Yoshida M. (1984). A unique photoreceptive structure in the arrowworms *Sagitta crassa* and *Spadella schizoptera* (Chaetognatha). **Cell Tissue Res. 235** 471-8

Götting K.-J. (1980*a*). Origin and relationships of the Mollusca. **Z. Zool. Syst. Evol. 18** 24-27

Götting K.-J. (1980*b*). Argumente für die Deszendenz der Mollusken von metameren Antezedenten. **Zool. Jb. Anat. 103** 211-8

Gould S. J. (1977). **Ontogeny and Phylogeny.** Belknap Press, Cambridge, Mass.

Gould S. J. (1985). **The Flamingo's Smile.** W. W. Norton & Co., New York & London.

Graham A. (1977). Introduction to the Symposium. In: **Comparative Biology of Skin,** (ed. R. I. C. Spearman), pp. 1-5. Academic Press, London.

Graham A. (1979). Gastropoda. In: **The Origin of Major Invertebrate Groups**, (ed. M. R. House), pp. 359-65. Academic Press, London & New York.

Green C. R. (1984). Intercellular junctions. In: **Biology of the Integument, Vol 1. Invertebrates,** (eds. J. Bereiter-Hahn, A. G. Matoltsy & K. S. Richards), pp. 5-16. Springer-Verlag, Berlin.

Green C. R. & Bergquist P. R. (1982). Phylogenetic relationships within the Invertebrata in relation to the structure of septate junctions and the development of occluding junctional types. **J. Cell. Sci. 53** 279-305

Greenberg M. (1959). Ancestors, embryos and symmetry. **Syst. Zool. 8** 212-21

Grell K. G. (1971). *Trichoplax adhaerens* (F. E. Schulze) und die Entstehung der Metazoen. **Naturw. Rundschau 24** 160-1

Grell K. G. (1981) *Trichoplax adhaerens* and the origin of the Metazoa. **Atti. Conv. Lincei 49** 107-21

Grobben K. (1908). Die systematische Einteilung des Tierreiches. **Verh. Zool. Bot. Ges. Wien. 58** 491-511

Grobben K. (1924). Theoretische Erörterungen betreffend die phylogenetische Ableitung der Echinodermen. **Sitzber. Akad. Wiss. Wien. 132** 262-90

Gupta A. P. (1979). Arthropod hemocytes and phylogeny. In: **Arthropod Phylogeny**, (ed. A. P. Gupta), pp. 669-735. Van Nostrand Reinhold, New York.

Gupta B. L. & Little C. (1975). Ultrastructure, phylogeny and Pogonophora. **Z. Zool. Syst. Evol.** Sonderheft **1** 45-63

Guthrie D. M. (1975). The physiology and structure of the nervous system of amphioxus (the lancelet) *Branchiostoma lanceolatum* Pallas. **Symp. Zool. Soc. Lond. 36** 43-80

Gutmann W. F. (1981). Relationship between invertebrate phyla based on functional-mechanical analysis of the hydrostatic skeleton. **Am. Zool. 21** 63-81

Gutmann W. F., Vogel K. & Zorn H. (1978). Brachiopods: biomechanical interdependences governing their origin and phylogeny. **Science 199** 890-3

Hadzi J. (1953). An attempt to reconstruct the system of animal classification. **Syst. Zool. 2** 145-54

Hadzi J. (1963). **The Evolution of the Metazoa.** Pergamon Press, Oxford.

Haeckel E. (1874). The gastraea-theory, the phylogenetic classification of the animal kingdom and the homology of the germ-lamellae. **Quart. J. Microsc. Sci. 14** 142-65

Haeckel E. (1875). Die Gastrula und die Entfurchung der Thiere. **Jena Z. Naturwiss. 9** 402-508

Hall A. K. & Ruttishauer U. (1985). Phylogeny of a neural cell adhesion molecule. **Developmental Biol. 110** 39-46

Hammarsten O. D. & Runnström J. (1925). Zur Embryologie von *Acanthochiton discrepans* Brown. **Zool. Jb. Anat. Ontog. 47** 261-318

Hand C. (1959). On the origin and phylogeny of the coelenterates. **Syst. Zool. 8** 191-202

Hand C. (1963). The early worm: a planula. In: **The Lower Metazoa**, (ed. E. C. Dougherty), pp. 33-39. University of California Press, Berkeley.

Hanson E. D. (1958). On the origin of the Eumetazoa. **Syst. Zool. 7** 16-47

Hanson E. D. (1963). Homologies and the ciliate origin of the Eumetazoa. In: **The Lower Metazoa,** (ed. E. C. Dougherty), pp. 7-22. University of California Press, Berkeley.

Hanson E. D. (1977). **The Origin and Early Evolution of Animals.** Pitman, London.

Harbison G. R. (1985). On the classification and evolution of the Ctenophora. In: **The Origin and Relationships of Lower Invertebrates**, (eds. S. Conway Morris, J. D. George, R. Gibson & H. M. Platt), pp. 78-100. Clarendon Press, Oxford.

Hardy A. (1953). On the origin of the Metazoa. **Quart. J. Microsc. Sci. 94** 441-3

Hartman W. (1963). A critique of the enterocoele theory. In: **The Lower Metazoa**, (ed. E. C. Dougherty), pp. 55-77. University of California Press, Berkeley.

Hatschek B. (1891). **Lehrbuch der Zoologie.** Fischer, Jena.

Hausmann K. (1978). Extrusive organelles in protists. **Int. Rev. Cytol. 52** 197-276

Hennig W. (1965). Phylogenetic systematics. **Ann. Rev. Ent. 10** 97-116

Hennig W. (1966). **Phylogenetic Systematics.** University of Illinois Press, Urbana.

Hentschel C. C. & Birnstiel M. L. (1981). The organisation and expression of histone gene families. **Cell 25** 301-13

Hernandez-Nicaise M.-L. (1984). Ctenophora. In: **Biology of the Integument, Vol 1. Invertebrates,** (eds. J. Bereiter-Hahn, A. G. Matoltsy & K. S. Richards), pp. 96-111. Springer-Verlag, Berlin.

Hessler R. R. & Newman W. A. (1975). A trilobitomorph origin for the Crustacea. **Fossils Strata 4** 437-59

Hibberd D. J. (1975). Observations on the ultrastructure of the choanoflagellate *Codosiga botrytis* (Ehr.) Saville-Kent with special reference to the flagellar apparatus. **J. Cell Sci. 17** 191-219

Holland C. H. (1979). Early Cephalopoda. In: **The Origin of Major Invertebrate Groups,** (ed. M. R. House), pp. 367-78. Academic Press, London & New York.

Holland P. W. H. & Hogan B. L. M. (1986). Phylogenetic distribution of Antennapedia-like homeo boxes. **Nature 321** 251-3

Holley M. C. (1984). The ciliary basal apparatus is adapted to the structure and mechanics of the epithelium. **Tissue & Cell 16** 287-310

Holley M. C. (1986). Cell shape, spatial patterns of cilia, and mucus net construction in the ascidian endostyle. **Tissue & Cell 18** 667-84

Holman E. W. (1985). Gaps in the fossil record. **Palaeobiology 11** 221-6

Horn H.S. *et al.* (group report) (1982). Adaptive aspects of development. In: **Evolution and Development,** (ed. J. T. Bonner), pp. 215-35. Springer-Verlag, Berlin.

Hoyle G. (1983). **Muscles and their Neural Control.** John Wiley & Sons, New York.

Hull D. L. (1979). The limits of cladism. **Syst. Zool. 28** 416-40

Hyman L. H. (1940). **The Invertebrates; Protozoa through Ctenophora.** McGraw Hill, New York.

Hyman L. H. (1951*a*). **The Invertebrates; Platyhelminthes and Rhynchocoela.** McGraw Hill, New York.

Hyman L. H. (1951*b*). **The Invertebrates; Acanthocephala, Aschelminthes, and Entoprocta.** McGraw Hill, New York.

Hyman L. H. (1955). **The Invertebrates; Echinodermata.** McGraw Hill, New York.

Hyman L. H. (1959). **The Invertebrates; Smaller Coelomate Groups.** McGraw Hill, New York.

Inglis W. G. (1983). The structure and operation of the obliquely striated supercontractile somatic muscles in nematodes. **Aust. J. Zool. 31** 677-93

Inglis W. G. (1985). Evolutionary waves; patterns in the origins of animal phyla. **Aust. J. Zool. 33** 153-78

Ishikawa H. (1977). Evolution of ribosomal RNA. **Comp. Biochem. Physiol. 58.B** 1-7

Ivanov A. V. (1968). (**The Origin of Multicellular Animals**). In Russian. Nauka, Leningrad.

Ivanov A. V. (1973). *Trichoplax adheerens* - a phygocytella-like animal. In Russian. **Zool. Zhurn. 52** 1117-31

Ivanov A. V. (1975). Embryonalentwicklung der Pogonophora und ihre systematische Stellung. **Z. Zool. Syst. Evol.** Sonderheft **1** 10-44

Iversen L. L. (1984). Amino acids and peptides; fast and slow chemical signals in the nervous system? **Proc. R. Soc. Lond. B 221** 245-60

Jacob F. (1983). Molecular tinkering in evolution. In: **Evolution from Molecules to Men.** (ed. D. S. Bendall), pp. 131-44. Cambridge University Press.

Jägersten G. (1955). On the early phylogeny of the Metazoa. The bilaterogastraea theory. **Zool. Bidr. Uppsala 30** 321-54

Jägersten G. (1959). Further remarks on the early phylogeny of the Metazoa. **Zool. Bidr. Uppsala 33** 79-108

Jägersten G. (1972). **Evolution of the Metazoan Life Cycle.** Academic Press, London.

Jefferies R. P. S. (1979). The origin of chordates - a methodological essay. In: **The Origin of Major Invertebrate Groups**, (ed. M. R. House), pp. 443-77. Academic Press, London & New York.

Jefferies R. P. S. (1981). In defence of the calcichordates. **Zool. J. Linn. Soc. 73** 351-96

Jefferies R. P. S. (1986). **The Ancestry of the Vertebrates**. British Museum (N.H.), London.

Jeffreys A. J., Harris S., Barrie P. A., Wood D., Blanchetot A. & Adams S. M. (1983). Evolution of gene families: the globin genes. In: **Evolution from Molecules to Men,** (ed. D. S. Bendall), pp. 175-95. Cambridge University Press.

Jensen D. D. (1963). Hoplonemertines, myxinoids, and vertebrate origins. In: **The Lower Metazoa,** (ed. E. C. Dougherty), pp. 113-26. University of California Press, Berkeley.

Jeuniaux C. (1963). **Chitine et Chitinolyse.** Masson, Paris.

Jeuniaux C. (1971). Chitinous structures. In: **Comprehensive Biochemistry,** vol 26C, (eds. M. Florkin & E. H. Stotz), pp. 595-632. Elsevier, Amsterdam.

Jeuniaux C. (1982). La chitine dans la regne animale. **Bull. Soc. Zool. France 107** 363-86

Joffe B. I. (1979). The comparative embryological analyses of the development of Nemathelminthes. **Dokl. Akad. Nauk SSSR 84** 39-62 (In Russian.)

Johansson J. (1952). On the phylogeny of Mollusca. **Zool. Bidr. Uppsala 29** 77-291

Jollie M. (1982). What are the 'calcichordates'; and the larger question of the origin of chordates. **Zool. J. Linn. Soc. 75** 167-88

Jones M. L. (1985). Vestimentiferan pogonophorans: their biology and affinities. In: **The Origins and Relationships of Lower Invertebrates,** (eds. S. Conway Morris, J. D. George, R. Gibson & H. M. Platt), pp. 327-42. Clarendon Press, Oxford.

Joysey K. A. & Friday A. E. (1982) (eds.). **Problems of Phylogenetic Reconstruction.** Academic Press, London & New York.

Karling T. G. (1966). On the defecation apparatus in the genus *Archimonocelis* (Turbellaria, Monocelididae). **Sarsia 24** 37-44

Katz M. J. (1983). Comparative anatomy of the tunicate tadpole, *Ciona intestinalis.* **Biol. Bull. 164** 1-27

Kerkut G. A. (1960). **Implications of Evolution.** Pergamon Press, Oxford & New York.

Kimura M. (1983). The neutral theory of molecular evolution. In: **Evolution of Genes and Proteins,** (eds. M. Nei & R. K. Koehn), pp. 208-33. Sinauer, Sunderland, Mass.

Knauss E. B. (1979). Indication of an anal pore in Gnathostomulida. **Zool. Scripta 8** 181-6

Kristensen N. P. (1979). (book review, of Boudreaux 1979). **Syst. Zool. 28** 638-43

Kristensen R. M. (1983). Loricifera, a new phylum with Aschelminthes characters from the meiobenthos. **Z. Zool. Syst. Evol. 21** 163-80

Lammert V. (1985). The fine structure of Gnathostomulid protonephridia and their comparison within the Bilateria. **Zoomorphology 105** 308-16

Lane N. J. & Skaer H. leB. (1980). Intercellular junctions in insect tissues. **Adv. Insect Physiol. 15** 35-213

Lang K. (1963). The relation between the Kinorhyncha and Priapulida and their connection with the Aschelminthes. In: **The Lower Metazoa.** (eds. E. C. Dougherty), pp. 256-62. University of California Press, Berkeley.

Lapan E. A. & Morowitz H. (1972). The Mesozoa. **Sci. Am. 227** 94-101

Larwood G. P. & Taylor P. D. (1979). Early structural and ecological diversification in the Bryozoa. In: **The Origin of Major Invertebrate Groups**, (ed. M. R. House), pp. 209-34. Academic Press, London & New York.

Lauterbach K.-E. (1972). Uber die sogenannte Gansbein-Mandibel der Tracheata insbesondere der Myriapoda. **Zool. Anz. 188** 145-54

Lauterbach K.-E. (1973). Schlussèlereignisse in der Evolution der Stammgruppe der Euarthropoda. **Zool. Beitr. 19** 251-99

Lauterbach K.-E. (1978). Gedanken zur Evolution der Euarthropoden- Extremität. **Zool. Jb. Anat. 99** 64-92

Lawrence J. M. (1987) **A Functional Biology of Echinoderms**. Johns Hopkins University Press.

Lehmann U. & Hillmer G. (1983). **Fossil Invertebrates**. Cambridge University Press.

Lemche H. (1959*a*). Molluscan phylogeny in the light of *Neopilina*.. **Proc. XV Int. Congr. Zool.** 380-1

Lemche H. (1959*b*). Protostomian interrelationships in the light of *Neopilina*. **Proc. XV Int. Congr. Zool.** 381-9

Lewin R. (1981). Seeds of change in embryonic development. **Science 214** 24-44

Linzen B. *et al.* (17 authors) (1985). The structure of arthropod haemocyanins. **Science 229** 519-24

Lorenzen S. (1985). Phylogenetic aspects of pseudocoelomate evolution. In: **The Origins and Relationships of Lower Invertebrates**, (eds. S. Conway Morris, J. D. George, R. Gibson & H. M. Platt), pp. 210-23. Clarendon Press, Oxford.

Løvtrup S. (1975). Validity of the Protostomia-Deuterostomia theory. **Syst. Zool. 24** 96-108

Løvtrup S. (1977). **The Phylogeny of Vertebrata.** John Wiley & Sons, London.

Løvtrup S. (1978). On von Baerian and Haeckelian recapitulation. **Syst. Zool. 27** 348-52

Lowenstam H. A. & Margulis L. (1980). Evolutionary prerequisites for Early Phanerozoic calcareous skeletons. **BioSystems 12** 27-41.

McFarland W. N., Pough F. H., Cade T. J. & Heiser J. B. (1979). **Vertebrate Life.** Collier McMillan, New York & London.

McGinnis W. (1985). Homeo-box sequences of the Antennapedia class are conserved only in higher animal genomes. **Cold Spring Harbour Symp. Quant. Biol. 50** 263-70

McGinnis W., Garber R. L., Wirz J., Kuroiwa A. & Gehring W. J. (1984). A homologous protein-coding sequence in *Drosophila* homeotic genes and its conservation in other metazoans. **Cell 37** 403-8

McLaughlin P. J. & Dayhoff M. O. (1973). Eukaryocyte evolution: a view based on cytochrome *c* sequence data. **J. Mol. Evol. 2** 99-116

Maggenti A. R. (1976). Taxonomic position of nematodes among the pseudocoelomate Bilateria. In: **The Organisation of Nematodes**, (ed. N. A. Croll), pp. 1-10. Academic Press, New York & London.

Malakhov V. V. (1980). Cephalorhyncha, a new type of animal kingdom uniting Priapulida, Kinorhyncha, Gordiacea and a system of Aschelminthes worms. **Zool. Zhurn. 59** 485-99 (In Russian.)

Manton S. M. (1949). Studies on the Onychophora VII. The early embryonic stages of *Peripatopsis* and some general considerations concerning the morphology and phylogeny of the Arthropoda. **Phil. Trans. R. Soc. B 233** 483-580

Manton S. M. (1964). Mandibular mechanisms and the evolution of arthropods. **Phil. Trans. R. Soc. B. 247** 1-183

Manton S. M. (1974). Arthropod phylogeny - a modern synthesis. **J. Zool. Lond. 171** 111-30

Manton S. M. (1977). **The Arthropods: Habits, Functional Morphology, and Evolution.** Oxford University Press.

Manton S. M. (1978). Habits, functional morphology and evolution of pycnogonids. **Zool. J. Linn. Soc. 63** 1-22

Manton S. M. & Anderson D. T. (1979). Polyphyly and the evolution of arthropods. In: **The Origin of Major Invertebrate Groups,** (ed. M. R. House), pp. 269-321. Academic Press, London & New York.

Marcus E. (1958). On the evolution of the animal phyla. **Quart. Rev. Biol. 33** 24-58

Margulis L. (1971). Whittaker's five kingdoms: minor modifications based on considerations of the origin of mitosis. **Evolution 25** 242-5

Margulis L. (1974). Five-kingdom classification and the origin and evolution of cells. **Evolutionary Biology 7** 45-78

Mariscal R. N. (1974). Nematocysts. In: **Coelenterate Biology, Reviews and New Perspectives,** (eds. L. Muscatine & H. M. Lenhoff), pp. 129-78. Academic Press, New York.

Mariscal R. N. (1984). Cnidaria: cnidae. In: **Biology of the Integument, Vol 1. Invertebrates,** (eds. J. Bereiter-Hahn, A.G. Matoltsy & K. S. S. Richards), pp. 57-68. Springer-Verlag, Berlin.

Mathews M. B. (1967). Macromolecular evolution of connective tissue. **Biol. Rev. 42** 499-551

Maxson R., Cohn R. & Kedes L. (1983). Expression and organisation of histone genes. **Ann. Rev. Genetics 17** 239-77

Maynard Smith J. (1983). Evolution and development. In: **Development and Evolution,** (eds. B. C. Goodwin, N. Holder and C. C. Wylie), pp. 33-45. Cambridge University Press.

Mettam C. (1985). Functional constraints in the evolution of the Annelida. In: **The Origins and Relationships of Lower Invertebrates,** (eds. S. Conway Morris, J. D. George, R. Gibson & H. M. Platt), pp. 297-309. Clarendon Press, Oxford.

Mileikovsky S. A. (1971). Types of larval development in marine bottom invertebrates, their distribution and ecological significance; a re-evaluation. **Mar. Biol. 10** 193-213

Mill P. J. (1982). **Comparative Neurobiology.** Edward Arnold, London.

Moore R. C. & Teichert C (eds.) (1952-79). **Treatise on Invertebrate Palaeontology**, Parts A-W, Geol. Soc. Am. & University of Kansas Press/ McGraw Hill, New York.

Morris N. J. (1979). On the origin of the Bivalvia. In: **The Origin of Major Invertebrate Groups,** (ed. M. R. House), pp. 381-413. Academic Press, London & New York.

Nei M. & Koehn R. K. (1983). **Evolution of Genes and Proteins**. Sinauer, Sunderland, Mass.

Nelson G. J. (1978). Ontogeny, phylogeny, palaeontology and the biogenetic law. **Syst. Zool. 27** 324-45

Nicholas W. L. & Hynes H. B. N. (1963). Embryology, post-embryonic development and phylogeny of the Acanthocephala. In: **The Lower Metazoa,** (ed. E. C. Dougherty), pp. 385-402. University of California Press, Berkeley.

Nichols D. (1967). The origin of echinoderms. **Symp. Zool. Soc. Lond. 20** 209-229

Nichols D. (1971). The phylogeny of the invertebrates. In: **The Invertebrate Panorama**, (eds. E. Smith, G. Chapman, R. B. Clark, D. Nichols & J. D. Carthy), pp. 362-81. Weidenfeld & Nicholson, London.

Nichols D. (1986). A new class of echinoderms. **Nature 321** 808

Nielsen C. (1971). Entoproct life cycles and the entoproct/ectoproct relationship. **Ophelia 9** 209-341

Nielsen C. (1977*a*). The relationships of Entoprocta, Ectoprocta, and Phoronida. **Am. Zool. 17** 149-50

Nielsen C. (1977*b*). Phylogenetic considerations: the Protostomian relationships. In: **Biology of Bryozoans**, (eds. R. M. Woollacott & R. L. Zimmer), pp. 519-34. Academic Press, New York & London.

Nielsen C. (1979). Larval ciliary bands and metazoan phylogeny. **Fortschr. zool. Syst. Evol. 1** 178-84

Nielsen C. (1985). Animal phylogeny in the light of the trochaea theory. **Biol. J. Linn. Soc. 25** 243-99

Nielsen C. & Nørrevang A. (1985). The trochaea theory: an example of life cycle phylogeny. In: **The Origins and Relationships of Lower Invertebrates,** (eds. S. Conway Morris, J. D. George, R. Gibson & H. M. Platt), pp. 28-41. Clarendon Press, Oxford.

Norenburg J. (1985). Structure of the nemertine integument with consideration of ecological and phylogenetic significance. **Am. Zool. 25** 37-51

Nørrevang A. (1970). The position of Pogonophora in the phylogenetic system. **Z. Zool Syst. Evol. 8** 161-72

Nursall J. R. (1962). On the origins of the major groups of animals. **Evolution 16** 118-23

Odselius R. & Elofsson R. (1981). The basement membrane of the insect and crustacean compound eye: definition, fine structure, and comparative morphology. **Cell & Tissue Res. 216** 205-14

Olive P. J. W. (1985). Covariability of reproductive traits in marine invertebrates, implications for the phylogeny of the lower invertebrates. In: **The Origins and Relationships of Lower Invertebrates,** (eds. S. Conway Morris, J. D. George, R. Gibson & H. M. Platt), pp. 42-59. Clarendon Press, Oxford.

Pantin C. (1960). Diploblastic animals. **Proc. Linn. Soc. London 171** 1-14

Patterson C. (1981). Significance of fossils in determining evolutionary relationships. **Ann. Rev. Ecol. Syst. 12** 195-223

Patterson C. (1982). Morphological characters and homology. In: **Problems of Phylogenetic Reconstruction**, (eds. K. A. Joysey & A. E. Friday), pp. 21-74. Academic Press, London & New York.

Patterson C. (1983). How does phylogeny differ from ontogeny? In: **Development and Evolution**, (eds. B. C. Goodwin, N. Holder & C. C. Wylie), pp. 1-31. Cambridge University Press.

Paul C. R. C. (1977). Evolution of primitive echinoderms. In: **Patterns of Evolution as Illustrated by the Fossil Record**, (ed. A. Hallam), pp. 123-58. Elsevier, Amsterdam.

Paul C. R. C. (1979). Early echinoderm radiation. In: **The Origin of Major Invertebrate Groups**, (ed. M. R. House), pp. 415-34. Academic Press. London & New York.

Paul C. R. C. (1982). The adequacy of the fossil record. In: **Problems of Phylogenetic Reconstruction**, (eds. K. A. Joysey & A. E. Friday), pp. 75-117. Academic Press, London & New York.

Paul C. R. C. & Smith A. B. (1984). The early radiation and phylogeny of echinoderms. **Biol. Rev. 59** 443-81

Paulus H. F. (1979). Eye structure and the monophyly of Arthropoda, In: **Arthropod Phylogeny,** (ed. A. P. Gupta), pp. 299-383. Van Nostrand Reinhold, New York.

Pennington J. T. & Chia F. S. (1985). Gastropod torsion: a test of Garstang's hypothesis. **Biol. Bull. 169** 391-6

Person P. & Philpott D. E. (1969). The nature and significance of invertebrate cartilages. **Biol. Rev. 44** 1-16

Pflug H. D. (1974). Vor- und Fruh-geschichte der Metazoen. **N. Jahrb. Geol. Palaeont. Abh. 145** 328-74

Philip G. M. (1979). Carpoids - echinoderms or chordates? **Biol. Rev. 54** 439-71

Phillips D. C., Sternberg M. J. E. & Sutton B. J. (1983). Intimations of evolution from the three-dimensional structures of proteins. In: **Evolution from Molecules to Men,** (ed. D. S. Bendall), pp. 145-73. Cambridge University Press.

Phillips D. M. (1974). Structural variants in invertebrate sperm flagella and their relation to motility. In: **Cilia and Flagella,** (ed. M. A. Sleigh), pp. 397-402. Academic Press, London.

Pilato G. (1981). The significance of musculature in the origin of the Annelida. **Boll. Zool. 48** 209-26

Pitelka D. R. (1974). Basal bodies and root structures. In: **Cilia and Flagella,** (ed. M. A. Sleigh), pp. 437-69. Academic Press, London.

Platt H. M. (1981). Meiofaunal dynamics and the origin of the Metazoa. In: **The Evolving Biosphere,** (ed. P. L. Forey), pp. 207-16. British Museum (NH) & Cambridge University Press.

Pojeta J. (1980). Molluscan phylogeny. **Tulane Studies in Geol. & Pal. 16** 55-80.

Por F. D. & Bromley H. J. (1974). Morphology and anatomy of *Maccabeus tentaculatus* (Priapulida, Seticoronaria). **J. Zool. Soc. Lond., 173** 173-97

Raff R. A. & Kaufmann T. C. (1983). **Embryos, Genes and Evolution.** MacMillan Publishing Co., New York.

Raup D. M. (1972). Taxonomic diversity during the Phanerozoic. **Science 177** 1065-71

Raup D. M. (1976*a*). Species diversity in the Phanerozoic: a tabulation. **Palaeobiology 2** 279-88

Raup D. M. (1976*b*). Species diversity in the Phanerozoic: an interpretation. **Palaeobiology 2** 289-97

Raup D. M. (1983). On the early origin of major biologic groups. **Palaeobiology 9** 107-15

Raven C. P. (1966). **Morphogenesis, the Analysis of Molluscan Development.** (2nd ed.) Pergamon Press, Oxford.

Rehkämper G. & Welsch U. (1985). On the fine structure of the cerebral ganglion of *Sagitta* (Chaetognatha). **Zoomorphology 105** 83-9

Reise K. & Ax P. (1979). A meiofaunal "Thiobios" limited to the anaerobic sulfide system does not exist. **Marine Biol. 54** 225-37

Reisinger E. (1970). Zur Problematik der Evolution der Coelomaten. **Z. Zool. Syst. Evol. 8** 81-109

Reisinger E. (1972). Die Evolution des Orthogons der Spiralier und das Archicölomatenproblem. **Z. Zool. Syst. Evol. 10** 1-43

Reisinger E. & Kelbetz S. (1964). Feinbau und Endlanguns-mechanismus der Rhabditen. **Z. Wiss. Mikroskopie und Mikrosk. Technik. 65** 472-508

Remane A. (1952). **Die Grundlagen des naturlichen Systems, der vergleichenden Anatomie ubd der Phylogenetik.** Reprinted 1971, Koeltz, Königstein.

Remane A. (1958). Zur Verwandtschaft und Ableitung der niederen Metazoen. **Zool. Anz. Suppl. 21** 179-96

Remane A. (1963*a*). The enterocoelic origin of the coelom. In: **The Lower Metazoa,** (ed. E. C. Dougherty), pp. 78-90. University of California Press, Berkeley.

Remane A. (1963*b*). The evolution of the Metazoa from colonial flagellates vs. plasmodial ciliates. In: **The Lower Metazoa,** (ed. E. C. Dougherty), pp. 23-32. University of California Press, Berkeley.

Remane A. (1963*c*). The systematic position and phylogeny of the pseudocoelomates. In: **The Lower Metazoa**, (ed. E. C. Dougherty), pp. 247-55. University of California Press, Berkeley.

Remane A. (1967). Die Geschichte der Tiere. In: **Die Evolution der Organismen,** (ed. G. Herberer), pp. 589-677. G. Fischer, Stuttgart.

Remane A., Storch V. & Welsch U. (1980). **Systematische Zoologie**. G. Fischer, Stuttgart.

Reutterer A. (1969). Zum Problem der Metazoenabstammung. **Z. Zool. Syst. Evol. 7** 30-53

Rice M. E. (1975). Sipuncula. In: **Reproduction of Marine Invertebrates Vol II,** (eds. A. Giese & J. Pearse), pp. 67-127. Academic Press, New York.

Rice M. E. (1985). Sipuncula: developmental evidence for phylogenetic inference. In: **The Origins and Relationships of Lower Invertebrates**, (eds. S. Conway Morris, J. D. George, R. Gibson & H. M. Platt), pp. 274-96. Clarendon Press, Oxford.

Richards K. S. (1984). Introduction. In: **Biology of the Integument, Vol. 1, Invertebrates**, (eds. J. Bereiter-Hahn, A. G. Matoltsy & K. S. Richards), pp. 1-4. Springer-Verlag, Berlin.

Richardson J. S. (1981). The anatomy and taxonomy of protein structure. **Adv. in Protein Chem. 34** 167-339

Rickards R. B. (1975). Palaeoecology of the Graptolithina, an extinct class of the phylum Hemichordata. **Biol. Rev. 50** 397-436

Rickards R. B. (1979). Early evolution of graptolites and related groups. In: **The Origin of Major Invertebrate Groups**, (ed. M. R. House), pp. 435-41. Academic Press, London & New York.

Riedl R. J. (1960). Beitrage zur Kenntnis der *Rhodope veranii*. **Z. wiss. Zool. 163** 76-81

Riedl R. J. (1969). Gnathostomulida from America. **Science 163** 445-52

Riedl R. J. (1979). **Order in Living Organisms: a Systems Analysis of Evolution.** John Wiley & Sons, Chichester.

Rieger G. E. & Rieger R. M. (1977). Comparative fine structure study of the gastrotrich cuticle and aspects of cuticle evolution within the Aschelminthes. **Z. Zool. Syst. Evol. 15** 81-124

Rieger R. M. (1976). Monociliated epidermal cells in Gastrotricha: significance for concepts of early metazoan evolution. **Z. Zool. Syst. Evol. 14** 198-226

Rieger R. M. (1980). A new group of interstitial worms, Lobatoceridae nov. fam. (Annelida) and its significance for metazoan phylogeny. **Zoomorphologie 95** 41-84

Rieger R. M. (1984). Evolution of the cuticle in the lower Metazoa. In: **Biology of the Integument, Vol 1, Invertebrates**, (eds. J. Bereiter-Hahn, A. G. Matoltsy & K. S. Richards), pp. 389-99. Springer-Verlag, Berlin.

Rieger R. M. (1985). The phylogenetic status of the acoelomate organisation within the Bilateria; a histological perspective. In: **The Origins and Relationships of Lower Invertebrates,** (eds. S. Conway Morris, J. D. George, R. Gibson & H. M. Platt), pp. 101-22. Clarendon Press, Oxford.

Rieger R. M. & Mainitz M. (1977). The body wall in Gnathostomulida, a comparative fine structure study. **Z. Zool. Syst. Evol. 15** 9-35

Rieger R. M. & Rieger G. E. (1976). Fine structure of the archiannelid cuticle and remarks on the evolution of the cuticle within the Spiralia. **Acta Zool. 57** 53-68

Rieger R. M., Ruppert E. E., Rieger G. E. & Schoepfer-Sterrer C. (1974). On the fine structure of gastrotrichs with description of *Chordodasys antennatus* sp.n. **Zool. Scripta. 3** 219-37

Rieger R. M. & Sterrer W. (1975). New spicular skeletons in Turbellaria and the occurrence of spicules in marine meiofauna. **Z. Zool. Syst. Evol. 13** 207-48

Rieger R. M. & Tyler S. (1979). The homology theorem in ultrastructural research. **Am. Zool. 19** 655-64

Riley J., Banaja A. A. & James J. L. (1978). The phylogenetic relationships of the Pentastomida; the case for their inclusion within the Crustacea. **Int. J. Parasitol. 8** 245-54

Roberts K. & Hyams J.S. (1979) (eds.). **Microtubules.** Academic Press, New York.

Roche J., Thoai N-V. & Robin Y. (1957). Sur la présence de la créatine chez les invertebres et sa signification biologique. **Biochim. Biophys. Acta 24** 514-9

Rockstein M. (1971). The distribution of phosphoarginine and phosphocreatine in marine invertebrates. **Biol. Bull. 141** 167-75

Rudall K. M. & Kenchington W. (1973). The chitin system. **Biol. Rev. 49** 597-636

Rudwick M. J. S. (1970). **Living and Fossil Brachiopods.** Hutchinson & Co., London.

Runnegar B. (1980). Mollusca: the first hundred million years. **J. Malac. Soc. Aust. 4** 223-4

Runnegar B. (1984). Derivation of the globins from type *b* cytochromes. **J. Mol. Evol. 21** 33-41

Runnegar B. (1985). Collagen gene construction and evolution. **J. Mol. Evol. 22** 141-9

Runnegar B. & Pojeta J. (1974). Molluscan phylogeny: the palaeontological viewpoint. **Science 186** 311-7

Runnegar B. & Pojeta J. (1985). Origin and diversification of the Mollusca. In: **The Mollusca, Vol.10, Evolution**, (eds. E. R. Trueman & M. R. Clarke), pp. 1-57. Academic Press, New York & London.

Ruppert E. E. (1982). Comparative ultrastructure of the gastrotrich pharynx and the evolution of myoepithelial foreguts in Aschelminthes. **Zoomorphologie 99** 181-220

Ruppert E. E. & Carle K. J. (1983). Morphology of metazoan circulatory systems. **Zoomorphologie 103** 193-208

Russell G. J. & Subak-Sharpe J. H. (1977). Similarity of the general designs of protochordates and invertebrates. **Nature 266** 533-6

Russell-Hunter W. D. (1979). **A Life of Invertebrates.** Macmillan Publishing Co., New York.

Ruttner-Kolisko A. (1963). The interrelationships of the Rotatoria, In: **The Lower Metazoa,** (ed. E.C. Dougherty), pp. 263-72. University of California Press, Berkeley.

Ryland J. S. (1970). **Bryozoans.** Hutchinson & Co., London.

Salvini-Plawen L. von (1969). Solenogastres und Caudofoveata (Mollusca, Aculifera); Organisation und phylogenetische Bedeutung. **Malacologia 9** 191-216

Salvini-Plawen L. von (1973). Zur Klärung des "Trochophora"-Begriffes. **Experientia 29** 1434-6

Salvini-Plawen L. von (1978). On the origin and evolution of the lower Metazoa. **Z. Zool Syst. Evol. 16** 40-88

Salvini-Plawen L. von (1980*a*). Phylogenetischer Status und Bedeutung der mesenchymaten Bilateria. **Zool. Jb. Anat. 103** 354-73

Salvini-Plawen L. von (1980*b*). Was ist eine Trochophora? Eine Analyse der Larventypen mariner Protostomier. **Zool. Jb. Anat. 103** 389-423

Salvini-Plawen L. von (1981). On the origin and evolution of the Mollusca. **Atti Conv. Lincei 49** 237-93

Salvini-Plawen L. von (1982*a*). A paedomorphic origin of the oligomerous animals? **Zool. Scripta 11** 77-81

Salvini-Plawen L. von (1982*b*). On the polyphyletic origin of photoreceptors. In: **Visual Cells in Evolution**, (ed. J. A. Westfall), pp. 137-54. Raven Press, New York.

Salvini-Plawen L. von (1985). Early evolution and the primitive groups. In: **The Mollusca, Vol 10, Evolution,** (eds. E. R. Trueman & M. R. Clarke), pp. 59-150. Academic Press, London & New York.

Salvini-Plawen L. von, & Mayr E. (1977). On the evolution of photoreceptors and eyes. In: **Evolutionary Biology, Vol 10**, (eds. M. K. Hecht, W. Steere & B. Wallace), pp. 207-63. Plenum Press, New York.

Salvini-Plawen L. von & Splechtna H. (1979). Zur Homologie der Keimblätter. **Z. Zool. Syst. Evol. 17** 10-30

Sandberg P. A. (1977). Ultrastructure, mineralogy and development of bryozoan skeletons. In: **Biology of Bryozoans**, (eds. R. M. Woollacott & R. L. Zimmer), pp. 143-81. Academic Press, New York & London.

Sanderson M. J. (1984). Cilia. In: **Biology of the Integument, Vol 1. Invertebrates,** (eds. J. Bereiter-Hahn, A. G. Matoltsy & K. S. Richards), pp. 17-42. Springer-Verlag, Berlin.

Schäfer W. (1973) (ed). **Das Archicoelomatenproblem**. Woldemar Kramer, Frankfurt-am-Main.

Schaller F. (1979). Significance of sperm transfer and formation of spermatophores in arthropod phylogeny. In: **Arthropod Phylogeny,** (ed. A. P. Gupta), pp. 587-608. Van Nostrand Reinhold, New York.

Scheltema A. H. (1978). Position of the class Aplacophora in the phylum Mollusca. **Malacologia 17** 99-109

Schopf J. W. (1975). Precambrian palaeobiology: problems and perspectives. **Ann. Rev. Earth Planet. Sci. 3** 213-49

Schopf J. W., Haugh B. N., Molnar R. E. & Satterthwait D. F. (1973). On the development of metaphytes and metazoans. **J. Palaeont. 47** 1-9

Schopf J. W. & Oehler D. Z. (1976). How old are the eukaryotes? **Science 193** 47-9

Schopf T. J. M. (1980). **Palaeooceanography.** Harvard University Press, Cambridge, Mass.

Schopf T. J. M. (1983). DNA structure: the fourth approach to comparative biology. **Cold Spring Harbour Symp. Quant. Biol. 47** 1159-64

Schram F. R. (1979). (Book review, of Boudreaux 1979). **Syst. Zool. 28** 635-8

Schram F. R. (1982). The fossil record and evolution of Crustacea. In: **The Biology of Crustacea, Vol. 1,** (ed. D. E. Bliss), pp. 93-147. Academic Press, London.

Scrutton C. T. (1979). Early fossil cnidarians. In: **The Origin of Major Invertebrate Groups**, (ed. M. R. House), pp. 161-207. Academic Press, London & New York.

Segler K., Rahmann H. & Rösner H. (1978). Chemotaxonomical investigations on the occurrence of sialic acids in Protostomia and Deuterostomia. **Biochem. Syst. Ecol. 6** 87-93

Seilacher A. (1977). Evolution of trace fossil communities. In: **Patterns of Evolution as Illustrated by the Fossil Record,** (ed. A. Hallam), pp. 357-76. Elsevier, Amsterdam.

Seilacher A. (1984). Late Precambrian and early Cambrian Metazoa: preservational or real extinctions? In: **Patterns of Change in Earth Evolution**, (eds. H. D. Holland & A. R. Trendall), pp. 159-68. Springer Verlag, Berlin.

Sepkoski J. J. (1978). A kinetic model of Phanerozoic taxonomic diversity. 1. Analysis of marine orders. **Palaeobiology 4** 223-51

Sepkoski J. J. (1979). A kinetic model of Phanerozoic taxonomic diversity. II. Early Phanerozoic families and multiple equilibria. **Palaeobiology 5** 222-51

Sepkoski J. J., Bambach R. K., Raup D. M. & Valentine J. W. (1981). Phanerozoic marine diversity and the fossil record. **Nature 293** 435

Shapeero W. L. (1961). Phylogeny of the Priapulida. **Science 133** 879-80

Sharov A. G. (1966). **Basic Arthropodan Stock with Special Reference to Insects.** Pergamon Press, Oxford.

Shelton G. A. B. (1982). **Electrical Conduction and Behaviour in 'Simple' Invertebrates.** Clarendon Press, Oxford.

Siewing R. (1969). **Lehrbuch der vergleichenden Entwicklungs- geschichte der Tiere.** Paul Parey, Hamburg.

Siewing R. (1975). Thoughts about the phylogenetic-systematic position of Pogonophora. **Z. Zool. Syst. Evol.** Sonderheft 1 127-38

Siewing R. (1976). Probleme und neuere Erkentnisse in der Gross-systematik der Wirbellosen. **Verhandl. Dtsch. Zool. Ges. Stuttgart 1976** 59-83

Siewing R. (1978). Zur mutmasslichen Phylogenie der Arthropoden- extremität. **Zool. Jb. Anat. 99** 93-8

Siewing R. (1979). Homology of cleavage types. **Fortschr. Zool. Syst. Evol. 1** 7-18

Siewing R. (1980). Das Archicoelomatenkonzept. **Zool. Jb. Syst. 103** 439-82

Signor P. W. (1978). Species richness in the Phanerozoic: an investigation of sampling effects. **Palaeobiology 4** 394-406

Simpson T. L. (1984). **The Cell Biology of Sponges.** Springer-Verlag, New York.

Sleigh M. A. (1974) (ed.). **Cilia and Flagella.** Academic Press, London.

Sleigh M. A. (1979). Radiation of the Eukaryote Protista. In: **The Origin of Major Invertebrate Groups,** (ed. M. R. House), pp. 23-53. Academic Press, London & New York.

Smith J. P. S. & Tyler S. (1985). The acoel turbellarians: kingpins of metazoan evolution or a specialised offshoot? In: **The Origins and Relationships of Lower Invertebrates,** (eds. S. Conway Morris, J. D. George, R. Gibson & H. M. Platt), pp. 123-42. Clarendon Press, Oxford.

Smith J. P. S., Tyler S. & Rieger R. M. (1985). Is the Turbellaria polyphyletic? **Hydrobiologia 132** 13-21

Smith J. P. S., Tyler S., Thomas M. B. & Rieger R. M. (1982). The morphology of turbellarian rhabdites: phylogenetic implications. **Trans. Am. Microsc. Soc. 101** 209-28

Snodgrass R. E. (1938). Evolution of the Annelida, Onychophora and Arthropoda. **Smithson. Misc. Coll. 97** 1-159

Solomon E. & Cheah K. S. E. (1981). Collagen evolution. **Nature 291** 450-1

Southward E. C. (1971). Recent researches on the Pogonophora. **Oceanogr. Mar. Biol. Ann. Rev. 9** 193-220

Southward E. C. (1975). Fine structure and phylogeny of the Pogonophora. **Symp. Zool. Soc. Lond. 36** 235-51

Sprinkle J. (1976). Classification and phylogeny of 'pelmatozoan' echinoderms. **Syst. Zool. 25** 83-91

Staehelin L. A. & Hull B. E. (1978). Junctions between living cells. **Sci. Am. 238** 140-52

Stanley S. M. (1975). Fossil data and the Precambrian-Cambrian evolutionary transition. **Am. J. Sci. 276** 56-76

Stanley S. M. (1976). Ideas on the timing of metazoan diversification. **Palaeobiology 2** 209-19

Stanley S. M. (1979). **Macroevolution: Pattern and Process.** W. H. Freeman & Co., San Francisco.

Starck D. & Siewing R. (1980). Zur Diskussion der Begriffe Mesenchym und Mesoderm. **Zool. Jb. Anat. 103** 374-88

Stasek C. (1972). The molluscan framework. In: **Chemical Zoology, Vol VII,** (eds. M. Florkin & B. T. Scheer), pp. 1-44. Academic Press, London & New York.

Stearns S. C. (1980). A new view of life-history evolution. **Oikos 35** 266-81

Steinböck O. (1958). Zur Phylogenie der Gastrotrichen. **Zool. Anz. Suppl. 21** 128-69

Steinböck O. (1963). Origin and affinities of the lower Metazoa. In: **The Lower Metazoa,** (ed. E. C. Dougherty), pp. 45-54. University of California Press, Berkeley.

Stephens R. E. (1977). Major membrane protein differences in cilia and flagella - evidence for a membrane associated tubulin. **Biochemistry 16** 2047-58

Sterrer W., Mainitz M & Rieger R. M. (1985). Gnathostomulida: enigmatic as ever. In: **The Origins and Relationships of Lower Invertebrates,** (eds. S. Conway Morris, J. D. George, R. Gibson & H. M. Platt), pp. 181-99. Clarendon Press, Oxford.

Storch V. (1979). Contributions of comparative ultrastructural research to problems of invertebrate evolution. **Am. Zool. 119** 637-45

Storch V. (1984). Minor pseudocoelomates. In: **Biology of the Integument Vol 1, Invertebrates,** (eds. J. Bereiter-Hahn, A. G. Matoltsy & K. S. Richards), pp. 242-68. Springer-Verlag, Berlin.

Størmer L. (1944). On the relationships and phylogeny of fossil and recent Arachnomorpha. A comparative study on Arachnida, Xiphosura, Eurypterida, Trilobita and other fossil arthropods. **Skrift. Vid. Akad. Oslo I. Math. Nat. Kl. 5** 1-158

Strathmann R. R. (1978). The evolution and loss of feeding larval stages of marine invertebrates. **Evolution 32** 894-906

Stunkard H. W. (1972). Clarification of taxonomy in Mesozoa. **Syst. Zool. 21** 210-4

Sylvester-Bradley P. C. (1975). The search for Protolife. **Proc. R. Soc. B. 189** 213-33

Sylvester-Bradley P. C. (1979). Precambrian prelude. In: **The Origin of Major Invertebrate Groups**, (ed. M. R. House), pp. 1-5. Academic Press, London & New York.

Tavener-Smith R. (1973). Some aspects of skeletal organisation in Bryozoa. In: **Living and Fossil Bryozoa**, (ed. G. P. Larwood), pp. 349-59. Academic Press, London.

Terwilliger R. C. (1980). Structures of invertebrate haemoglobins. **Am. Zool. 20** 53-67

Teuchert G. (1968). Zur Fortpflanzung und Entwicklung der Macrodasyoidea (Gastrotricha). **Z. Morph. Tiere 63** 343-418

Teuchert G. (1977). The ultrastructure of the marine gastrotrich *Turbanella cornuta* Remane (Macrodasyoidea) and its functional and phylogenetic importance. **Zoomorphologie 88** 189-246

Thorpe A. & Thorndyke M. C. (1975). The endostyle in relation to iodine binding. **Symp. Zool. Soc. Lond. 36** 159-77

Tiegs O. W. & Manton S. M. (1958). The evolution of the Arthropoda. **Biol. Rev. 33** 255-337

Tombes A. S. (1979). Comparison of arthropod neuroendocrine structures and their evolutionary significance. In: **Arthropod Phylogeny**, (ed. A. P. Gupta), pp. 645-67. Van Nostrand Reinhold, New York.

Towe K. M. (1981). Biochemical keys to the emergence of complex life. In: **Life in the Universe**, (ed. J. Billingham), pp. 297-306. MIT Press, Cambridge, Mass.

Turbeville J. M. & Ruppert E. E. (1985). Comparative ultrastructure and the evolution of nemertines. **Am. Zool. 25** 53-71

Tuzet O. (1973). Introduction et place des spongiaires dans la classification. In: **Traité de Zoologie, Vol 3.1,** (ed. P. Grassé), pp. 1-26. Masson, Paris.

Tyler S. (1979). Distinctive features of cilia in metazoans and their significance for systematics. **Tissue & Cell 11** 385-400

Tyler S. (1984). Turbellarian platyhelminths. In: **Biology of the Integument, Vol 1, Invertebrates,** (eds. J. Bereiter-Hahn, A. G. Matoltsy & K. S. Richards), pp. 112-31. Springer-Verlag, Berlin.

Tyler S. & Rieger R. M. (1975). Uniflagellate spermatozoa in *Nemertoderma* (Turbellaria), and their phylogenetic significance. **Science 188** 730-2

Ubaghs G. (1969). General characteristics of the echinoderms. In: **Chemical Zoology, Vol III,** (eds. M. Florkin & B. T. Scheer), pp. 3-45. Academic Press, New York & London.

Ubaghs G. (1979). Classification of the echinoderms. In: **Treatise on Invertebrate Palaeontology,** (ed. R. A. Robison), pp. 359-401. Geol. Soc. Am., New York.

Uchida T. (1963). On the interrelationships of the Coelenterata, with remarks on their symmetry. In: **The Lower Metazoa,** (ed. E. C. Dougherty), pp. 169-77. University of California Press, Berkeley.

Vagvolgyi J. (1967). On the origin of molluscs, the coelom, and coelomic segmentation. **Syst. Zool. 16** 153-68

Valentine J. W. (1973*a*). Coelomate superphyla. **Syst. Zool. 22** 97-102

Valentine J. W. (1973*b*). **Evolutionary Palaeoecology of the Marine Biosphere.** Prentice-Hall, Englewood Cliffs, N.J.

van der Land J. & Nørrevang A. (1975). The systematic position of *Lamellibrachia* (sic) (Annelida, Vestimentifera). **Z. Zool. Syst. Evol.** Sonderheft **1** 86-101

van der Land J. & Nørrevang A. (1985). Affinities and intraphyletic relationships of the Priapulida. In: **The Origins and Relationships of Lower Invertebrates,** (eds. S. Conway Morris, J. D. George, R. Gibson & H. M. Platt), pp. 261-273. Clarendon Press, Oxford.

Vanfleteren J. R. & Coomans A. (1976). Photoreceptor evolution and phylogeny. **Z. Zool. Syst. Evol. 14** 157-69

Verdonk N. H. & van den Biggelaar J. A. M. (1983). Early development and the formation of the germ layers. In: **The Mollusca, Vol 3, Development,** (eds. N. H. Verdonk & J. A. M. van den Biggelaar), pp. 91-122. Academic Press, New York & London.

Vogel K. & Gutmann W. F. (1981). Zur Entstehung von Metazoen-Skeletten an der Wende vom Präkambrium zum Kambrium. In: **Festschrift d. wissensch. Ges. J. W. Goethe Univ.**, pp. 517-37. Steiner Verlag, Wiesbaden.

Warren L. (1963). The distribution of sialic acids in nature. **Comp. Biochem. Physiol. 10** 153-71

Watts D. C. (1975). Evolution of phosphagen kinases in the chordate line. **Symp. Zool. Soc. Lond. 36** 105-27

Webb M. (1964). The posterior extremity of *Siboglinum fiordicum* (Pogonophora). **Sarsia 15** 33-6

Welsch L. T. & Welsch U. (1978). Histologische und elektronmikroskopische Untersuchungen an der präoralen Wimpergrube von *Saccoglossus horsti* (Hemichordata) und der Hatschekschen Grube von *Branchiostoma lanceolatum*

(Acrania). Ein Beitrag zur phylogenetischen Entwicklung der Adenohypophyse. **Zool. Jb. Anat. 100** 564

Welsch U. & Storch V. (1976). **Comparative Animal Cytology And Histology.** Sidgewick & Jackson, London.

Wessels N. K. (1982). A catalogue of processes responsible for metazoan morphogenesis. In: **Evolution and Development,** (ed. J. T. Bonner), pp. 115-54. Cambridge University Press.

Westfall J. A. (1982) (ed.). **Visual Cells in Evolution.** Raven Press, New York.

Westheide W. (1985). The systematic position of the Dinophilidae and the archiannelid problem. In: **The Origins and Relationships of Lower Invertebrates,** (eds. S. Conway Morris, J. D. George, R. Gibson & H. M. Platt), pp. 310-26. Clarendon Press, Oxford.

Weygoldt P. (1979). Significance of later embryonic stages and head development in arthropod phylogeny. In: **Arthropod Phylogeny**, (ed. A. P. Gupta), pp. 107-35. Van Nostrand Reinhold, New York.

Whitfield P. J. (1971). Phylogenetic affinities of Acanthocephala: an assessment of ultrastructural evidence. **Parasitology 63** 49-58

Whittington H. B. (1966). Phylogeny and distribution of Ordovician trilobites. **J. Palaeontol. 40** 696-737

Whittington H. B. (1979). Early arthropods, their appendages and relationships. In: **The Origin of Major Invertebrate Groups,** (ed. M. R. House), pp. 253-68. Academic Press, London & New York.

Whittington H. B. (1980). The significance of the fauna of the Burgess Shale, Middle Cambrian, British Columbia. **Proc. Geol. Ass. 91** 127-48

Whittington H. B. (1985). **The Burgess Shale.** Yale University Press, New Haven & London.

Wiley E. O. (1981). **Phylogenetics: The Theory and Practice of Phylogenetic Systematics.** John Wiley & Sons, New York.

Williams A. (1984). Lophophorates. In: **Biology of the Integument, Vol 1, Invertebrates,** (eds. J. Bereiter-Hahn, A. G. Matoltsy & K. S. Richards), pp. 728-45. Springer-Verlag, Berlin.

Williams A. & Hurst J. M. (1977). Brachiopod evolution. In: **Patterns of Evolution as Illustrated by the Fossil Record**, (ed. A. Hallam), pp. 79-121. Elsevier, Amsterdam.

Williams J. B. (1960). Mouth and blastopore. **Nature 187** 1132

Wilson A. C., Carlson S. C. & White T. J. (1977). Biochemical evolution. **Ann. Rev. Biochem. 46** 573-639

Wilson R. A. & Webster A. (1974). Protonephridia. **Biol. Rev. 49** 127-60

Winston J. E. (1977). Feeding in marine bryozoans. In: **Biology of Bryozoans,** (eds. R. M. Woollacott & R. L. Zimmer), pp. 233-71. Academic Press, New York & London.

Woollacott R. M. & Zimmer R. L. (1977) (eds.). **Biology of Bryozoans**. Academic Press, New York & London.

Wright A. D. (1979). Brachiopod radiation. In: **The Origin of Major Invertebrate Groups**, (ed. M. R. House), pp. 235-52. Academic Press, London & New York.

Yochelson E. L. (1978). An alternative approach to the interpretation of the phylogeny of ancient molluscs. **Malacologia 17** 165-91

Yochelson E. L. (1979). Early radiation of Mollusca and mollusc-like groups. In: **The Origin of Major Invertebrate Groups**, (ed. M. R. House), pp. 323-58. Academic Press, London & New York.

Zilch R. (1979). Cell lineage in arthropods? **Fortschr. Zool. Syst. Evol. 1** 19-41

Zimmer R. L. (1973). Morphological and developmental affinities of the Lophophorates. In: **Living and Fossil Bryozoa,** (ed. G. P. Larwood), pp. 593-9. Academic Press, London.

Zuckerkandl E. & Pauling L. (1962). Molecular disease, evolution and genic heterogeneity. In: **Biochemistry,** (eds. M. Kasha & B. Pullman), pp. 189-225. Academic Press, New York.

Zuckerkandl E. & Pauling L. (1965). Molecules as documents of evolutionary history. **J. Theor. Biol. 8** 357-66

Index